Symbol	Written As	Read As	
		Relations	
$=$	$A = B$	A is equal to B	7
$\neq$	$A \neq B$	A is not equal to B	7
$//$	$\overleftrightarrow{AB} // \overleftrightarrow{CD}$	$\overleftrightarrow{AB}$ is parallel to $\overleftrightarrow{CD}$	15
$\perp$	$\overleftrightarrow{AB} \perp \overleftrightarrow{CD}$	$\overleftrightarrow{AB}$ is perpendicular to $\overleftrightarrow{CD}$	108
∟	used in diagrams to indicate perpendicularity and right angles		108
$\stackrel{\circ}{=}$	$m(\angle A) \stackrel{\circ}{=} 40$	the degree measure of $\angle A$ is 40	111
$T(\ )$	$T(a) = b$	the image of element a is b in set T	125
$\sim$	$\triangle ABC \sim \triangle DEF$	$\triangle ABC$ is similar to $\triangle DEF$	127
$\cong$	$\triangle ABC \cong \triangle DEF$	$\triangle ABC$ is congruent to $\triangle DEF$	130
$\stackrel{m}{=}$	$\angle A \stackrel{m}{=} \widehat{BC}$	the measure of $\angle A$ is equal to the measure of $\widehat{BC}$	156
$\approx$	$m(\angle A) \approx 40$	$m(\angle A)$ is approximately equal to 40	177
$\rightarrow$	$p \rightarrow q$	p implies q	256
$\{\ \}$	$\{a, b, c\}$	the set with elements a, b, and c	266
$\in$	$a \in A$	a is an element of set A	267
$\notin$	$a \notin A$	a is not an element of set A	267
$\leftrightarrow$	$a \leftrightarrow b$	a corresponds to b, and b corresponds to a	269
$\sim$	$A \sim B$	set A is equivalent to set B	270
$\subset$	$A \subset B$	set A is a subset of set B	271
$\supset$	$A \supset B$	set A is a superset of set B	271
$'$	A'	the complement of set A	272

Operations

Symbol	Written As	Read As	
$\cup$	$A \cup B$	A union B	272
$\cap$	$A \cap B$	A intersection B	273
$\times$	$A \times B$	The Cartesian product of A and B	274

INTRODUCTION TO CONCEPTS OF GEOMETRY

PRENTICE-HALL INTERNATIONAL, INC., *London*
PRENTICE-HALL OF AUSTRALIA, PTY. LTD., *Sydney*
PRENTICE-HALL OF CANADA, LTD., *Toronto*
PRENTICE-HALL OF INDIA PRIVATE LIMITED, *New Delhi*
PRENTICE-HALL OF JAPAN, INC., *Tokyo*

INTRODUCTION TO CONCEPTS OF GEOMETRY

Carl A. Backman
The University of West Florida

Robert G. Cromie
St. Lawrence University

Prentice-Hall, Inc.
Englewood Cliffs, New Jersey

Current printing (last digit):
10 9 8 7 6 5 4 3 2 1

13-479683-7

Library of Congress Catalog Card Number 77-142430

Printed in the United States of America

To
Dolores, David, and Diane
Ingrid, Carla, and Linda

Foreword

The revolution in mathematics education during the past ten years has influenced the content of the elementary and junior high school textbook programs in two very significant ways. It has presented the real-number system in a more mathematically correct form, and it has greatly increased the quality and quantity of geometry teaching. Of these two changes, the former has received the greater attention in teacher education programs through pre-service courses, in-service courses, and workshops. Today, in some mathematics textbooks for the middle elementary grades, as many as one-third to one-half of the pages may be concerned with geometry.

The purpose of this textbook by Professors Backman and Cromie is to provide the professionalized knowledge of geometry that is essential to the mathematics programs for children of ages 6 to 14. The content of the book is presented in a mathematically correct manner, yet in a literary style that is highly readable. It is well illustrated to assist the student in "seeing" abstract geometric ideas. The content was selected and organized with careful consideration to the CUPM recommendations for a one-semester course in geometry for elementary school teachers. From a study of this text, the student will obtain increased competence in the subject matter of geometry that will lead to increased confidence in teaching mathematics in the elementary school, middle school, and junior high school.

Storrs, Connecticut VINCENT J. GLENNON

Preface

In writing this book we have attempted to present an informal, readable introduction to basic geometric concepts. We feel this approach is appropriate for those who have little or no background in geometry and for those who have neither time nor desire for a full course in formal geometry. On the basis of our use of the manuscript with groups of pre-service and in-service teachers, we feel that the book can be used with these groups as a resource tool for geometric concepts and as a textbook in a one-semester course in informal geometry.

The geometric content of the book is similar to that suggested in the *Course Guides for the Training of Teachers of Elementary School Mathematics* prepared for the Mathematical Association of America by the Committee on the Undergraduate Program in Mathematics.

Chapters 1–4 are primarily concerned with nonmetric concepts, 5–10 with metric geometry. Chapter 11 provides an introduction to coordinate geometry and the intrinsic geometry of the surface of a sphere. Chapter 12 presents some familiar problems associated with topology, and Chapter 13 provides both a framework for a reorganization of the content of the book and an invitation to study further. We emphasize the importance of the deductive nature of formal geometry by describing it in the transitional position at the end of the book.

The discourse throughout is informal, although both intuitive and deductive approaches have been used to develop geometric concepts and relationships. On several occasions a particular relationship is first generalized from experimental situations and is later shown to be a logical consequence of the assumptions and relationships already accepted. In several cases concepts are developed from more than one viewpoint. We have tried to do this particularly for those situations in which there is some controversy in school mathematics programs.

The degree of formalization of the geometric concepts and the logical interrelationships is left to the instructor. Not all students desire or are ready for a highly formalized approach. If a more formal approach than ours is preferred, we suggest that the sections in Chapter 13 on logic and structure, definition, elementary logic and proof, and consistency and independence be used early in the course, perhaps as early as Chapter 3 or 4. The descriptions, assumptions, and derived relationships can then be very easily transformed into the definitions, axioms, and theorems of a geometric structure.

There are additional features of the book that the reader and instructor should consider:

1. The style is conversational and gives detailed developmental discussions. This will allow a gradual assimilation of material by the mathematically inexperienced reader.

2. In keeping with the style, and in an attempt to increase reader participation in the process of concept development, we have included questions within the expository discourse. The reader should consciously answer these questions as he comes to them. The questions have been written so that the reader will have no difficulty answering them if he has understood the material studied earlier.

3. The sets of exercises following chapter sections and those at the ends of chapters are an important part of the book. To keep the book from becoming excessively long, we have used some of the exercises to develop important geometric concepts.

4. Several exercises suggest that the reader investigate entries in the References. The readings have been selected to extend topics presented in the text, to illustrate different viewpoints on geometric topics, to indicate opinions on the geometric content appropriate at various grade levels, and to illustrate methods appropriate for teaching geometry (especially in the elementary school). Reference citations are explained in problem 17 of the Exercises at the end of Chapter 1.

5. Appendix C contains patterns for making models of several important geometric figures.

6. Appendix A provides a review of basic set concepts.

We wish to express our gratitude to Dr. Vincent J. Glennon for his kind and persistent encouragement of our writing. We are indebted to Jerry Becker, Richard Diedrich, Vincent Naramore, Augusta Schurrer, George Spooner, and Frederic Warner for their critical reading of the manuscript; to our students at Syracuse University with whom we used various preliminary versions of the book; to the editors and staff of Prentice-Hall for their many hours of assistance; and to our many friends who have read and reread portions of the manuscript. And last but certainly not least, we owe a special measure of thanks to our families, without whose unending patience and support we could never have completed this work.

C.A.B.
R.G.C.

Contents

CHAPTER 1

Geometry Related to Familiar Concepts

1-1 INTRODUCTION

Through the centuries men have constantly asked questions about themselves, about what they see, and about their ideas. Certain of these questions and the methods of investigating them have been grouped because of their relationships to each other. For example, questions about the nature of man or the nature of knowledge are usually classified as philosophical. Questions about events involving men are often called historical questions. Questions and answers about matter and energy, their laws and processes, their states and properties, are considered to be in the realm of physics. Questions and answers about the composition of substances and the various elementary forms of matter are in the realm of chemistry.

In this book we shall investigate one of these categories—the branch of knowledge called mathematics. What kinds of questions, answers, and methods have been considered mathematical in nature? We would be oversimplifying if we said that mathematicians have considered questions about numbers and their relationships, about shapes and relative positions, and about methods appropriate for investigating these questions. This simplified description, however, will do for now; later we shall expand and refine it.

Geometry—the part of mathematics that investigates questions about shapes and relative positions—is the concern of this book. As you will soon see, "shapes and relative postions" is a considerable simplification, but this is a good place from which to start. Once we begin asking a few questions about shapes and relative positions you will be surprised how many more questions you can suggest, how many you can answer, and how many you can begin investigating. Moreover, you will find that geometry is related to the other branches of mathematics both in content and in method.

1-2 IMPRECISION OF COMMON LANGUAGE

We shall now show that some of the questions asked by geometers are already familiar to you and our common language can often be used to give partial answers to these questions.

Let us start with a question about relative position. What word do you commonly use to indicate the relative positions of the two rails that make up a track? Does the same word apply to wires on a telephone pole? (See Fig. 1-1.)

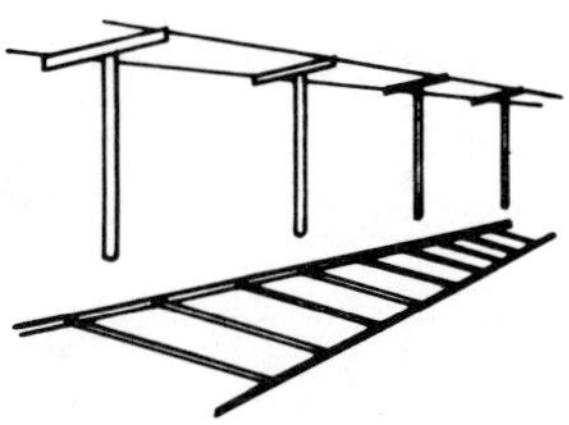

Figure 1-1

The word "parallel" is used frequently. The gymnast uses parallel bars; the grammarian speaks of parallel sentence structure; pilots may fly their planes in parallel formation; people may express parallel ideas. Given "parallel" as the answer to the questions above about relative position, we might go on to ask some further questions. Does the use of this word in geometry correspond to its use in ordinary discourse? How are the uses the same? How different? What are we saying when we describe two physical objects as being parallel?

To this last question you may give answers such as "the tracks are everywhere the same distance apart" or "the wires, if extended, will never intersect or cross each other." These answers in turn lead to other questions, such as: "What is meant by 'same distance' or 'never intersect'?" Suppose you want to be more precise about the meaning of these ideas. Geometry gives us a way of doing this.

Aside from the achievement of precision of thought, geometric ideas have other aspects. Some people might ask why parallelism is so important a notion that it warrants the attention of geometers. Some might want to know when parallelism began to be recognized as an important relation.

Did prehistoric man know of it? Did the ancient Greeks have the concept? Is it of modern origin? There is also the perplexing psychological question of how the mind singles out and differentiates a concept or property, such as parallelism, which we see existing between so many objects. How and why are the qualities referred to by the word "parallelism" separated from the other qualities of objects, such as color, substance, or function?

Now let us consider a question about shape. What is the shape of a photograph produced by a camera that uses size 127 film? What is the shape of a saltine cracker? (See Fig. 1-2.)

Figure 1-2

Like the word "parallel," the word "square" is often used. Once we realize we are using such a common word to indicate a concept of shape, we might want to ask what criteria we use to determine when the concept square applies to a given object. Some might say that an object is square "if all its sides are equal" or "if it has square corners."

Are these conditions sufficient to determine whether an object is to be classified as a square? In other words, do you need to specify more restrictions on the shape of an object in order to call it square? Can an object have five sides and be square? Is an object like a jack-in-the-box considered square? (See Fig. 1-3.) Is any part of the box considered square? The cover?

Figure 1-3

Is it necessary that a square have all sides equal? Can you think of objects referred to as square that do not have all sides equal?

Now that we have briefly discussed the two concepts parallel and square, we could ask whether they are related to each other. Does parallelism have anything to do with squares? Suppose you notice that in the square represented in Fig. 1-4 side AB is parallel to side CD. Would this be true of all squares? How do you know? How can you convince someone of this?

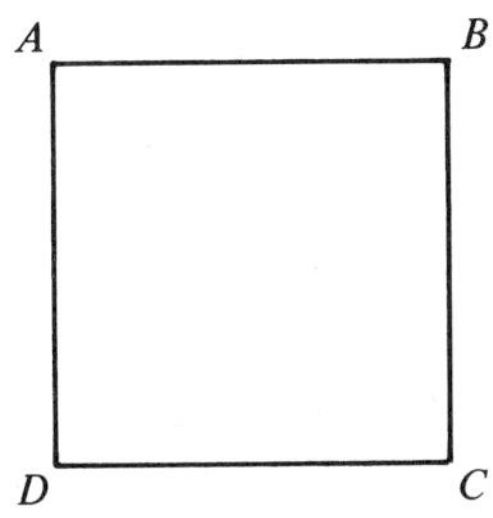

Figure 1-4

Again we see that a common notion, like square, can lead to a multitude of questions about the concept, its meaning and its properties.

As a third example consider a teacher who plans to have the children make full-length silhouette cutouts of themselves. How much paper should she buy? Obviously, she will need to know each child's height. How do you determine how tall a child is? Do you need to know width to get height? Will the teacher need to know the width of each child in order to determine how much paper she needs?

What is meant by "measuring"? How is measuring height like measuring area? Why do we use inches or feet for measuring length? Are there other units available for measuring length? Does measurement have any relationship to squares and parallels?

This questioning procedure is one way in which mathematical knowledge grows. Each new question suggests many, many more. Thus far we have taken but a few geometric concepts (parallel, square, measurement), illustrated how they are used in the common language, and suggested some questions that could be asked about each of them. Of course, there are many more concepts, questions, and answers. It is our plan to discuss with you the nature of the concepts that constitute geometry, to discover the nature of the questions the geometer attempts to answer, and to consider the ways in which he attempts to answer his questions.

EXERCISE 1-2

For Exercises 1–4 identify the geometric concept and then try to describe what the concept means.

1. What shape-word do we use to describe the top rim of an orange-juice can or the tire on a bicycle?

2. What do we say we draw with the straight edge of a ruler?

3. What do we say is formed by the positions of the two antennas of a set of "rabbit ears"?

4. Of what do we obtain a measure when determining how much shellac is needed to cover a floor or when deciding how many ceiling tiles are needed to cover a ceiling?

5. Using Fig. 1-5, indicate a part of the box that suggests:
(a) A point. (c) A plane.
(b) A line. (d) A space.

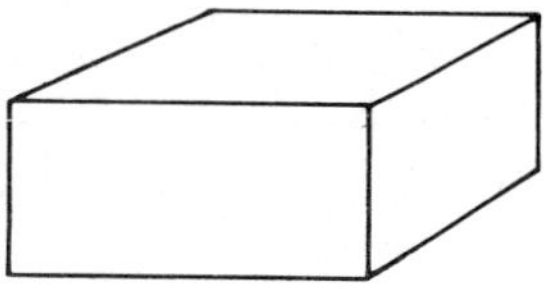

Figure 1-5

1-3 OUR STRATEGY

The time has come to be somewhat more technical. We shall investigate the concepts of geometry, the special ways of writing about geometric concepts, the relationships between geometric concepts, and the ways of justifying or verifying these relationships.

We would not hear so many complaints about the difficulties of studying geometry if all that there was to know about it were conveyed in the common language. The content and methods of geometry, although derived from and supported by common ideas, require a more technical, more precise language. For the most part, we shall not be as precise as the mathematician in our development of topics. We shall develop many important geometric relationships without giving the supporting arguments that a mathematician would. The development in this book is informal and intuitive. We shall claim to know some concepts from geometry and claim to be sure that certain relationships are true, but we shall not generally produce the sort of grounds or reasons that the mathematician would accept.

We shall produce some illustrations and arguments for the truth of relationships that are very convincing without being proofs in the strict mathematical sense. However, we shall also be studying to some extent the grounds for proof that the mathematician would accept as adequate, so that where appropriate we can justify the truth of our claims using the mathematician's reasoning. This approach to the study of mathematics resembles, perhaps, one of the ways in which mathematics is created. The mathematician discovers or intuitively knows a relationship between mathematical concepts. He then attempts to justify the truth of, or prove, the relationship by using the formal methods of mathematical reasoning. (See Chapter 13.)

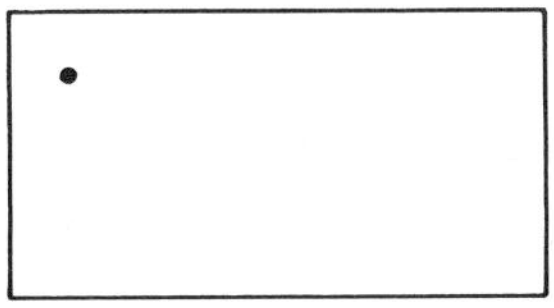

Figure 1-6

1-4 SPACE

The geometric concepts discussed in this book are concerned principally with the space around us. This space is most often described in terms of the positions that objects occupy. In our geometry discussion we shall use the term **space** when we are referring to the set of all possible positions.

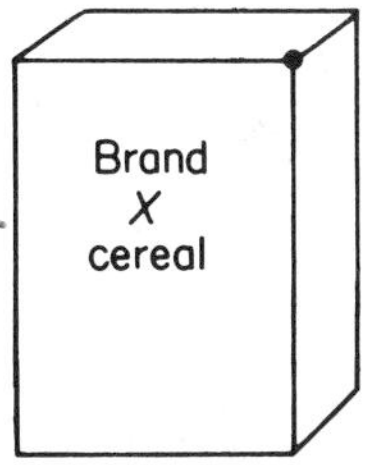

Figure 1-7

1-5 POINTS

The most elemental concept of position in geometry is that of **point**. Indeed, space is often characterized as the set of all possible points.

Most people have a fairly good idea of the notion of point and know that points are usually represented by small dots. For example, if a small bead were dropped on a piece of paper and we were then asked where the bead landed, we could "point" to the position and indicate it with a small dot. (See Fig. 1-6.) If we wanted to indicate the point in the extreme upper front right corner of a box, we could draw a diagram of the box as in Fig. 1-7 and place a dot in the appropriate position.

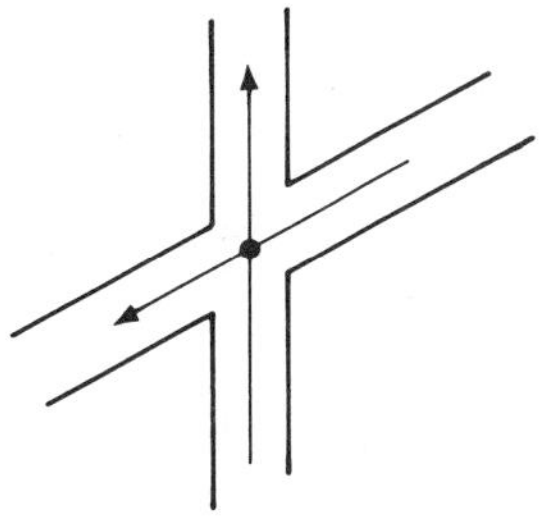

Figure 1-8

Assuming that the box is a normal cereal box, how many other corner positions are there? Are there other positions or points that could be located on the box? Is there a point that is at the same time on the front, on the top, and on the left side of the box?

If we draw a diagram to indicate the paths of two cars, as in Fig. 1-8, the place, or position, where the two paths cross is called the point of intersection. Suppose, however, that we sketch the intersection of two streets. (See Fig. 1-9.) The shaded region indicates what is commonly referred to as the intersection of the streets. Do we usually call this intersection a point? If we consider the same streets as diagrammed on the map in Fig. 1-10, the intersection does seem more closely related to the concept of point. Actually, the dot printed on the map is not the point itself; it is

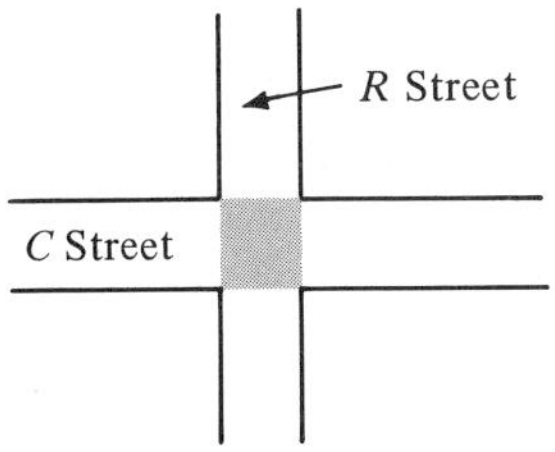

Figure 1-9

but a representation intended to convey the idea of a singular position or point.

The uses of the two kinds of maps may suggest why one is closely associated with the notion of point while the other is not. In Fig. 1-9 the shape or size of the intersection is important, as it might be in an accident report for an insurance company. In Fig. 1-10, however, the actual size of the intersection is of no consequence; only its position is important.

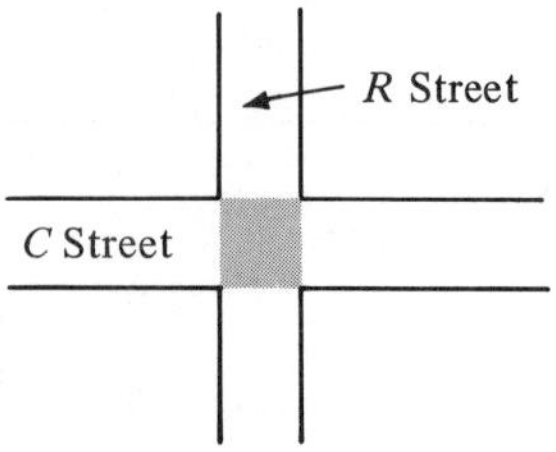

Figure 1-9

Often much is made of the idea that points have no size, no dimension. As Figs. 1-9 and 1-10 show, the reason is simply that if the notion of a point is applicable to a given situation, then the notion of size or shape is not. If size is a quality of the situation that is being discussed, then whatever is being discussed goes beyond our notion of a single point.

A notion often mentioned is that points do not move. A common illustration involves holding a pencil in front of you. The tip of the pencil occupies a position or point. Now, if the pencil is moved to the side, as in Fig. 1-11, the tip of the pencil has changed position, but the point originally occupied by the tip has not moved. This is sometimes referred to as the static interpretation of point.

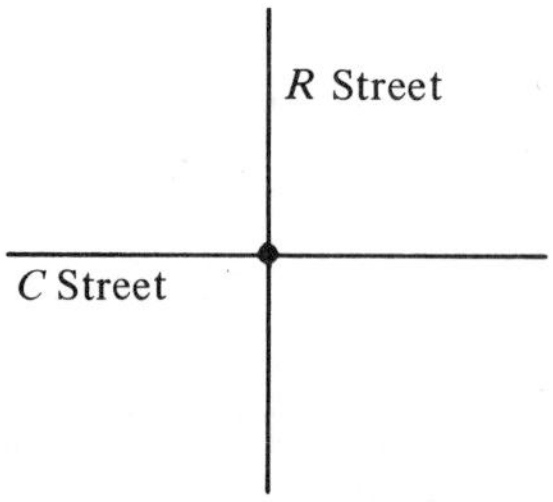

Figure 1-10

At times, however, it may be useful to think of points as capable of motion. For example, using the pencil situation again, suppose you want to describe the path that the tip of the pencil took in going from one point to the other. This situation is often described as the path of a moving point, and the approach is sometimes referred to as the dynamic interpretation of point.

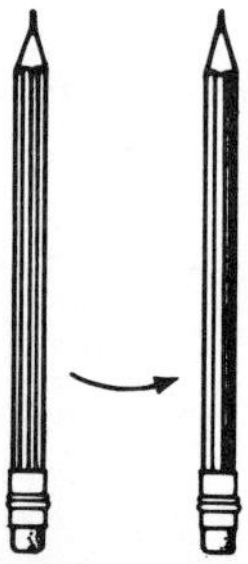

Figure 1-11

You are probably wondering which interpretation is correct. For clearly, according to normal understanding, that which moves does not stand still. (See Ref. 9, pp. 86–89, for an interesting discussion of this paradox attributed to Zeno.[1]) Among people who use mathematics, some defend the static interpretation while others hold to the dynamic. We introduce the controversy here to show that it is quite possible to have more than one viewpoint on some very fundamental ideas in mathematics. In order to determine which view to take, you will need to analyze the immediate situation to decide which interpretation will be most relevant and useful. This is, of course, not without parallel in other disciplines. For example, in physics there are two interpretations of the nature of light—the wave theory and the particle theory. Physicists tend to use that theory which best explains the situation at hand.

A third idea about points often occurs in elementary textbooks. This is the notion that you cannot see a point because "a point is an idea." This issue is raised to emphasize two things about the nature of geometry in

[1] Bibliographic citations are explained in problem 17 in the Exercises at the end of Chapter 1.

general and of points in particular. The first is that the dot that is drawn on paper is but a representation of a point. The dot is not to be construed as a point in and of itself. The problem is that the diagram introduces a quality that is not a quality of points: a dot can actually be measured; it has a certain size. The notion of size is not applicable to points. Just now, the distinction between the representation of a point and the point probably seems of little importance. But if diagrams can introduce qualities that we do not associate with a given concept, we must be very careful not to associate an unwanted property of a diagram with the properties of the idea we are really talking about. For example, since one could reasonably ask how big to draw the dot, one might want to know how big the point is.

The second idea being emphasized in saying that you cannot see a point is that *point* and other geometric concepts are abstractions. That is, it is possible to consider geometry as a system of ideas entirely independent of physical objects.

Now we return to the example of the position of a bead on a piece of paper. Suppose a red bead and a yellow bead are dropped on a paper and we make a diagram again. [See Fig. 1-12(a).] In order to distinguish the two points, the convention is to label each dot with a capital letter. In Fig. 1-12(b) the dot associated with the letter R represents the position occupied by the red bead. R is said to be the name of the point. What does the dot associated with the letter Y represent? What does Y name?

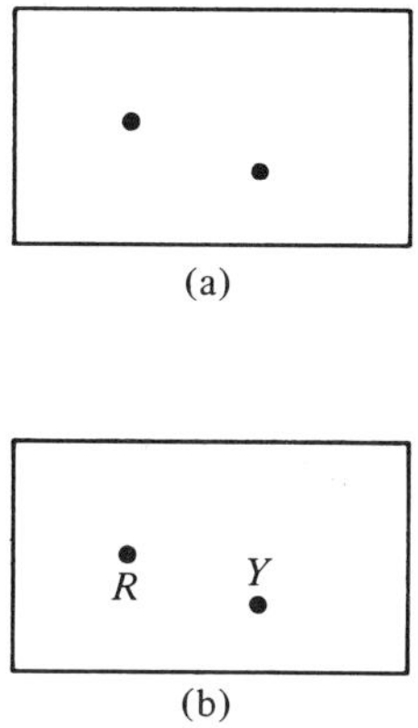

Figure 1-12

Within one discussion, if two references are made to "point R," then we know that the same point is being talked about.

Sometimes two different letters are used to refer to the same point. For example, in Fig. 1-13 F and R are names for the same point. What is another name for the point Y? Perhaps you recall from other work in mathematics that when two names refer to the same thing, this may be indicated by an equal sign between the two names. For this example, then, we may write point F = point R, or, more simply, $F = R$. To indicate that two names do not refer to the same thing, we use the notation $\neq$. Thus: $F \neq Y$.

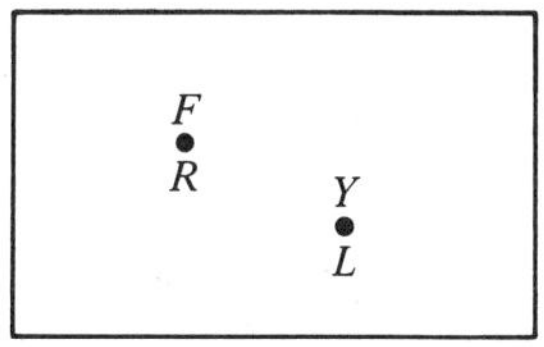

Figure 1-13

EXERCISE 1-5

1. Draw a diagram of a cube and, using capital letters, label the points at each of the corners, a point in each edge, and a point in each face.

2. Using Fig. 1-14, indicate whether the mathematical sentence should be completed with "=" or "$\neq$."

(a) Point A _____ point L.
(b) Point A _____ point M.
(c) Point X _____ point C.
(d) Point M _____ point X.
(e) Point M _____ point G.

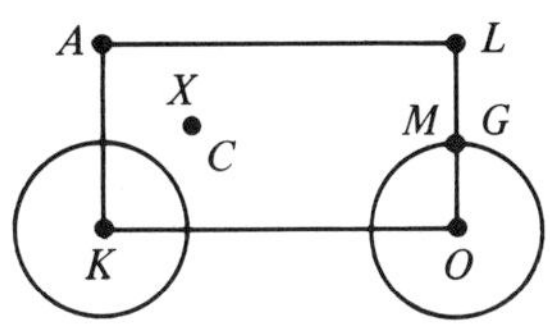

Figure 1-14

3. Indicate whether the notion of point is applicable in each situation below. Give a reason for your answer.
   (a) The portion of the border common to the states of Utah, Colorado, Arizona, and New Mexico.
   (b) A footprint.
   (c) The place where a thumbtack enters a bulletin board.
   (d) The size of the hole made by the thumbtack.

4. For each situation below, indicate whether a static or dynamic interpretation of point is applicable.
   (a) The position of a nail in a wall.
   (b) The location of a rivet on the wing of an airborne jet.
   (c) The position of a nail in an automobile tire.
   (d) The position of a period at the end of a sentence.

1-6 LINES AND LINE SEGMENTS

Another position concept of fundamental importance is the **line**. We want to know here what is meant by "line" and also how the concept of line is related to the concept of point.

The notion of line has different interpretations in the various geometries; in fact, the interpretation of line in the geometry we are studying has not been altogether consistent over the centuries. A clear distinction has not always been made between the two concepts of line and line segment. What follows is an attempt to clarify how we will be using the terms "line" and "line segment."

"Point" was never actually defined in the previous section. Later you will understand the reason. (See Secs. 13-2 and 13-3.) For the present it is enough to know when and how we communicate about points. We shall approach lines in the same way. "Line" will not be defined mathematically, but we can gain an understanding of the concept by using appropriate interpretations.

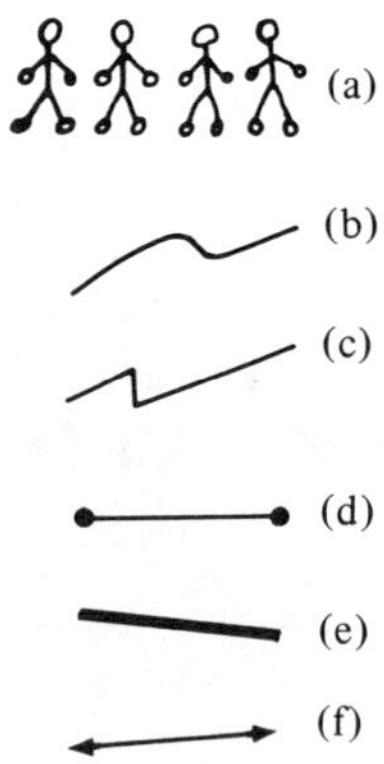

Figure 1-15

The use of the word "line" in geometry needs to be carefully distinguished from its common uses. If people were asked to draw a line, we might get various responses, as suggested in Fig. 1-15. Not all of these are acceptable interpretations of line in the geometry we are now considering.

Drawings (b) and (c) are not acceptable as lines because they are not straight (although in some geometries lines need not be straight). The queue pictured in (a) is inappropriate because it consists of objects separated from one another; the line is incomplete. The diagram in (d) does not fit the concept since it implies that a line has fixed length; it is more appropriate to the concept of a line segment. Drawing (e) does not reflect a line because it indicates that the figure has thickness. Even (f), the representation of line that is used by many students of mathematics, is not actually a line—for a line, like a point, is an abstraction. The diagram is used only to call attention to certain qualities of the concept. As with diagrams for points, we must be careful not to assume that all properties of the diagram are properties of the concept.

What then are the qualities of the abstraction of position called line? We shall use the idea of a line of people to investigate some important ideas about lines, although lines of people lack some important properties of the abstraction.

Each of the people in a line occupies a distinct position in the line. This observation is useful for two reasons. First, it illustrates that "a line is a continuous set of points"—a set of positions. Second, it shows that thickness is not a part of the concept of line. When you ask students to line up, you do not care about the width of the line; all you care about is how the positions of the students in the line relate to one another. One reason that no drawing can be a line is that a drawn line has thickness, which is not a relevant property of lines.

Figure 1-16

The line of people we have in mind is a straight line. In Fig. 1-16 Ann has stepped "out of line." The teacher asks her to get back "into line." What the teacher means is that Ann is to get directly between the other two students. This illustrates that, for our kind of line, each point must be directly between two other points of the line—the line is straight.

How many people can be in a line? Or, how many positions are there in a line? Two questions are involved here: (1) What is the extent of a line? (2) Given two positions on a line, how many positions are there between these two positions? For any line of people the actual length of the line is limited by the number of people available. However, we can always imagine that there are more people, and our line must contain positions for all of them. Any line in our geometry extends indefinitely, without limit. We simply recognize the set of all possible positions that "fall in line" with one another, often indicating the limitless extent of a line by placing arrowheads as in Fig. 1-17.

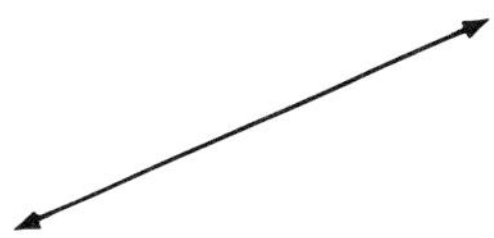

Figure 1-17

The description "all possible positions," used above, also shows why lines of people lack another important quality of geometric lines. Using people, it is physically possible to have a queue in which the people are packed so tightly that it is physically impossible to squeeze another person between two others. The obvious reason is that people have size. But the line abstraction refers to the set of all logically possible positions that are in "line with one another." Since size is not relevant to position or point as we use these terms, it appears that if you are given two points in a line, no matter how close they are, it is always logically possible to think of a third position that is directly between the other two. In fact, would there not be many such positions?

You say, "Fine! I can think about it, but I certainly can draw two dots so close that I cannot fit another dot between them." This is entirely true, and it sums up the whole problem of drawing points and lines. The irrelevant quality of size is introduced by the drawing. Again we must remind ourselves of the difference between a concept and its physical representation.

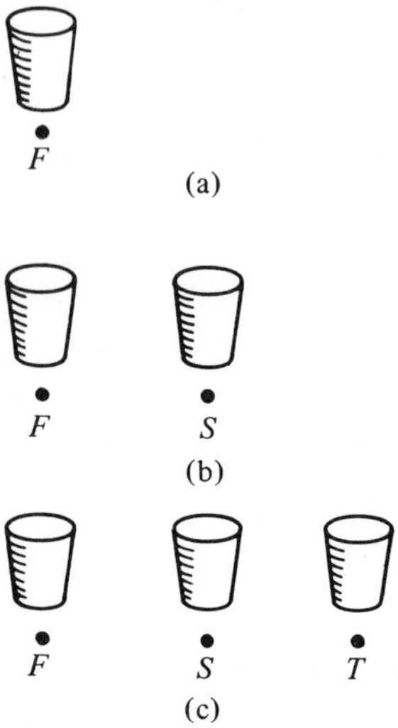

Figure 1-18

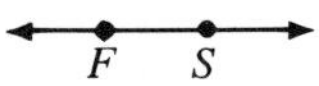

Figure 1-19

The notion of betweenness is also involved when we say that a line is *continuous*, having no breaks. This means simply that you can always conceive of a point in the line that falls directly between any other two points in the line that you might select. In representing a line, we draw a continuous stroke to indicate this.

Thus far, we have talked about qualities of lines in general. How can you distinguish one line from another? Notice that we are assuming there is more than one line. How can you describe the position of a line? Suppose that you are making a row of glasses. You set one glass at position *F*, as shown in Fig. 1-18(a). This represents one point in the line. Where can you put a second glass so that the two glasses will be in line? Clearly, *F* is a point on many lines, so that any position will do for the second glass except to put it on top of the first glass. Call this position *S*, as in (b). Where should the third glass go to be in line with the other two? Now the choice of positions is not so free. There are still many positions at which the glass may be placed, but if it is to be in line with the other two, as in (c), the positions of the first two glasses limit the range of positions for the third glass or any other glasses. Figure 1-19 represents the set of all points that fall in line with points *F* and *S*—that is, the set[2] of all points on the same line as *F* and *S*.

How many lines could contain the position *F*? How many lines could contain the two positions *F* and *S*? Is it always possible to find a line that contains any three given positions? We shall assume that there is one and only one line that contains two distinct points. We say that two points *determine* a line. Notice how the word "determine" is used here. It means that there is a line that contains the two points and, furthermore, there is only one such line.

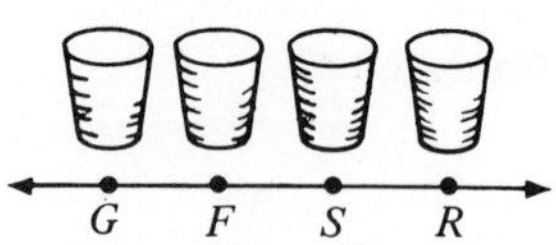

Figure 1-20

Notice that once the glasses have been lined up, any two glasses in the set determine the line. In Fig. 1-20 the glasses at *G* and *R* could be used as reference points just as well as those at *F* and *S*.

That two points determine a line is the basis for a procedure for naming lines. Two different points known to be on the line, such as *G* and *S* in Fig. 1-20, are selected as reference points. We use the names of these two points in naming the line by writing the two point names side by side and

[2] A review of set concepts is included in Appendix A.

placing over them the symbol for line: $\leftrightarrow$. For example, $\overleftrightarrow{SG}$. Note that $\overleftrightarrow{SG} = \overleftrightarrow{GS}$. Does $\overleftrightarrow{GS} = \overleftrightarrow{FR}$? Sometimes the symbol is omitted and the word "line" used: line *SG*. Another convention for naming lines uses simply a lowercase letter. For example, in Fig. 1-21 the line may be called line *r*. Thus, this same line may be named $\overleftrightarrow{JK}$, line *JK*, or line *r*.

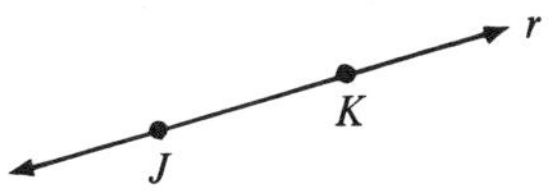

Figure 1-21

Figure 1-22

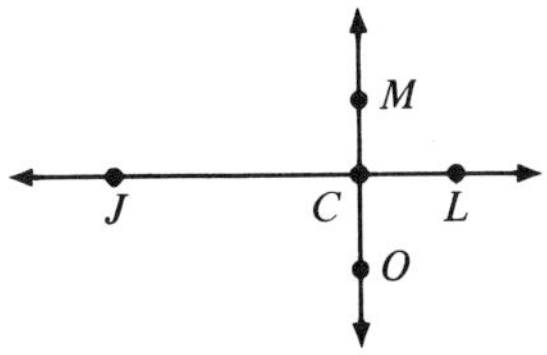

Figure 1-23

Is it possible for two lines to have a point in common? Interpreting this question in terms of rows of glasses, is it possible for one glass to be in two different rows? The answer is, of course, yes, as Fig. 1-22 shows. The glass at position *C* is common to the two rows. Figure 1-23 provides an analysis of the situation using line representations. The intersection of the two lines is the point *C*. Since we can think of lines as sets of points, the situation may also be expressed in set notation: $\overleftrightarrow{JL} \cap \overleftrightarrow{MO} = \{C\}$.[3]

Is it possible for two different lines to have more than one point in common? Try to arrange two straight rows of glasses so that two of the glasses occupy positions in each of the two rows. Let us try to think this one out. We have said that two points determine one line; that is, through two points there is one and only one line. The two glasses occupy two positions. Can both of these positions be contained on one line? Can both of these positions be contained on more than one line? To say yes to this question would be to contradict our earlier assumption about the number of lines containing two different points.

Although we find it useful to think of a line as a set of all possible positions that has certain characteristics, one of which is unlimited extent, few, if any, physical objects have this characteristic. Most physical objects that are classified as straight have definite endpoints. (See Fig. 1-24.) A special subset of a line is often used to represent such objects. Two points in a line are selected to represent the endpoints. The object then corresponds

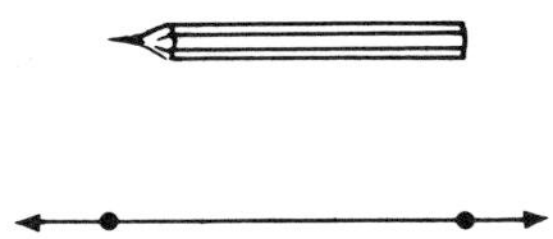

Figure 1-24

[3] For a comment on this notation see Appendix B.

to the union of the two endpoints and all the points of the line between the two endpoints. Such sets of points are called **line segments**. Figure 1-25 pictures a line segment whose **endpoints** are X and Y. What is the name of the line of which the segment is a subset?

Line segments are named by placing a bar over the two letters that name its endpoints. For example, $\overline{XY}$. Sometimes the bar is omitted and the name is given: line segment XY. Note that $\overline{XY} = \overline{YX}$.

Figure 1-25

EXERCISE 1-6

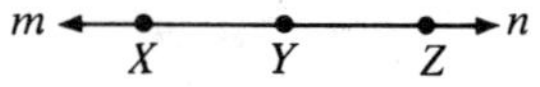

Figure 1-26

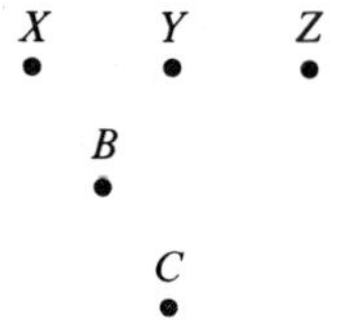

Figure 1-27

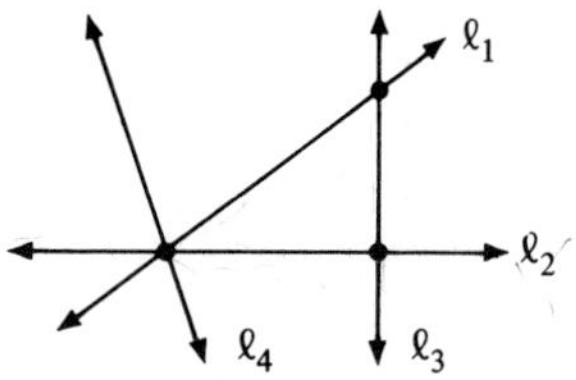

Figure 1-28

1. Write four different names for the line represented in Fig. 1-26.
2. Write the names of three segments whose endpoints are contained in the line shown in Fig. 1-26.
3. What are the endpoints of (a) $\overleftrightarrow{EF}$, (b) $\overline{CD}$?
4. Draw a diagram to illustrate two lines whose intersection contains: (a) no points, (b) one point, (c) two points.
5. A set of points is said to be **collinear** if all points in the set are in the same line. What sets of points appear to be collinear in Fig. 1-27?
6. A set of lines is said to be **concurrent** if a single point is common to all of them. Name the sets of concurrent lines in Fig. 1-28.
7. Draw diagrams to illustrate the possible intersections of three distinct lines. (There are four possibilities.)

1-7 PLANES

A third fundamental abstraction of position in space is called a **plane**. The word "plane" is not as common as point and line in ordinary discussions; however, the idea is often used when we note that something is smooth and flat—for example, the top of a table or the surface of a pane of glass. But we have to be careful that the word "flat" does not intrude on what we mean by plane. Many times "flat" describes only those things which have a horizontal orientation. In our geometry, planes can have any directional orientation whatsoever. On the house sketched in Fig. 1-29, planes are suggested by the pitched side of the roof, the face of a wall, and the surface of a floor.

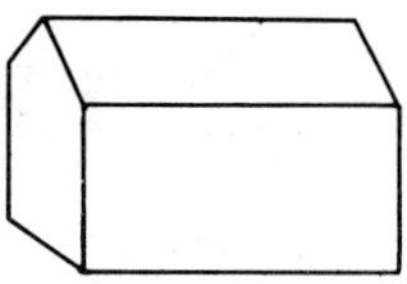
Figure 1-29

How would you describe the orientation of the plane referred to in the physical concept of inclined plane?

The concept of plane is very important in sectional drawing, which attempts to show all the features of an object that lie in one plane. For example, Fig. 1-30 shows various sections of a carrot. Part (a) shows the features in a plane that corresponds to a vertical cut of the carrot. Part

(b) shows a cross section, while (c) shows features in a plane corresponding to an oblique cut.

The word "plane" has been used in press releases describing attempts to join two orbiting satellites. Before the two satellites can dock, their orbits must be in the same plane.

How are the position concepts of line and point related to the concept of plane? First, a plane may be considered as a set of positions—in other words, a set of points. Second, there is no limit to the number of points in a plane. This would mean that we could select any two of the points. From Sec. 1-6 we recall that two points determine a line. Are all the points of a line contained in a given plane if two of its points are? Think of two dots drawn to represent two points. The two dots are on a flat piece of paper. Now think of the line that contains the two points. Is there any point of this line not in the plane suggested by the paper? Remember that the set of points in the plane extends beyond the edges of the paper. We shall assume that if two points of a line are contained in a plane, then every point of the line is contained in the plane.

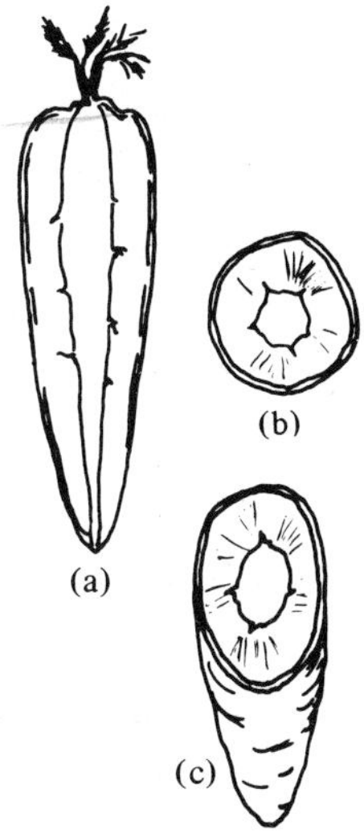

Figure 1-30

Since a line can be contained in a plane, a plane must have limitless extent. This is one reason why we say that a roof, a table top, or any other physical object only suggests the concept of plane. The notion of plane represented by the table top extends beyond the limits of the table to include every position at the *same level* as the top. However, thickness is not relevant to the concept of plane. Plates of glass are only suggestive of planes, because plate glass has thickness. It might be better to say that plate glass is a model of two planes: the top surface is a model of one plane while the bottom surface represents another plane. There are actually six planes. Can you find them?

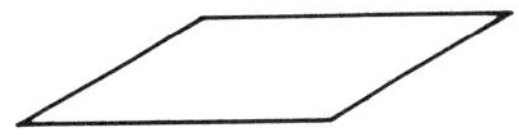

Figure 1-31

These two characteristics of planes make it difficult to draw adequate representations of planes. Typically, only a portion of a plane is sketched, the understanding being that the plane is not limited by the four edges of the diagram. Thus, the part of the plane shown in Fig. 1-31 depicts the whole plane.

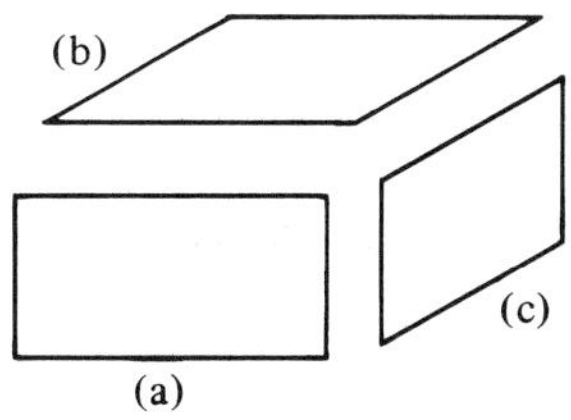

Figure 1-32

Sometimes it is quite difficult to draw a diagram to show a plane's orientation. Figure 1-32(a) represents a face view of part of a vertical plane; (b) represents an oblique view of part of a horizontal plane; and (c) represents part of a vertical plane not seen full face. Putting these three diagrams together in Fig. 1-33, we have a drawing of a box that makes the orientation of the planes clearer.

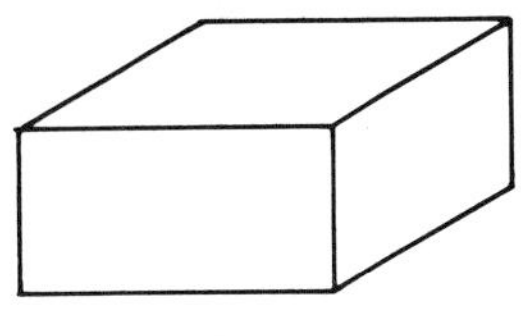

Figure 1-33

We can now continue discussing the relationship of points, lines, and planes. In Sec. 1-6 we assumed that a unique line is determined by two points. Is there a similar assumption about planes and points? That is, even though there is an unlimited number of points in a plane, is it possible to find a minimum number of points all of which are in one plane and in no other plane? The answer is yes. We shall assume that any three noncollinear points determine a plane.

To see that one point does not determine a plane, simply find a point that is contained in more than one plane. Figure 1-34 shows that any point in the edge of a box is contained in at least two planes. The same diagram shows that two points may lie in two planes and therefore do not determine a unique plane. It also shows that if we pick the correct three points, then they may lie in more than one plane. But notice that three points do not always have to be collinear, whereas two points always are. We shall return to this idea in a moment. Four noncollinear points may lie in one plane, as do the four corners of one side of the box in Fig. 1-34. But this does not always hold. Think of a table standing on a flat floor. The floor represents a plane. The feet of the table legs represent four points. If one of the legs is shorter than the other three, will all four points lie in the plane of the floor? For this case there is no plane that contains all four points. The situation is similar for tables with five, six, seven or more legs.

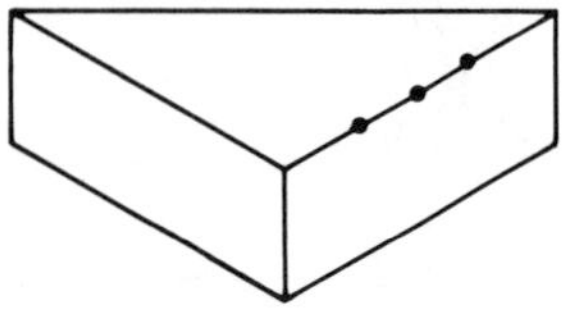

Figure 1-34

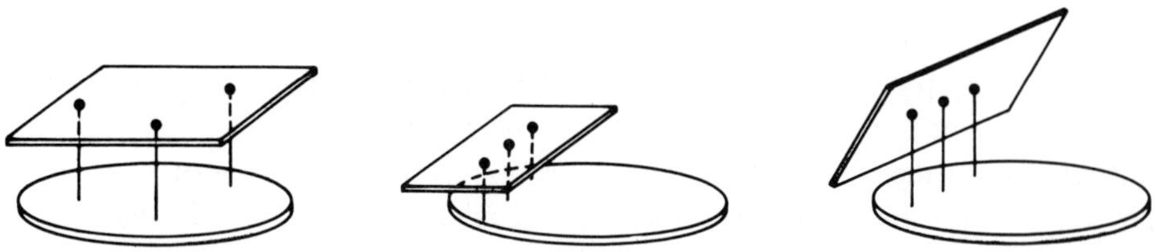

(a) (b) (c)

Figure 1-35

Let us return to the three-point case. Figure 1-35(a) shows a piece of plastic set on the tips of three toothpicks stuck in styrofoam. If the tips of the toothpicks are noncollinear, will the plastic lie flat without wobbling? Is there any other orientation of the plastic (aside from turning the plastic over) for which the plastic touches each of the three tips? Notice in Fig. 1-35(b) and (c) that several orientations of the plastic are possible if the three tips are collinear. This, then, is the basis of our assumption that any set of three noncollinear points is contained in one and only one plane.

This assumption provides the rationale for one of the conventions used in naming planes. We simply select names for three noncollinear points of the plane—for example, plane *ABC*. Planes are sometimes named with two capital letters: plane *XY*; and they are often named with a single lowercase letter: plane *n*.

We have already assumed that if two points of a line are in a plane, then all points of the line are in the plane. Since every point of the line is a point of the plane, we may say that the line is a subset of the plane. If m is the plane and r is the line, we might write this using set notation: line $r \subset$ plane m.

If no points of a line are contained in a plane, we say that the line and the plane are **parallel.** In Fig. 1-36, $\overleftrightarrow{ST}$ is above plane m. $\overleftrightarrow{ST}$ does not intersect[4] plane m; that is, the line has no points in common with the plane. Using set notation, we may write: $\overleftrightarrow{ST} \cap m = \varnothing$. $\overleftrightarrow{ST}$ and plane m are parallel. The geometric symbol for parallelism is $//$. Thus, we write: $\overleftrightarrow{ST} \; // \; m$. A ceiling beam and the floor beneath it are suggestive of this situation.

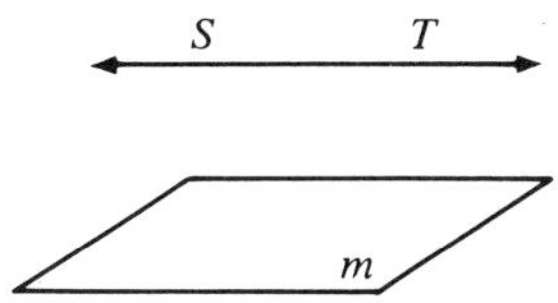

Figure 1-36

It is possible for a line and a plane to have but one point in common. A dowel piercing a board is a model of this relationship. [See Fig. 1-37(a).] In this case we say that the line intersects the plane in a point. From the line-and-plane representation in Fig. 1-37(b) we see that $r \cap \overleftrightarrow{AB} = \{C\}$.

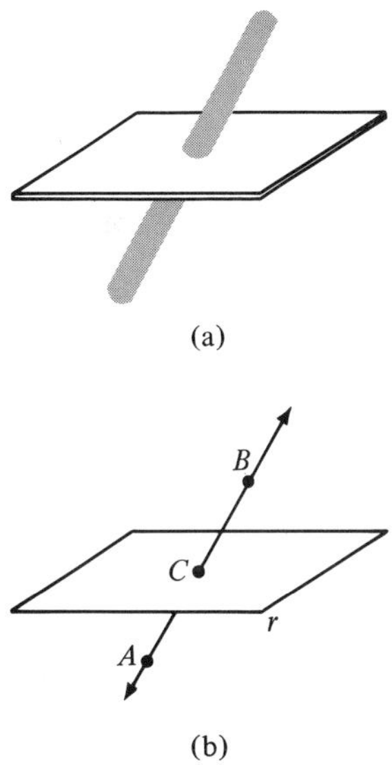

Figure 1-37

If two distinct lines lie in one plane, it is possible that the intersection of the two lines is empty. For example, in Fig. 1-38, $r \cap t = \varnothing$. In such cases we say that the lines are **parallel.** Using the symbol for parallelism, we can write: $r \; // \; t$. Does this remind you of the railroad track situation in Sec. 1-2?

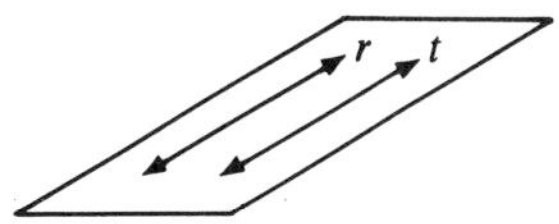

Figure 1-38

Similarly, if two planes have no points in common, we say that they are parallel. If the two planes are not parallel, what is the intersection? The front end and side of the pup tent shown in Fig. 1-39 suggest that the intersection may be a line. The line is suggested by the seam. Can you think of a situation for which the intersection of two planes is a single point? Don't forget that any plane has unlimited extent. In fact, we shall now agree that if the intersection of two planes is not empty, then the intersection is a line.

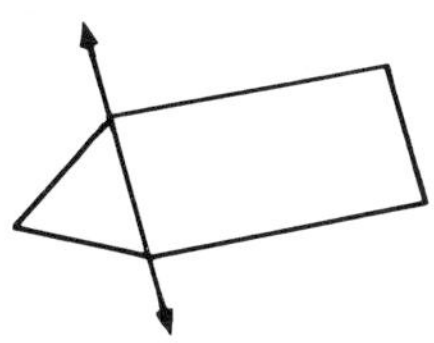

Figure 1-39

If space is considered the set of all possible points, then any set of points is a subset of space. Thus, planes, lines, and other sets of points are subsets of space. We have also noted that lines and other sets of points are subsets of planes. To complete the analysis we have considered that segments and other sets of points are subsets of a line. In subsequent chapters we shall use points, lines, planes, and space as the basic concepts from which to build other geometric figures.

EXERCISE 1-7

1. Name the intersection of planes m and n in Fig. 1-40.

[4] With reference to sets of points, the words "intersect" and "intersection" are often used ambiguously. Any two sets of points have an *intersection*; it may be empty or it may be nonempty. However, when we say that two sets *intersect*, we imply that the intersection is *nonempty*. When we say that two sets do *not intersect*, we imply that the intersection is the *empty set*.

2. Name the intersection of planes a, b, and c in Fig. 1-41.

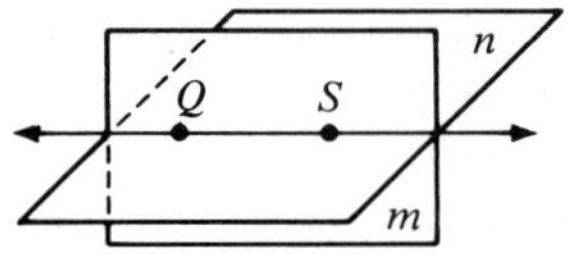

Figure 1-40

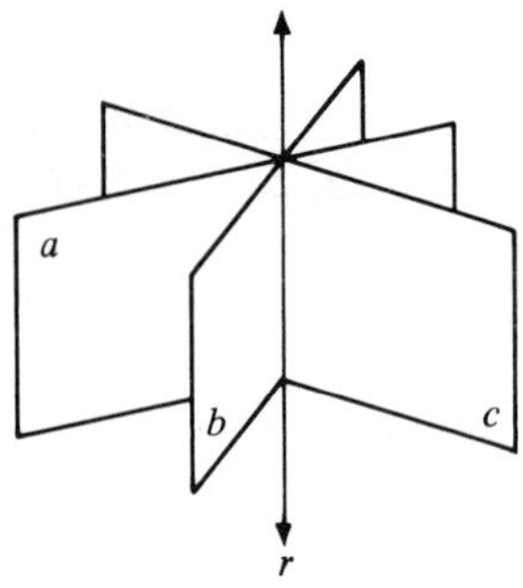

Figure 1-41

3. Using Fig. 1-42, name each of the following intersections:
   (a) Plane j and plane k.
   (b) Plane k and plane ℓ
   (c) Planes j, k, and ℓ.
   (d) Plane k and line XY.
4. Answer the following questions about Fig. 1-43.
   (a) What is the intersection of planes q and r?
   (b) Line AB is the intersection of which planes?
   (c) What is the intersection of planes q, r, and s?

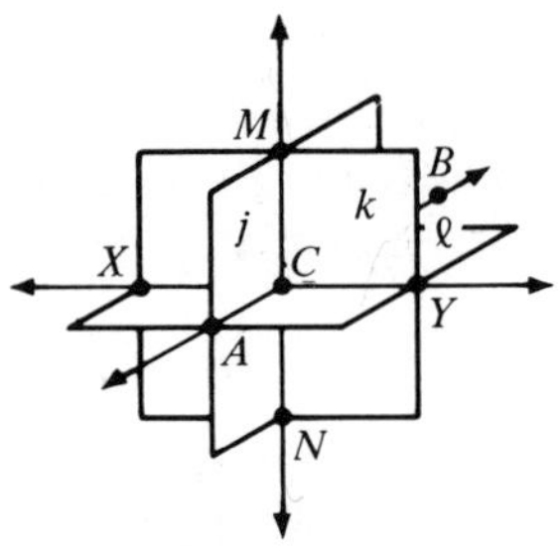

Figure 1-42

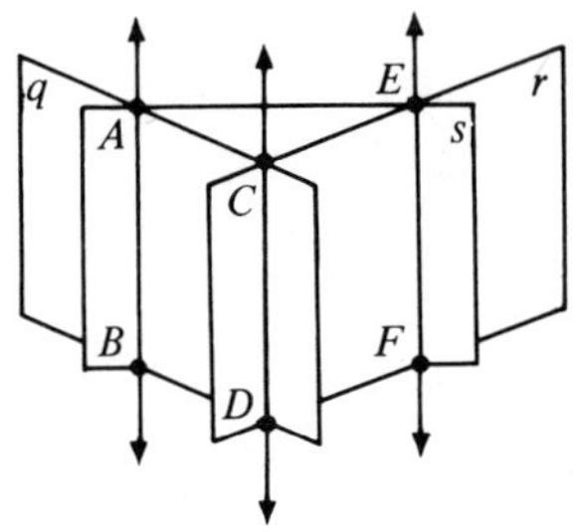

Figure 1-43

5. If all the points in a given set are contained in one plane, the set of points is said to be **coplanar**. Why are two lines whose intersection is a point always coplanar in our geometry? [*Hint:* How many points determine a plane?]
6. On several occasions we have said about a geometric relationship that such and such is true in "our" geometry. This may seem peculiar, if the relationship obviously agrees with normal experience. You might well ask, "But, what other geometry is there?"

   Our discussion of lines and points may, in fact, seem a little overdone, since lines obviously are straight, are contained in a plane, and so on. But suppose you are restricted to points on the surface of a sphere; that is, suppose the universal set for the following discussion is the set of all points on the surface of a sphere. With this in mind, consider the following:

(a) What is a "line" on a sphere; that is, what on a sphere is analogous to our line in a plane or space?
(b) Do two points determine one and only one of these lines?
(c) Will such lines be straight?
(d) Are such lines infinite in extent?
(e) Are lines that do not intersect parallel?
(f) Do lines intersect in one and only one point?

Some of the answers to these questions agree with the descriptions of lines given in the text. Some do not. This is why we have qualified our remarks by saying, "In our geometry. . . ." In other geometries the relation being considered may not hold.

Because of the importance of the plane and the relationship of the line to the plane, this aspect of our geometry has often been called *plane geometry.*

SUMMARY OF PRINCIPLES

Given two distinct points, there is one and only one line containing the two points.

The intersection of two distinct lines is either (1) a single point, or (2) empty, in which case the lines are parallel only if they lie in the same plane.

If two points of a line are contained in a plane, then every point of the line is contained in the plane.

Given three distinct noncollinear points, there is one and only one plane containing the three points.

The intersection of two distinct planes is either (1) a line, or (2) empty, in which case the planes are parallel.

The intersection of a line and a plane is either (1) empty, in which case the line is parallel to the plane, or (2) a single point, or (3) a line, in which case the line lies in the plane.

EXERCISES

The end-of-chapter exercises are designed to help you review concepts and principles discussed in the chapter, to extend these ideas, and to relate the ideas in the various chapters. As you answer each question, try to think in terms of explaining your answer to someone else—a student or instructor, perhaps.

The choices A, S, and N in some of the exercises correspond to Always, Sometimes, and Never true.

1. In a dictionary, look up the etymology (origin) of the word "geometry."

2. For each of the following concepts or principles, (a) name a physical model, and (b) make a sketch of the situation using representations of points, lines, line segments, and planes.
   (1) Point.
   (2) Line segment.
   (3) The intersection of two lines is empty but the lines are not parallel.
   (4) The intersection of two lines is a point.
   (5) Three planes whose intersection is a line.
   (6) Three planes whose intersection is empty such that no pair of the planes are parallel.

3. For each of the following physical models (a) name a geometrical abstraction of the model in terms of points, lines, line segments, and planes, and (b) make a sketch.
   (1) A pair of rays of sunlight each traveling in the direction opposite to that of the other.
   (2) All positions in the universe.
   (3) All positions on a floor.
   (4) Opposite edges of a piece of writing paper.
   (5) A revolving door.
   (6) The covers of a closed book.
   (7) The floors in a three-story house.
   (8) Three sides that form a corner of a box.

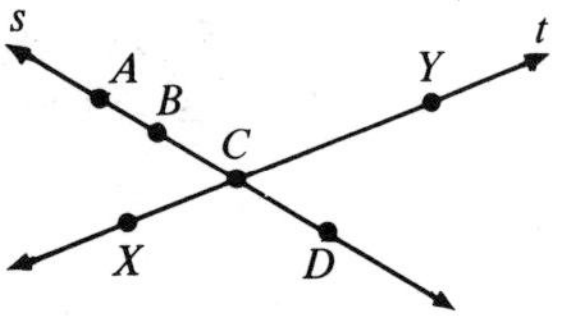

Figure 1-44

4. For each of the following statements encircle A, S, or N.
   (a) Any set of points that is collinear is also coplanar. A S N
   (b) If the intersection of two lines is empty, then the lines are parallel. A S N
   (c) The intersection of a line segment and a line is a line segment. A S N
   (d) A set of concurrent lines lies in one plane. A S N

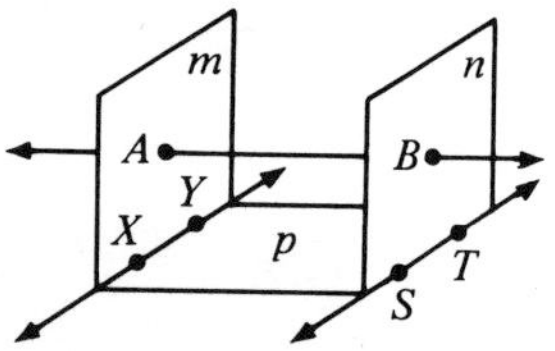

Figure 1-45

5. Using Fig. 1-44, find each of the following unions and intersections.
   (a) Line $s \cap$ line t.
   (b) $\overline{AB} \cap \overline{BC}$.
   (c) $\overline{AB} \cap \overleftrightarrow{CD}$.
   (d) $\overline{AB} \cap \overline{CD}$.
   (e) $\overline{AB} \cup \overline{BC}$.
   (f) $\overline{AC} \cap \overline{BD}$.
   (g) $\overline{XD} \cap \overline{AB}$.
   (h) $s \cup t$.
   (i) $\overleftrightarrow{AC} \cap \overleftrightarrow{BD}$.

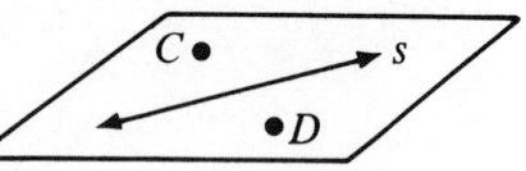

Figure 1-46

6. Using Fig. 1-45, find each of the following intersections.
   (a) Plane $m \cap$ plane p.
   (b) Line $AB \cap$ plane m.
   (c) $\overleftrightarrow{ST} \cap$ plane n.
   (d) If $\overleftrightarrow{XY}$ and $\overleftrightarrow{ST}$ are parallel, then $\overleftrightarrow{XY} \cap \overleftrightarrow{ST} =$ _____.

7. Points C and D and line s lie in a plane as in Fig. 1-46. Describe a path from C to D.

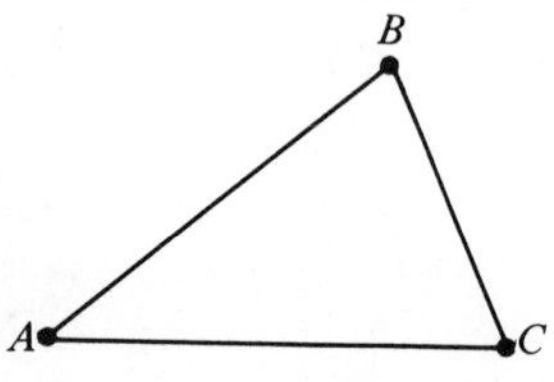

Figure 1-47

8. Describe Fig. 1-47 in terms of the union of geometric sets.

9. Using the notion of betweenness, describe the possible locations of a third glass G if the glass is to be in line with glasses E and F.

10. How might each of the following be interpreted using a dynamic viewpoint (that is, in terms of positions that are capable of moving)?

(a) Line. [*Hint:* Think of describing a path.]
(b) Plane. [*Hint:* Think of moving a rod across a table top.]
(c) Space. [*Hint:* Think of the rising floor of an elevator.]

11. Using a static viewpoint of geometry, we noted earlier that $\overleftrightarrow{AB} = \overleftrightarrow{BA}$ and $\overline{XY} = \overline{YX}$. Why might these agreements be inappropriate if we used a dynamic viewpoint of geometry?

12. (a) Complete the following table.

Number of points	*Maximum number of lines determined by these points*
2	1
3	3
4	6
5	10
6	15
n	

(b) Read Ref. 70, p. 247.

13. Complete the following table.

Number of points	*Maximum number of planes determined by these points*
3	1
4	4
5	10
6	20
n	

14. In Sec. 1-7 it was agreed that three noncollinear points determine a plane. When does each of the following determine a plane? Explain.
(a) The union of a line and a point not on the line.
(b) The union of two lines that intersect in a point.
(c) The union of two segments that intersect in a point.
(d) Two parallel lines.

15. The text indicates three different ways to name a plane, two different ways to name a line, and one way to name segments. Just as lines and planes are sometimes named with lowercase letters, the convention is sometimes used for naming line segments. Draw diagrams of a line, a plane, and a segment. Write names for each figure, giving an example of each of the various conventions available.

16. Make models from cardboard or stiff posterboard to illustrate each of the following:
(a) Two parallel planes.
(b) Three planes whose intersection is a line.
(c) Three planes whose intersection is a point.

17. In the References at the end of this book are relevant articles and books. Some extend material presented in the text; others give suggestions for teaching the content of geometry in the schools. From time to time we shall recommend that you read some of these. Each entry in the References has a number, which we will use as follows: "Ref. 12, p. 9" means that you should read page 9 of the article listed as 12 in the References. Our selection merely suggests the literature available; it is by no means exhaustive of the many rich sources.
   (a) For three views on the geometry appropriate for children in the elementary schools read Ref. 57, pp. 441–447; Ref. 58, pp. 460–467; and Ref. 84, pp. 453–459.
   (b) For two views on the geometry appropriate for students in the high schools read Ref. 90, pp. 226–238, and Ref. 96, pp. 466–471. (You may just want to skim these at this point and return to them after you have completed this book.)
   (c) For views on the nature of abstraction in mathematics and geometry read Ref. 3, pp. 4.1–4.6, and Ref. 54, pp. 1–6, 176–182.
   (d) For a research study on the concept of betweenness read Ref. 60, pp. 683–686.

CHAPTER 2

Building Blocks of Geometry

2-1 INTRODUCTION

Chapter 1 has reminded you of some geometric ideas already familiar to you. In this chapter you will encounter many of these ideas again, using them to extend your vocabulary and general understanding in your quest to discover the geometry that is applicable to the physical world.

Points, lines, planes, and space are basic units for our construction; indeed, the point has been used to build each of the others. Let us return for a moment to the point and line and develop other basic concepts.

2-2 LINE SEPARATION

You recall that if the intersection of two lines is not empty, this intersection consists of one point. Consider the points that comprise m in Fig. 2-1. Clearly, $m \cap n = \{P\}$. (See Appendix B.) We can now speak of those points in m on the A side of P, or those points in m on the B side of P. Point P **separates** m into three subsets—P itself and two others. Each of the

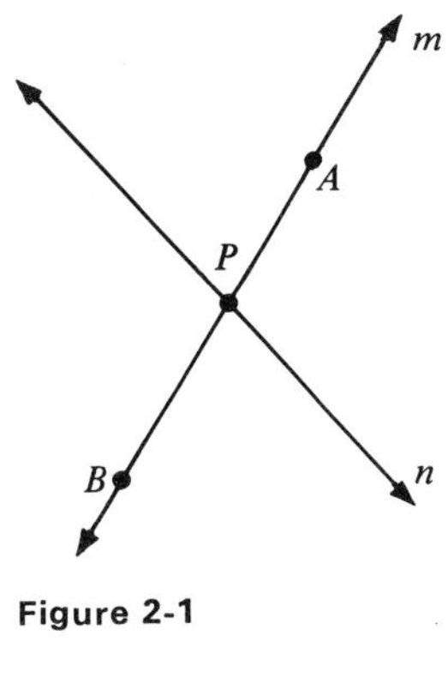

Figure 2-1

latter two we call a **half-line**. Therefore we say that any point in a line separates the line into three disjoint sets: the two half-lines and the point.

Now think of a piece of string cut with a pair of scissors. Clearly, the scissors separate the string into two pieces. What point of the string is the separating point that plays the role of P in Fig. 2-1? Surely it is not in the right-hand piece, nor is it in the left-hand piece. The **separator** of a line is contained in neither half-line, which should remind us that a point is merely a position and has no physical embodiment as does a dot.

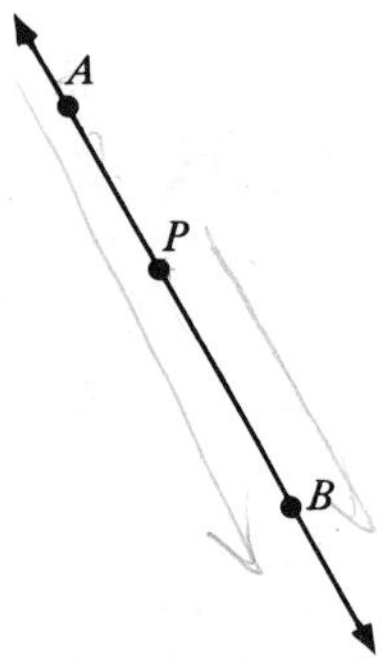

Figure 2-2

Thus we say that any point of a line determines three disjoint subsets of the given line: the separator and two half-lines. We also say that any point of a line separates the line into two half-lines.

In Sec. 2-7 we shall want to refer to a particular half-line and its separator. Hence in Fig. 2-2 we refer to the union of P and the points in the A side of P as **ray** PA. Notice that a ray consists of one point, which is the separator, and an endless set of other points. The point that is the separator is called the **endpoint** of the ray, the endless set of other points is the half-line. Thus in Fig. 2-2 we have rays PA and PB, each with the same endpoint, P. In ray PA the endless set of points is the half-line, the A side of P.

When we name a ray we name the endpoint first, and we symbolize this by drawing an arrow over this letter and a letter naming any other point of the ray. Notice that, as with the line, we use two points, but with the ray one of these is quite special, the endpoint. Thus it is much like a combination of the segment and line symbols.

The ray PA of Fig. 2-2 is symbolized $\overrightarrow{PA}$, and ray PB is written $\overrightarrow{PB}$. In Fig. 2-2 is there a ray AP? Is there such a geometric figure as $\overrightarrow{AB}$ in this figure? If so, what point is the endpoint? What about $\overrightarrow{PB}$? Is $\overrightarrow{AP}$ the same ray as $\overrightarrow{PA}$? It is generally agreed that the name of the endpoint is written on the left. We use this convention here. However, no ambiguity should arise if $\overrightarrow{PA}$ were symbolized $\overleftarrow{AP}$, the arrow direction indicating which point is the endpoint.

EXERCISE 2-2

1. In Fig. 2-2 find $\overleftrightarrow{AB}$, $\overrightarrow{AB}$, $\overline{AB}$, $\overleftrightarrow{PB}$, $\overrightarrow{AP}$, $\overrightarrow{PA}$, $\overline{PB}$, $\overrightarrow{BP}$, and $\overrightarrow{BA}$. Is P in the half-line that is the B side of $\overleftrightarrow{AB}$? A symbol that is not universal but occasionally convenient represents a half-line. Thus the B side of $\overleftrightarrow{AB} = \overset{\circ\rightarrow}{PB}$. How would the A side of $\overleftrightarrow{AB}$ be written?

2. In Fig. 2-2 what is $\overrightarrow{PA} \cup \overrightarrow{PB}$? $\overrightarrow{AB} \cup \overrightarrow{PB}$? $\overrightarrow{AB} \cap \overrightarrow{PB}$? $\overrightarrow{AB} \cap \overrightarrow{PA}$? $\overset{\circ\rightarrow}{PA} \cap \overset{\circ\rightarrow}{PB}$?

2-3 PLANE AND SPACE SEPARATION

In plane p of Fig. 2-3 how far does line m extend? Would it be clear to speak of points in p and on the R side of m? Would you be willing to say that m separates the plane p into two **half-planes**? Would you say that m lies in the R half of p? Would you say that m lies in the S side of p? Here we have the two-dimensional counterpart of the one-dimensional line separation. Think of cutting a flat piece of cardboard with a pair of scissors. What is the three-dimensional counterpart?

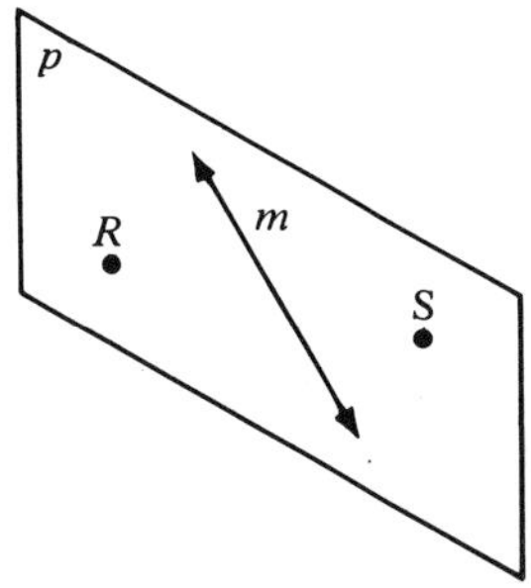

Figure 2-3

Suppose we think of space as represented by a rectangular room. If we build a partition across the room from one wall to an opposite wall, as in Fig. 2-4, would this represent a plane in space? Suppose you in one half-room represent a point in space on one side of the plane represented by the partition. If you are otherwise allowed to roam freely and take up any position in this half-room, what one limitation is being imposed on you in terms of the entire room? How do you decide whether you have selected a proper position? Could you move across or through the partition? (No mention was made of a door, arch or window.) We have, therefore, the room (space) separated by a partition (plane) into two half-rooms (**half-spaces**).

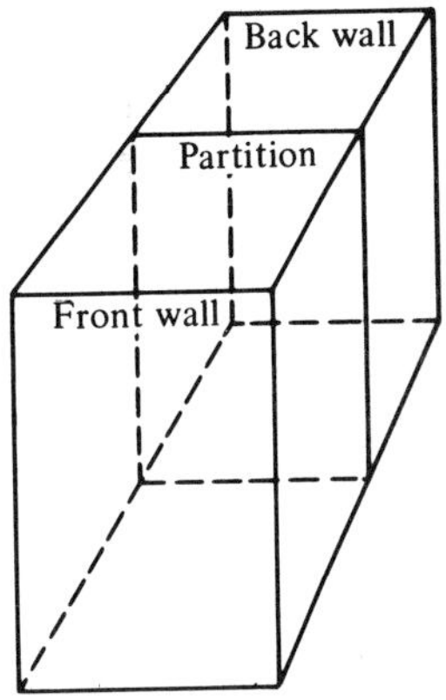

Figure 2-4

2-4 BOUNDARY, REGION

The previous analogy also alludes to another term frequently employed in geometry: **boundary**. A plane is considered to be the boundary of a half-space. Likewise the line m in plane p of Fig. 2-3 is spoken of as the boundary of either of the half-planes. What would be the boundary of a half-line? When used in this sense, boundary merely refers to the separator relative to separated sets; thus, the boundary contains no point of the set. However, this is not always the case (as with the ray), and the reader must proceed cautiously when boundary points are involved.

In this text we shall assume that the boundary does not contain any point of the set being bounded, unless we clearly indicate otherwise.[1] We have already followed this convention; that is, a point of separation is the boundary of a half-line, and we have agreed it is not in the half-line. However, a point of separation is surely the boundary of a ray and is a point of the ray. But we pointed out this fact, even giving it a name. What is the name of the boundary of a ray?

You are apt to encounter references to regions, usually modified by an adjective such as *plane*, *triangular*, or *closed*, but with no explanation of

[1] For a more formal discussion of boundary see Exercise 12-3, 9.

what region means. One possible interpretation of a region is that it is a subset of some previously mentioned set of points. In Fig. 2-3, for example, since the points on the S side of m in plane p and the points on the R side of m are each subsets of plane p, we might say that m separates plane p into two plane regions. Some might even say that plane p has been divided into three regions, meaning line m and the two half-planes. However, even though line m is a subset of plane p, we shall not speak of the line as being a plane region. We shall apply the term **region** only to subsets that are more in character with the superset. However, a continuous portion of a line could well be called a region of the line; a continuous planelike portion (not a curve) of a plane might be referred to as a region of the plane; a continuous spacelike portion (not a plane region or a curve) of space might be spoken of as a region of space. In order to avoid confusion, we shall speak of plane regions or space regions.

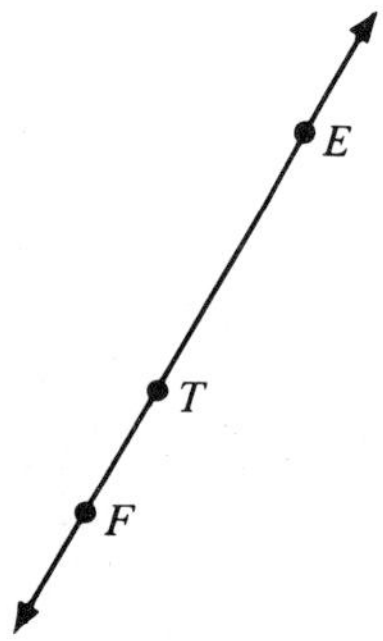

Figure 2-5

2-5 CARRIER

We have used the term collinear to indicate points that lie in the same line. Notice that all of the points, segments, and rays in Fig. 2-5 are subsets of $\overleftrightarrow{EF}$. A line that contains all the points of a segment, ray, or half-line is called the **carrier** of that segment, ray, or half-line. (See Ref. 41, p. 38.) We may also speak of the carrier of collinear points. If a set of points is collinear, any and all of the segments and rays determined by these points have the same carrier.

2-6 BETWEENNESS

Another property of collinear points is that of **betweenness.** If three points are collinear, one point lies **between** the other two. This statement then is more limiting than general usage. If three towns lie along a winding road, general usage would say that the second lies between the other two. In Chapter 12 you will find that in some geometries betweenness with respect to points along a path is not unlike that of towns along a road. In the present geometry, however, we reserve the term for points that are collinear.

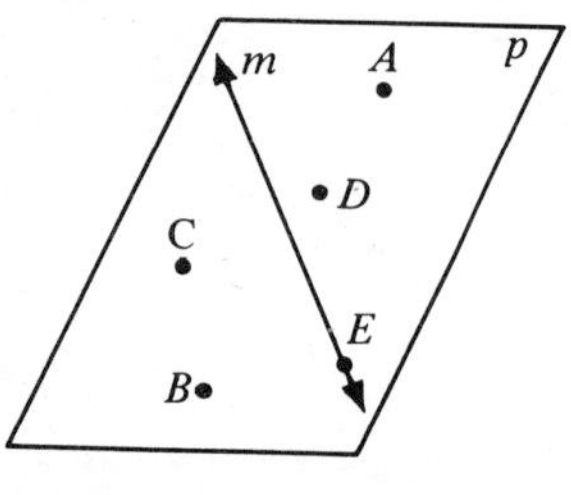

Figure 2-6

EXERCISE 2-6

1. Pictured in plane p in Fig. 2-6 are the points A, B, C, D, and E, and the line m. Point E is in m.
   (a) Is the half-plane that is the A side of m the same as the half-plane that is the D side of m?

(b) Is the A side of m the same as the C side of m?
(c) What can you say of $m \cap \overline{AD}$?
(d) What can you say of $m \cap \overline{BC}$?
(e) What can you say of $m \cap \overline{BD}$?
(f) Is m a separator of p?
(g) Now look at parts (a)–(f) and try to compose a statement that uses a line segment to test points with respect to the boundary of a set.
(h) Is $\overline{CD}$ a separator of p? Is $\overrightarrow{EA}$? Is E a separator of m?
(i) What is the boundary of $\overset{\circ\!\!\rightarrow}{ED}$? of $\overrightarrow{ED}$? of $\overleftrightarrow{ED}$?
(j) Describe the regions of p formed by m.
(k) Is E in the A side of m or the B side of m?
(l) As pictured, is C between B and D?

2. Draw a sketch to illustrate $\overleftrightarrow{AB} \cap \overrightarrow{PQ} = \{P\}$.

3. (a) From our discussion of carrier might we say that a plane is a carrier of coplanar lines? Coplanar points?
(b) What is the carrier of A, B, C, D, and E in Fig. 2-6? What is the carrier of m and D?
(c) Name some other geometric objects that might be considered carriers of lines or line segments. They need not be among those we have discussed.

4. If point K is between points P and Q, then decide whether the following are always, sometimes, or never true. (Encircle A, S, or N.)
(a) $\overrightarrow{KP}$ and $\overrightarrow{KQ}$ have the same carrier. A S N
(b) $\overrightarrow{QK}$ and $\overrightarrow{KQ}$ name the same ray. A S N
(c) Every point of $\overrightarrow{QK}$ is also a point of $\overrightarrow{QP}$. A S N
(d) Point P is between points K and Q. A S N
(e) Q is a separator of $\overleftrightarrow{KP}$. A S N

5. In Fig. 2-7, points D, E, and F lie in $\overleftrightarrow{DF}$.
(a) Describe a **path** of a point traveling from D to F. Does this statement allude to the static or dynamic point of view? Does more than one path satisfy this statement? How many? Is one path shorter than all the others? (Use your concept of shorter to answer this, since we have not yet told you ours.) If you take the path that you consider the shortest, is there anything special you might say about it? If so, what?
(b) Using the concepts of betweenness and carrier, make a statement about this shortest path.

6. Use the concept of carrier to describe parallelism for line segments, rays, lines, and combinations thereof.

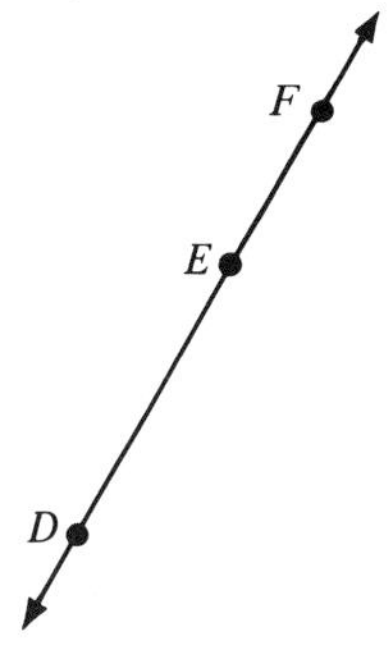

Figure 2-7

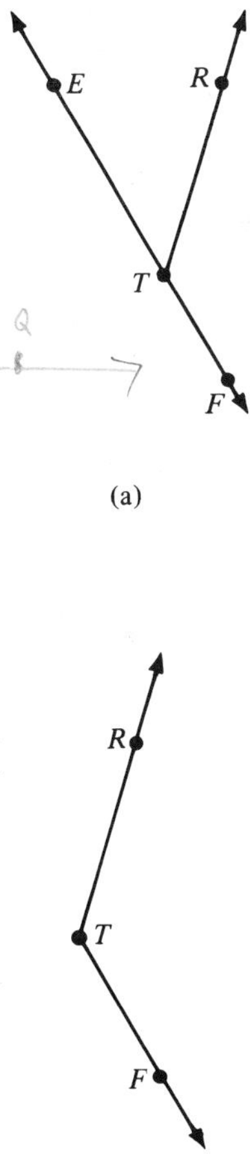

(a)
(b)
Figure 2-8

2-7 ANGLE

In Fig. 2-8(a) we have $\overleftrightarrow{EF}$, with separator T. We also have $\overrightarrow{TR}$. Therefore, we have three rays, all with the common endpoint T. What is the carrier of $\overrightarrow{TF}$? of $\overrightarrow{TE}$? of $\overrightarrow{TR}$? The figure comprised of $\overrightarrow{TR}$ and $\overrightarrow{TF}$ is the basis for another fundamental concept, that of angle.

Recall that the union of two sets is a set composed of all points that are members of either set (or both). Using this, we describe an **angle** as the union of two rays with a common endpoint. This common endpoint is called the **vertex** (plural: vertices, or vertexes), and the rays are called **sides.** Thus in Fig. 2-8(b) the angle formed by rays *TR* and *TF* has *T* as a vertex, and $\overrightarrow{TR}$ and $\overrightarrow{TF}$ as sides. It is named angle *RTF*. Notice that the middle letter names the vertex. Now we must decide in more detail what an angle is and is not. Figure 2-9 illustrates some physical models of angles. See how many you can find.

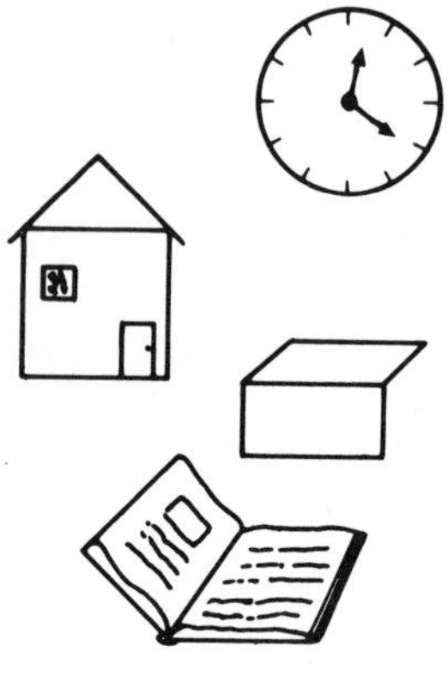

Figure 2-9

Authorities differ on whether the sides of an angle can have the same carrier. Notice that in the previous description of an angle nothing was said about the carrier or carriers of the rays. Figures called "straight angles" are sometimes permitted. Several groups require that the rays have different carriers, or in the words of one text, "[are] not both on the same straight line." (Ref. 44, p. 18.) They deny the existence of a straight angle. Proponents of this limitation feel that it allows simplicity, serves useful purposes, and eliminates some difficulties inherent in other definitions of angle—namely, the possible confusion between the concepts of angle and line if the rays have the same carrier. From a static and introductory point of view, these arguments may be persuasive. However, from a dynamic viewpoint, such a description is restrictive, perhaps too much so.

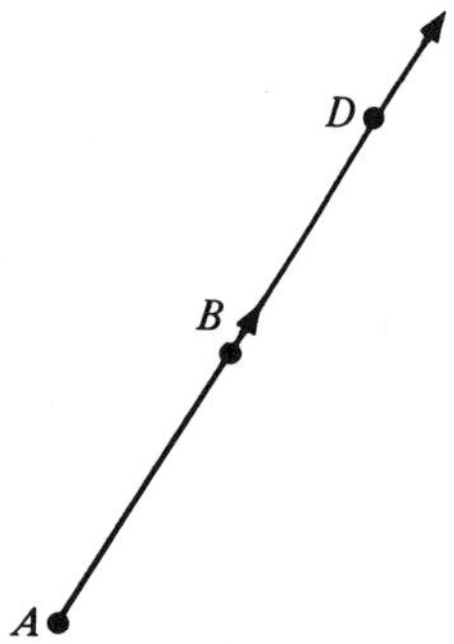

Figure 2-10

From the static viewpoint an angle is composed of two rays that are stationary. If we think dynamically, the rays are also fixed for a given angle, but we think of them as having been rotated into position. That is, one of the rays is fixed, while the other is rotated from it. In Figs. 2-10 and 2-11, $\overrightarrow{AB}$ and $\overrightarrow{AD}$ are shown in the "before" and "after" positions. Ray *AD* has been rotated counterclockwise from the position in Fig. 2-10 to that suggested in Fig. 2-11 to form the angle *DAB*. When we think dynamically,

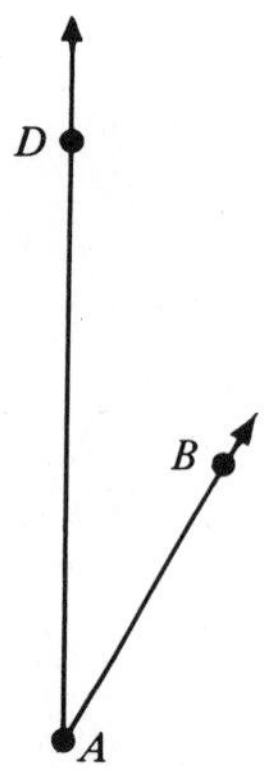

Figure 2-11

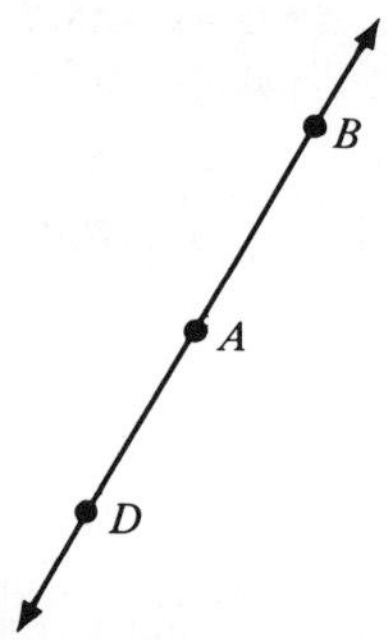

Figure 2-12

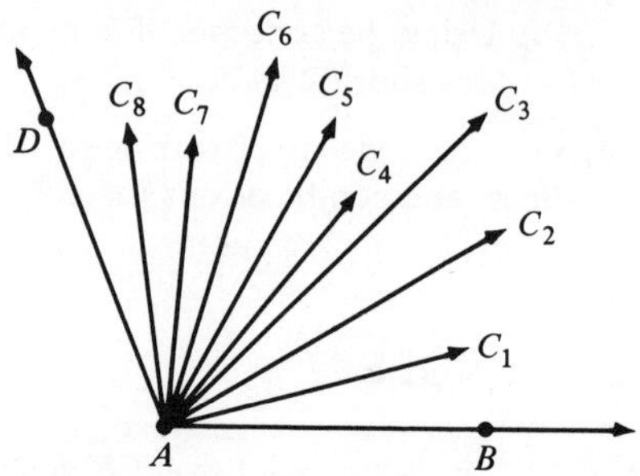

Figure 2-13

the fixed ray is called the **initial side**, the other ray the **terminal side**. Thus in Figs. 2-10 through 2-13 we have angles *DAB* with vertices at *A*, initial sides $\overrightarrow{AB}$, and terminal sides $\overrightarrow{AD}$. Notice that within this concept of angle (dynamic), each of the four figures is a good example of an angle. However, under the above mentioned limitation with respect to different carriers, Figs. 2-10 and 2-12 would not be angles. Why? These are "limiting situations" and cause some trouble later on. We prefer to include them as angles.

Another question is raised (Ref. 114, p. 18) concerning whether the angle is comprised of two rays, as in the dynamic and static concepts above, or of a set of rays that would include a region of the plane between the initial and terminal rays of the dynamic viewpoint. (We are using the term "between" here in the more general, nonmathematical sense.) Dynamically there is a good case for this viewpoint. The angle would include all of those points on the moving ray in any of its positions between the initial and terminal positions. In Fig. 2-13 all of the points of $\overrightarrow{AC_1}$, $\overrightarrow{AC_2}$, $\overrightarrow{AC_3}$, and so on, would be points between rays *AB* and *AD* as well as all the points of the rays between these rays. And all would then be points of angle *DAB*. There are other reasons why this view might be desirable, but we feel that an angle is best described as the union of two rays with a common endpoint. If we wish to speak of the points between the rays, we prefer to speak of points in the interior.

2-8 PLANE SEPARATION—ANGLE

Figure 2-14 illustrates $\overrightarrow{AB}$ and its carrier $\overleftrightarrow{AB}$ with its *D* side scored. Figure 2-15 depicts $\overrightarrow{AD}$ and its carrier $\overleftrightarrow{AD}$ with its *B* side scored. When these two figures are superimposed, as in Fig. 2-16, the crosshatched region clearly pictures that portion of the plane which we refer to as the **interior** *of the angle*, the intersection of the two half-planes. Since *A* is in neither half-plane, is the vertex of an angle in the interior of the angle? Figure 2-16 pictures the interior of angle *DAB* as the crosshatched region.

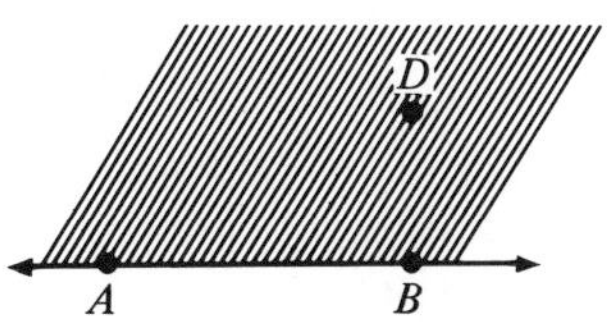

Figure 2-14

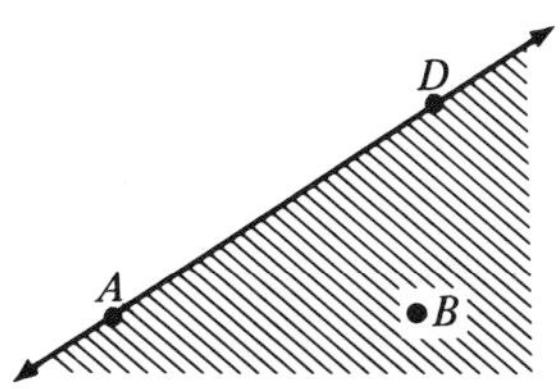

Figure 2-15

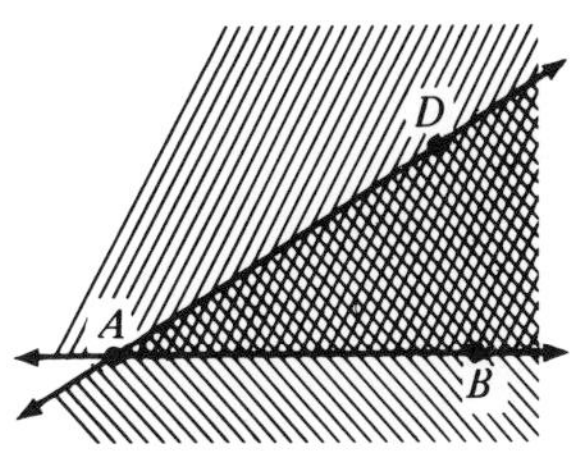

Figure 2-16

In Fig. 2-17 you see the interior of a different angle indicated by the cross-hatching. We form this angle using the same lines, $\overleftrightarrow{AB}$ and $\overleftrightarrow{DE}$, but we make use of $\overrightarrow{AB}$ and $\overrightarrow{AE}$. Hence, Fig. 2-17 pictures the interior of angle *BAE*. The interior of angle *BAE* is the intersection of which two half-planes?

Recall what we have said about a half-plane, then consider the question: Does an angle separate a plane? The answer is a conditional yes.

Consider the comments in Sec. 1-7 related to the conditions that determine a plane. Selecting one side and a point on the other side, other than the vertex, we then have a point and a line not containing this point; therefore, we have conditions that determine a plane. How long are the sides of an angle? How long is a ray? Then an angle is composed of two rays that, as indicated in Fig. 2-18, would separate the plane. Suppose, however, angle *RPS* appears as in Fig. 2-19 or 2-20? Perhaps these situations illustrate why some of the texts mentioned earlier do not call these angles. How can we reconcile Fig. 2-20 with the statement that an angle separates a plane? There would seem to be several possible choices. Let us discuss two of these.

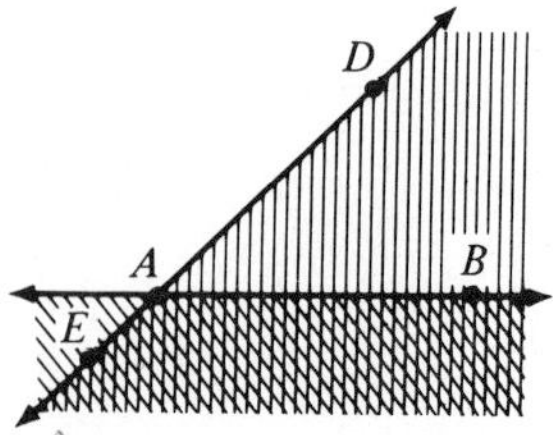

Figure 2-17

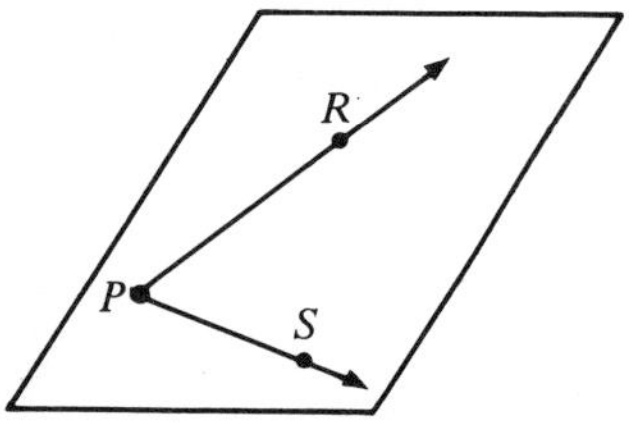

Figure 2-18

(a) We could select the statement that would not allow the angle in Fig. 2-20, or we could say that in this case the interior is the null set. And, in this case we would not say that the plane has been separated, since we do not have the two regions, one on each side of the boundary. What boundary?

(b) We might claim that those angles formed by rays that do not have the same carriers separate a plane. This statement would be correct, but it would not include the angle in Fig. 2-19. How do we reconcile this? Notice that the other description of an angle that excludes lines and rays as angles would contain this same error of incompleteness. How does all of this fit into our present total picture of separation?

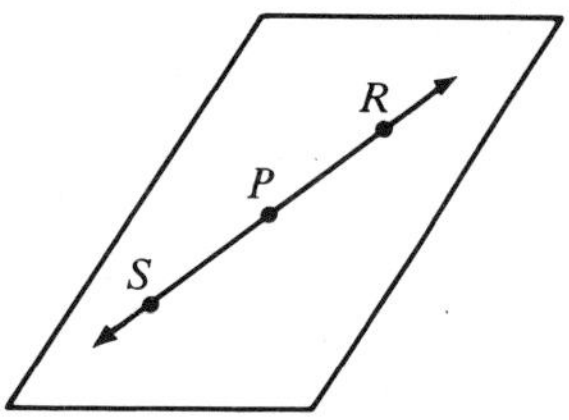

Figure 2-19

Suppose we say that an angle composed of rays with different carriers separates a plane. Then Figs. 2-19 and 2-20 would be disallowed. We feel that Fig. 2-20 does not illustrate separation, so we find this agreeable. But the situation in Fig. 2-19 would not be permitted, and we feel that it should be allowed as a case of separation. Notice in Fig. 2-19 that the carrier of rays *PR* and *PS* is $\overleftrightarrow{SR}$, and we have said that a line separates a plane. Therefore, although this case is not allowed under angle separation, our concept of line separation in a plane would include it. Perhaps this discussion helps you see how authorities can disagree concerning certain developments.

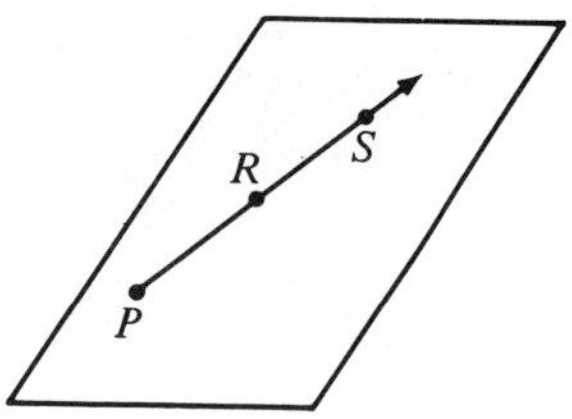

Figure 2-20

We now agree that (1) an angle is the union of two rays with a common endpoint, and (2) an angle whose sides have different carriers determines two regions, which we refer to as the interior and exterior of the angle. Thus, the **exterior** *of an angle* is the region of the plane that contains neither the angle nor interior points of the angle.

EXERCISE 2-8

1. List two reasons why the one-carrier limitation in the angle definition is considered too restrictive in this text.
2. List two reasons why the one-carrier limitation is desirable.
3. Locate four or more geometry texts (perhaps high school level), two that were written before 1958 and two or more that have been written since 1966. Compare their definitions and discussions of angle. See also Ref. 118, pp. 856–857.

2-9 NAMING AN ANGLE

Let us return to the problem of naming an angle. When there is no ambiguity, an angle can be named with a single letter, either upper or lowercase, preceded by the angle symbol: $\angle$. In the angles of Fig. 2-21 no confusion arises when we refer to $\angle a$, $\angle B$, $\angle c$, $\angle d$, $\angle e$, or $\angle f$. But suppose we refer to $\angle g$ or $\angle R$? Are you sure which angle is meant? Perhaps there would be more general agreement concerning $\angle g$ than $\angle R$. Our point is that there is a method that is free from ambiguity. An angle can be named with the angle symbol and three capital letters. The middle letter names the vertex; each of the other letters names a point, different from the vertex, on each of the sides.

Thus in Fig. 2-21 $\angle SRT$ and $\angle SRV$ are clearly distinguished, as is $\angle VRT$. Incidentally, general agreement would be that $\angle SRT = \angle g$, and we shall follow this convention. There would be some agreement that $\angle R = \angle SRV$, but we shall avoid this usage since it is not universal.

How about $\angle TRV$ and $\angle VRT$? Situations do arise in which it is desirable to distinguish between $\angle TRV$ and $\angle VRT$. For instance, in a dynamic appraisal, it may be desirable to distinguish between clockwise and counterclockwise rotation. Should the reader be interested in this, he may consult almost any high school geometry or trigonometry text. Look under positive, negative, and general angles. In this introductory treatment, however, we shall agree that $\angle VRT = \angle TRV$.

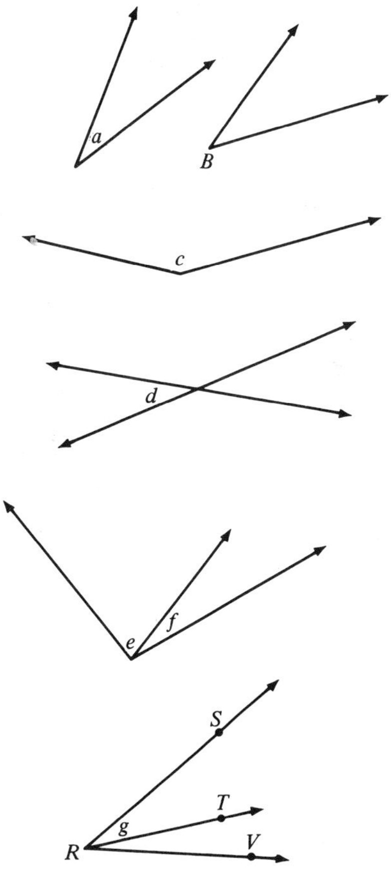

Figure 2-21

Another advantage of this notation is that it enables us to quickly distinguish the interior of an angle. Angle ABC in Fig. 2-22 has its interior shaded, its exterior not shaded. At this point you may wish to refer to Figs. 2-14 through 2-17, where we first discussed interior of an angle. Angle *ABC* in Fig. 2-22, and the symbol "$\angle ABC$" in general, tells us that the rays (sides) are $\overrightarrow{BA}$ and $\overrightarrow{BC}$. This in turn tells us that the interior consists of the intersection of two half-planes: the half-plane that is the *C* side of $\overleftrightarrow{AB}$ and the half-plane that is the *A* side of $\overleftrightarrow{BC}$.

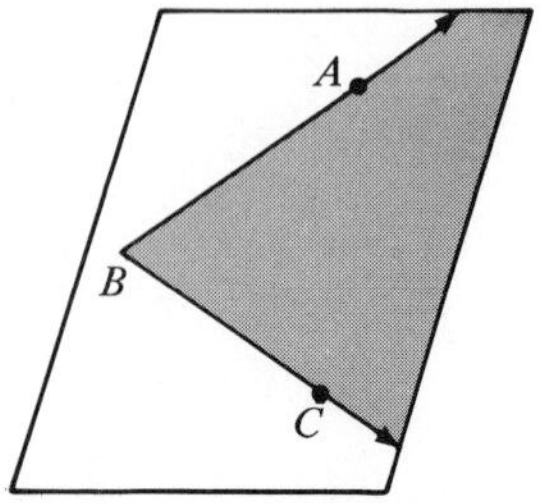

Figure 2-22

2-10 ADJACENT ANGLES

There are, of course, many kinds of angles, but for now let us limit ourselves to the discussion of only two kinds of plane angles, adjacent and vertical. We shall then look briefly at the three-dimensional counterpart.

Adjacent angles are those with a common ray and with no common interior points. In Fig. 2-23 do you think $\angle RSQ$ and $\angle QST$ are adjacent? How do you decide? Do they have a common ray? Shade the interior of angle *RSQ*. Without shading, check the interior of $\angle QST$. Are there any common interior points? If your answer to this last question was yes, then you had better check again, because $\angle RSQ$ is adjacent to $\angle QST$—and this means they have no interior points in common. Now check $\angle RSQ$ and $\angle RST$ using the same technique. What ray have they in common? How about their interiors? Since all of the interior of $\angle RSQ$ lies in the interior of $\angle RST$ (and it takes only one common interior point to disallow the adjacent concept), these are not a pair of adjacent angles.

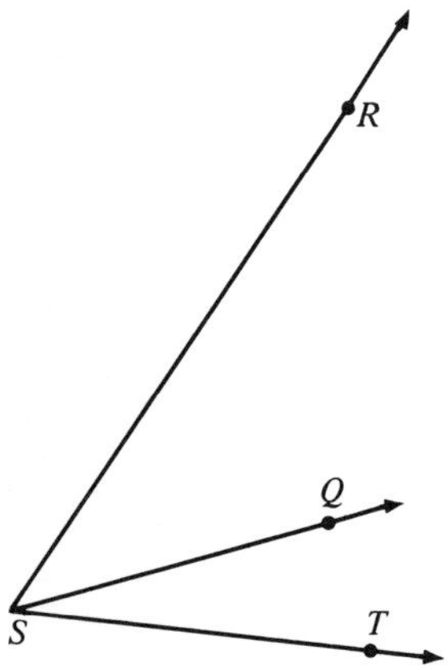

Figure 2-23

The use of the word "pair" in the previous sentence raises an interesting question: How many angles can we speak of as being adjacent? Our description did not specify the number of angles to which we can refer. Can we have one adjacent angle? Two? Three? More than three? Rereading our description and looking at the adjacent angles in Fig. 2-23, we see that when we speak of adjacent angles we are limited to two at a time.

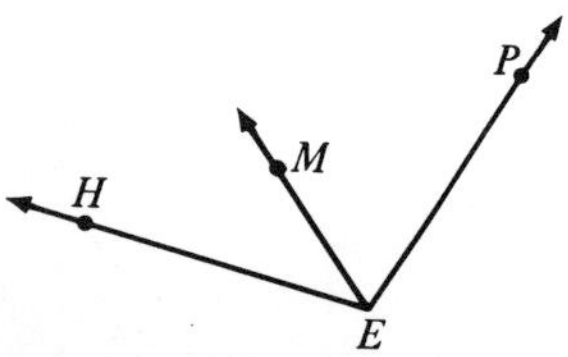

Figure 2-24

In Sec. 13-3 we list as a characteristic of a definition the fact that the statement contain no unnecessary information. It is not our purpose here to give formal definitions; nonetheless, situations such as the one above should stimulate you to think things through. In this case you should be able to decide that you can only consider two angles at a time.

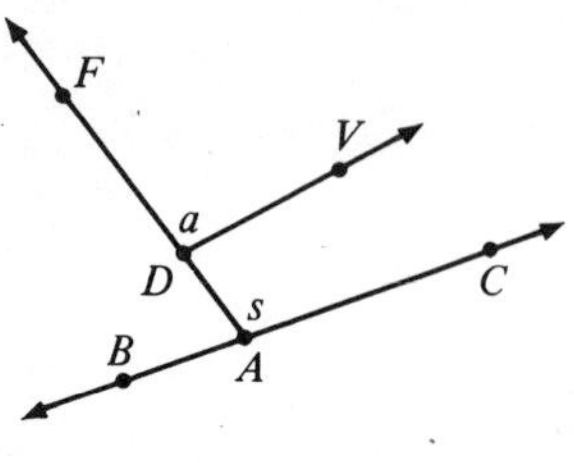

Figure 2-25

EXERCISE 2-10

1. Name the adjacent angles in Fig. 2-24.
2. In Fig. 2-25:
   (a) Are angles *FDV* and *FAC* adjacent?
   (b) Do they have a ray in common? If so, name it.
   (c) Do they have any interior points in common?

3. If there are any pairs of adjacent angles in Fig. 2-25 or 2-26, name them.
4. In Fig. 2-25 name $\angle FDV$ in another unambiguous way.
5. Should we refer to $\angle NOG$ as $\angle t$? Name $\angle t$ using a different convention. Which angle is $\angle O$?
6. In Fig. 2-27:
   (a) Name the sides of $\angle APD$. Of $\angle BPC$.
   (b) Name the vertex of $\angle PBC$. Of $\angle BPC$.
   (c) True or false: The sides of $\angle APD$ are not as long as the sides of $\angle BPC$.
   (d) Shade the interior of $\angle APC$ in pencil.
   (e) Shade the interior of $\angle CPD$ in some other conspicuous manner, perhaps ink or crayon.
   (f) What can you say of {interior $\angle APC$} $\cap$ {interior $\angle CPD$}? Are the angles adjacent?
   (g) Suppose $\angle APC$ is considered to be formed (generated) by rotating $\overrightarrow{PC}$ *counterclockwise* from $\overrightarrow{PA}$. What do you think should be called the interior of this $\angle APC$? Does our convention about naming angles and the resulting interior yield this? What, then, should we conclude with respect to "our" concept of angle interior? Would this $\angle APC$ be adjacent to $\angle CPD$? Incidentally, if we wished to refer to this angle, we would speak of *reflex angle APC*.
7. Draw a sketch to illustrate $\angle RST$ and $\angle USR$ such that $\angle UST$ and $\angle TSR$ are adjacent.
8. In Fig. 2-28 name an angle adjacent to $\angle ARB$. Name an angle not adjacent to $\angle BRC$.

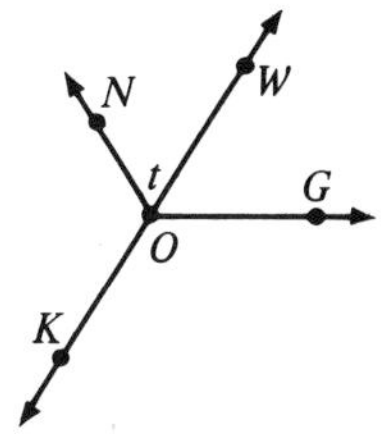

Figure 2-26

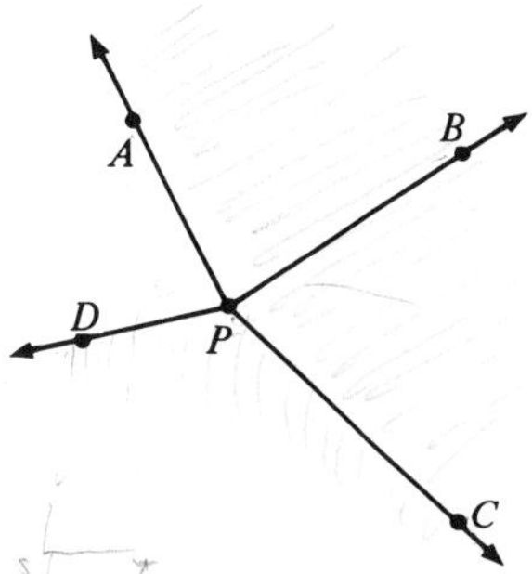

Figure 2-27

2-11 VERTICAL ANGLES

We have noted that although the description of adjacent angles did not specify that they must be a pair, it turns out that this must be the case. This situation resembles that for those angles called vertical angles. Vertical angles are actually a pair of angles, but we need not say so in the description that follows.

Vertical angles are the nonadjacent angles formed at the intersection of two lines. (Reminder: We are not allowing reflex angles unless we specify that we are.) Do any of the angles represented in Figs. 2-23 through 2-26 satisfy the description of vertical angles? Which ones? Do $\angle RPS$ and $\angle VPT$ in Fig. 2-29? Do $\angle ARD$ and $\angle BRC$ in Fig. 2-28? Do $\angle RPV$ and $\angle SPT$ in Fig. 2-29? Notice that for the three pairs of angles just named, only the pair $\angle ARD$ and $\angle BRC$ fails to satisfy the description. The angles in each of the other two pairs are called vertical angles. In the case of the vertical angles $\angle RPV$ and $\angle SPT$, the two intersecting lines are $\overleftrightarrow{RT}$ and $\overleftrightarrow{SV}$. Only four angles are formed (excluding the straight

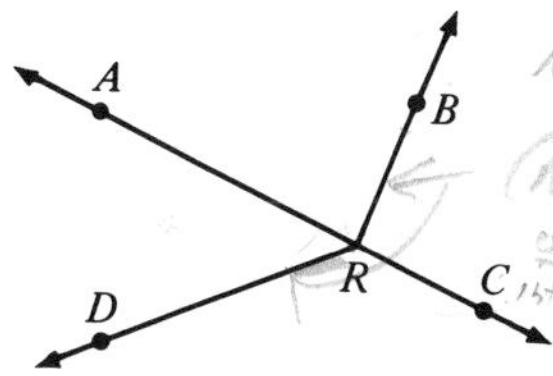

Figure 2-28

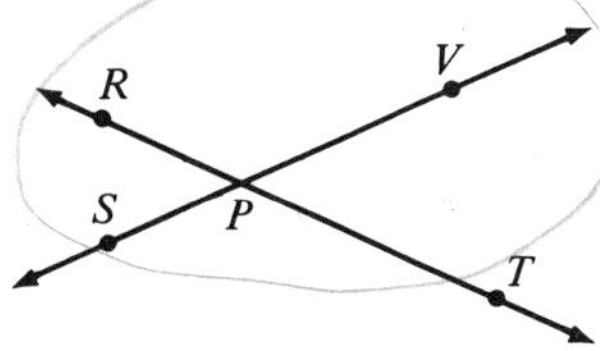

Figure 2-29

angles). What is the only angle of the four that is not adjacent to $\angle RPV$?

2-12 SPACE ANGLES

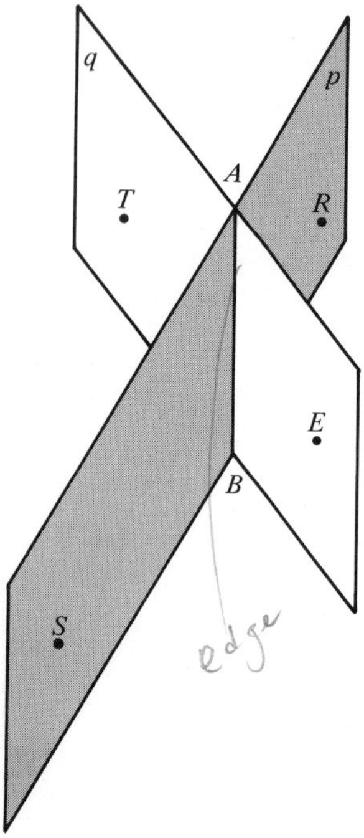

Figure 2-30

We mentioned earlier that an angle is a plane figure. What do you suppose is the three-dimensional counterpart of our two-dimensional plane angle? If two lines intersect, they form plane angles. What happens if two planes intersect? We know from Sec. 1-7 that their intersection is a straight line.

Suppose we consider two half-planes with a common separator. In Figs. 2-30 and 2-31, planes p and q have $\overleftrightarrow{AB}$ in common. Find the half-plane that is the E side of q in Fig. 2-30. Find the half-plane that is the S side of p. The union of these two half-planes and their common boundary ($\overleftrightarrow{AB}$) is called a **dihedral angle**. The line is called the **edge**, and the half-planes with the edge are called **faces**. In Fig. 2-30 the angle whose edge is $\overleftrightarrow{AB}$, one of whose faces contains point E and the other of which contains point S, is named dihedral angle $E\text{-}AB\text{-}S$ and symbolized: $\angle_{\text{dih}}\, E\text{-}AB\text{-}S$. The hyphens, which are not read, indicate two things: that the reader should pause, and that the letters between pauses name the edge. Thus the two middle letters name the edge. Name the dihedral angle in Fig. 2-31. Name another dihedral angle in Fig. 2-30.

The description of adjacent and vertical dihedral angles is analogous to that for adjacent and vertical angles in a plane. Name the adjacent dihedral angles and the vertical dihedral angles suggested by Fig. 2-30.

Figure 2-31

Other types of space angles can arise; at this point we will merely mention and illustrate one more. A **polyhedral angle** is the union of regions of distinct planes that are concurrent. In each of Figs. 2-32 and 2-33, the union of the angular regions suggests the polyhedral angle $P\text{-}ABC$. The point of concurrence is P and is called the **vertex** of the polyhedral angle. A good model of this type of space angle is illustrated in Fig. 2-32. In the classroom this would be suggested by the front wall, right-hand wall, and floor, meeting at the lower right-hand front corner of the room. This corner corresponds to the vertex P of the polyhedral angle in Fig. 2-33. The corner angles in each of the two walls and floor correspond to the plane angles APB, BPC, and CPA. Each of the plane regions and its boundary is called a **face**. The intersections of the faces are called **edges**. Are there any dihedral angles suggested by either figure? If so, name some.

Polyhedral angles are classified according to the number of faces that form them:

Number of faces	*Name*
3	trihedral angle
4	tetrahedral angle
5	pentahedral angle
6	hexahedral angle
8	octahedral angle

A different development of polyhedral angles is given in Sec. 4-6.

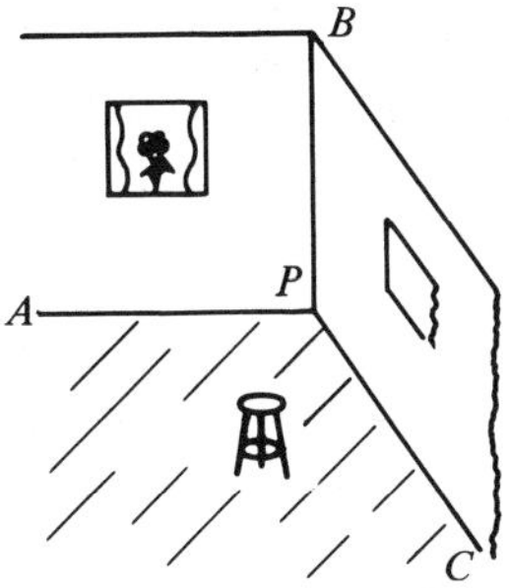

Figure 2-32

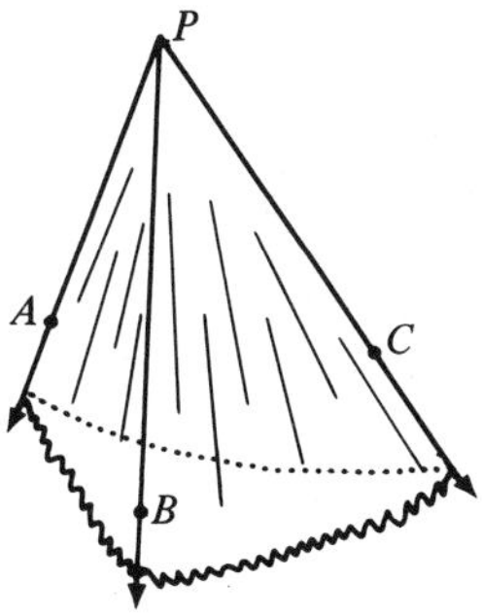

Figure 2-33

SUMMARY OF PRINCIPLES

Any point in a line separates the line and determines three disjoint subsets: the point and two half-lines.

Any line in a plane separates the plane and determines three disjoint subsets: the line and two half-planes.

Any plane separates space and determines three disjoint subsets: the plane and two half-spaces.

The boundary of a set contains no points of the set unless stated otherwise.

If a set of points is collinear, all the segments and rays determined by these points have the same carrier.

Any angle whose sides have different carriers separates a plane and determines three disjoint subsets: the angle, a plane region called the interior, and a plane region called the exterior.

EXERCISES

Encircle A, S, or N, where appropriate, to indicate always, sometimes, or never true.

1. The period at the end of this sentence is a point. A S N
2. The empty set may be symbolized {0}. A S N
3. You cannot see a line. A S N
4. The points of a line are collinear. A S N
5. If two lines do not intersect, then they are parallel. A S N
6. If two lines intersect, then they are not coplanar. A S N
7. Three points are coplanar. A S N
8. $\overleftrightarrow{GH} = \overleftrightarrow{HI}$. A S N
9. Given $\overleftrightarrow{RTS}$. Then the intersection of the two half-lines, one on the R side of T, the other on the S side of T—that is, $\overset{\circ\!\rightarrow}{TR} \cap \overset{\circ\!\rightarrow}{TS}$—is $\{T\}$. A S N
10. Suppose A is between B and C. Then:
    (a) $\overrightarrow{BA}$ and $\overrightarrow{CA}$ name the same ray. A S N
    (b) Every point of $\overrightarrow{CB}$ is a point of $\overrightarrow{AB}$. A S N
    (c) $\overrightarrow{BA} \cap \overrightarrow{AC} =$ _____.
11. $\overrightarrow{GH} = \overrightarrow{GI}$. A S N
12. Pencils placed flat on a desk would be a model for (a) lines in space, (b) concurrent lines, (c) coplanar lines.
13. Given two distinct lines, m and n, and plane p such that $m \in p$, $n \notin p$, and $m \cap n = \varnothing$. Then (a) $n \cap p = m$, (b) $m \cap p = \varnothing$, (c) $n \cap p = \varnothing$, (d) none of these necessarily follows.
14. If three lines are concurrent, they have how many points in common? (a) 1, (b) 2, (c) 3, (d) 0.
15. $\overline{MN}$ has how many endpoints?
16. $\overrightarrow{MN}$ has how many endpoints?
17. $\overleftrightarrow{MN}$ has how many endpoints?
18. In Fig. 2-34:
    (a) $\overleftrightarrow{RS} \cap m =$ _____.
    (b) $\overleftrightarrow{EF} \cap \overleftrightarrow{RS} =$ _____.
    (c) $m \cap p =$ _____.
    (d) $\overrightarrow{FE} \cap \overrightarrow{AB} =$ _____.
    (e) $\overrightarrow{RS} \cap p =$ _____.
    (f) $\angle R\text{-}AB\text{-}E$ is called _____.

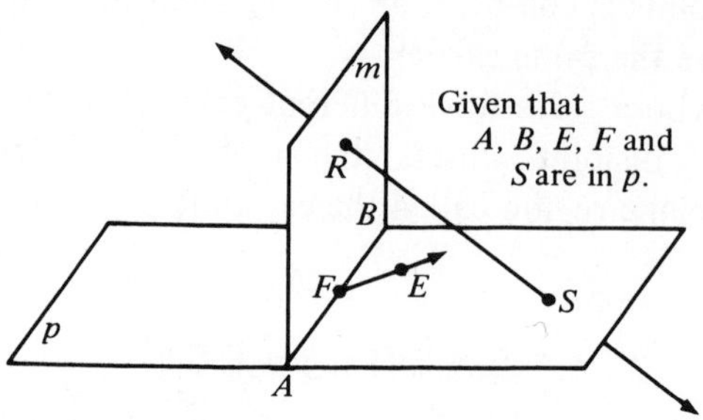

Figure 2-34

19. Draw a sketch to illustrate:

(a) $\overleftrightarrow{AB} \cap \overleftrightarrow{RS} = \{C\}$.

(b) Two intersecting planes with lines, one in each plane, not intersecting the other plane.

20. In Fig. 2-35:

(a) $\angle VRB \cap \overleftrightarrow{AB} =$ _____. (b) $\angle ASB \cap \overrightarrow{RT} =$ _____.

(c) $\overline{TV} \cap \overline{AB} =$ _____. (d) $\overrightarrow{TR} \cap \overrightarrow{BV} =$ _____.

(e) $\overrightarrow{TA} \cup \overrightarrow{TV} =$ _____. (f) $\angle RTS \cup \angle RTA =$ _____.

(g) $\overrightarrow{RV} \cap \overline{TV} =$ _____. (h) Name a pair of adjacent angles.

(i) Name a pair of vertical angles.

(j) Are angles RAU and TAB vertical angles? Explain.

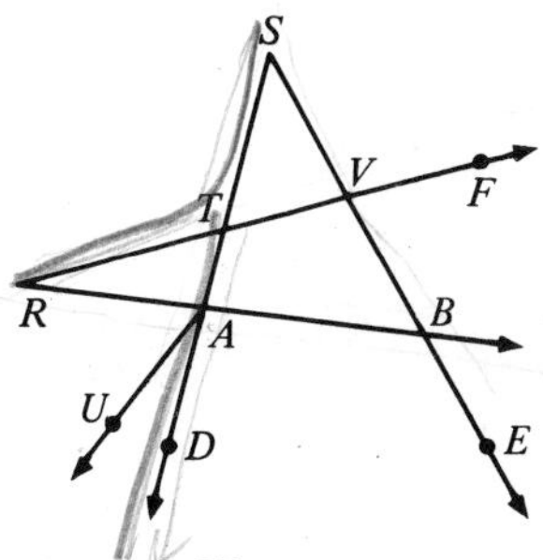

Figure 2-35

21. Think carefully now. Can the union of two half-planes be a plane? Explain, and be prepared to illustrate, perhaps with a model.

22. Draw a picture of a line. Label four points of the line, *R*, *S*, *T*, and *W*, in that order from left to right. Then:

(a) Name two rays whose union is the line.

(b) Name two rays whose union is not the line but contains *R*, *S*, and *T*,

(c) Name two rays whose union does not contain *R*.

(d) Name two rays whose intersection is $\{S\}$.

(e) Name two rays whose intersection is empty.

(f) Select a point *P* between *S* and *T*. Now name a pair of rays whose union does not contain *P* and whose intersection is empty.

23. Explain each of the following, using principles, concepts, or relationships discussed in Chapters 1 and 2.

(a) A gun sight consists of two guides rather than three.

(b) A three-legged table is less apt to wobble than a four-legged table.

(c) A sidewalk superintendent will "shuffle" a board along the frame of a newly poured cement sidewalk.

(d) A line segment has only one name, but a line and a ray may have many names.

24. Consider how you might explain the relationships between each of the following pairs, using a dynamic viewpoint: point—line, line—plane, plane—space.

25. Read Ref. 46, pp. 11–20, and Ref. 44, pp. 1–18.

CHAPTER 3

Building in a Plane: Curves, Polygons

3-1 INTRODUCTION

In previous chapters you were led from familiar ideas to the understanding that particular sets of points comprise a line, particular sets of lines comprise a plane, and space may be thought of as the union of a set of planes. These fundamental ideas were used to develop building blocks, which we shall continue to use in this intuitive approach to build further concepts. In this chapter, however, we shall limit ourselves to space figures that are contained in a plane in space. At times we shall remove this restriction, but at all other times it must be assumed we are working in a plane.

3-2 GEOMETRIC CURVES

Several sets of points are sketched in Fig. 3-1. Each is referred to as a geometric **curve** or **path.** Notice, first, that all lie in the plane suggested by this page. Think of a piece of string dangling from a tack in a side wall, with some of the string lying on the floor. Or think of a piece of wire wrapped around a box, so that it crosses each side of the box. Would these two situations be models of sets of points that would be called curves? There really is no reason why they could not be called curves; however, we have just agreed to exclude such curves from our discussion.

Notice two other things about the curves in Fig. 3-1. The single point, *P*, is a curve—a somewhat trivial case, but nonetheless it is a set of point(s) in a plane. It may seem strange to apply the term "curve" to the segment, ray, and line in Fig. 3-1, but this follows general usage in other branches of mathematics.

Thus we might be led to describe a curve in geometry as any set of points in a plane. This description, however, would allow compound curves, as in Fig. 3-2. Although some persons would include such sets of points as curves, we prefer to require that the points be "joined" to form what some call a continuous set of points.

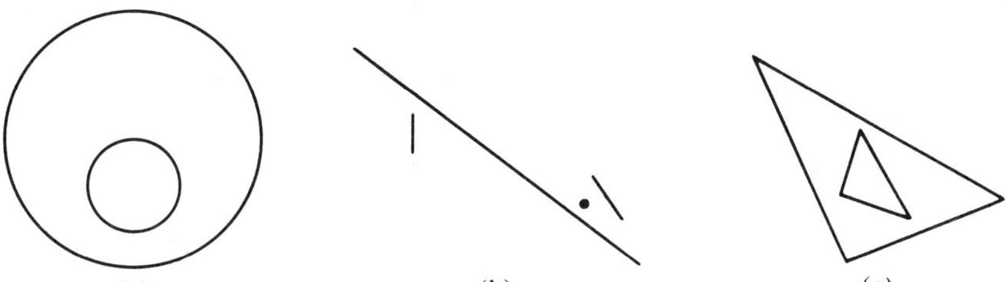

Figure 3-2

To decide whether a given set of points is a curve (to us), you must decide:

(1) Are all points coplanar?

(2) Could I trace out the curve with a pencil and reach all points without lifting the pencil off the page?

If your answer is yes to both questions, then you have a planar curve, which is what we are calling a curve.

Figure 3-3 illustrates three geometric configurations, each of which is called a **broken line** or a **polygonal path**. A curve of this type can be described as one composed of line segments, each sharing at least one of its endpoints with another segment. We prefer the term "polygonal path,"

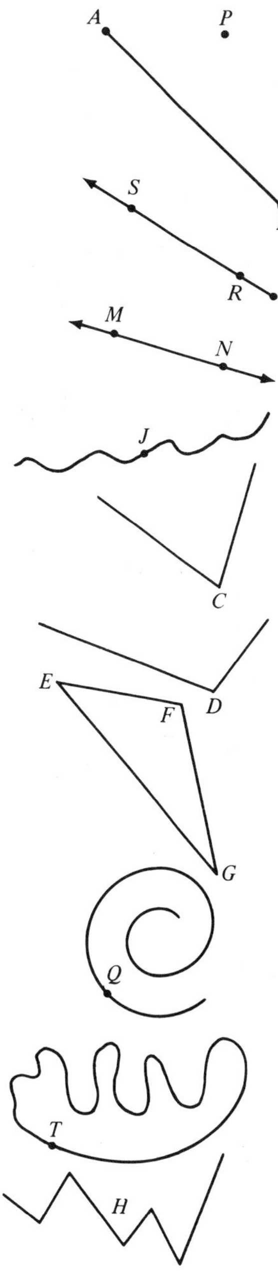

Figure 3-1

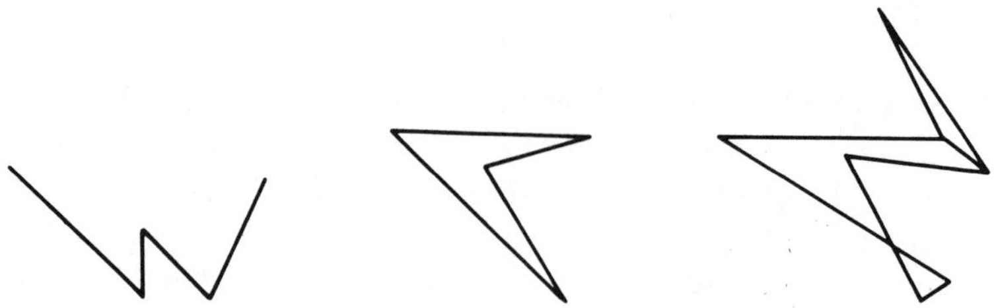

(a) (b) (c)

Figure 3-3

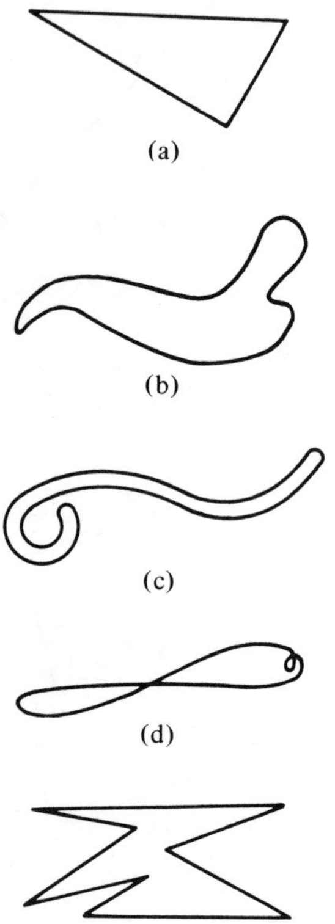

(a) (b) (c) (d) (e)

Figure 3-4

since "broken line" implies endless extent. Pick out the polygonal paths pictured in Fig. 3-1.

3-3 SIMPLE CLOSED CURVES

Figure 3-4 illustrates a class of special geometric curves called **closed curves or closed paths**. Other illustrations are the outline maps of Nevada or Utah. A closed curve may be traced by starting and stopping at the same point. Retracing is allowed only if any portion retraced is traversed in the same direction as originally traced.

The closed curves that seem to be of most interest are those called **simple closed curves**. We can distinguish these by adding two qualifying questions to those listed above:

(3) Did I cover all points without crossing or retracing?

(4) Did I return to the original point?

Again we must answer yes to both questions to refer to the curve as a simple closed curve.

We see in Fig. 3-4(a), (b), (c), and (e) an illustration of plane separation that embodies a different notion from that of Secs. 2-3 and 2-10. Notice that the boundary curve is not of infinite extent in the same sense as are the line and ray. But notice that this type of separation, like the previous type, does distinguish three sets: the boundary and two regions, one on either side of this boundary. Have we, in our vocabulary, words that might apply to these two regions? What type of curves display this type of separation? In Figs. 3-1 through 3-4 which of the pictures illustrate simple closed curves?

In this type of separation the union of the simple closed curve and its interior is referred to as a **closed region**.

EXERCISE 3-3

1. Which of the sketches in Fig. 3-5 do not illustrate curves?
2. Which of the sketches in Fig. 3-5 illustrate:

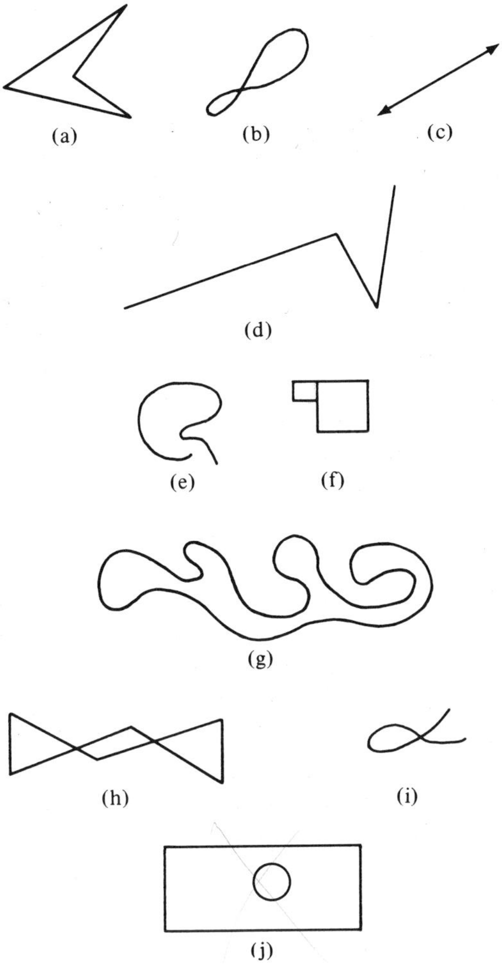

Figure 3-5

(a) Closed curves. (b) Simple closed curves.
(c) Polygonal paths. (d) Closed polygonal paths.
(e) Simple closed polygonal paths.

3. Does the term "polygonal path" refer to the static or the dynamic viewpoint?

3-4 POLYGON

We are now ready to point out how several elementary concepts can be used quickly and smoothly to build another concept. Consider a **simple closed polygonal path**. What are the distinguishing characteristics of such a figure? Surely if we are to encounter this type of figure rather often, it would be convenient to have a single-word name for it. Have you ever heard of the word "polygon"? Literally, this term means "many angles." Look over Figs. 3-1 through 3-5 and see whether you can find any that possess all of the properties you have listed. Any figure that possesses all of these properties is called a **polygon.**

Some people maintain that a polygon is a closed curve that is the union of segments. Notice that the property of simpleness has not been included here. There are even occasions when the curve is not limited to points in a plane. Consider Fig. 3-6 as illustrating a sheet of paper folded along the line joining two opposite corners and placed in a standing position on a table. The figure formed by $\overline{AB}$, $\overline{BC}$, $\overline{CD}$ and $\overline{AD}$ would represent the original four edges of the sheet of paper (before folding) and would be a space polygon. Notice that in this case the four endpoints are noncoplanar points. However, we shall follow the convention of requiring a polygon to be a planar curve, as stated above.

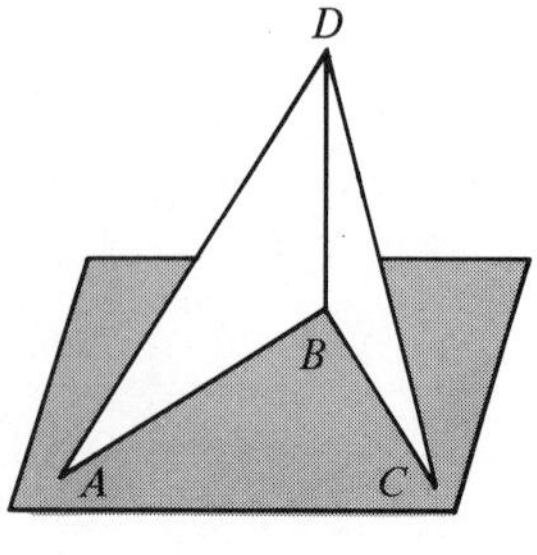

Figure 3-6

Figure 3-7 illustrates a polygon. Each segment is called a **side**. Any common endpoint of two sides is a **vertex**. We would name this polygon $ABCDE$. We may use any vertex as the first letter of the name, proceeding either clockwise or counterclockwise, naming each vertex once. Thus polygon $ABCDE$ = polygon $BCDEA$ = polygon $CBAED$, and so on.

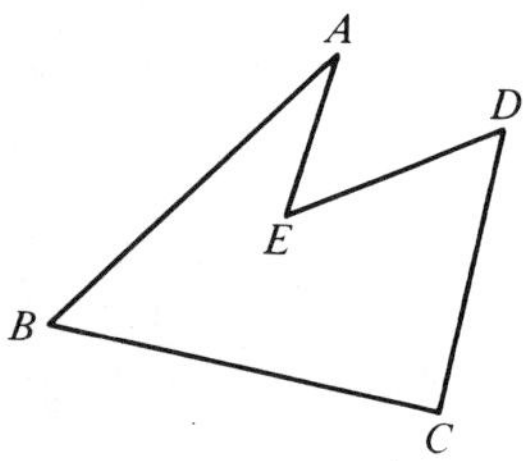

Figure 3-7

What figures have we encountered that have a vertex and sides? In Fig. 3-7 what is the union of $\overline{AB}$ and $\overline{BC}$? Before calling it an angle, perhaps you should recall our description of an angle. Since the sides of an angle are rays, and the sides of a polygon are segments, perhaps we really should not speak of $\angle ABC$. On the other hand, surely the points A, B, and C determine $\overrightarrow{BA}$ and $\overrightarrow{BC}$, which in turn form $\angle ABC$; therefore, we say $\angle ABC$ *of* polygon $ABCDE$, or $\angle ABC$ *determined by* polygon $ABCDE$.

Recall the discussion of separation in Sec. 3-3. Would you say that a polygon separates a plane? We say the shaded region of the plane in Fig. 3-8 is the interior of polygon $RSWVT$. How does this relate to the problem of naming the interior of $\angle SRT$ of polygon $RTVWS$? (You might refer to Exercise 2-10, 6(g), then consider $\angle TVW$ of the polygon.)

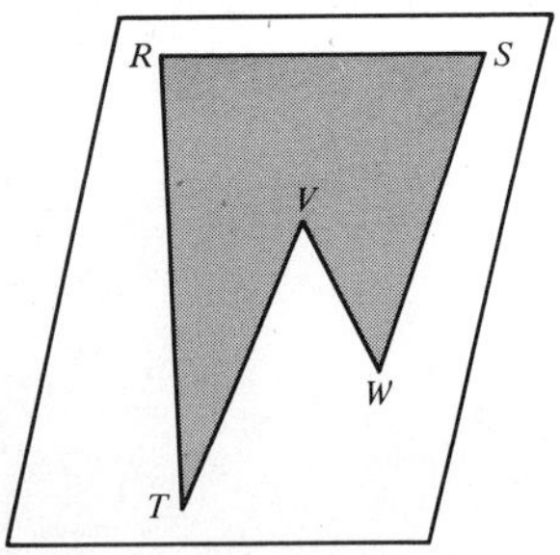

Figure 3-8

Suppose we consider the carriers of the sides of polygon $ABCDE$ of Fig. 3-9. Would any of these lines contain points of the interior of polygon $ABCDE$? A polygon for which the answer is no for all of the carriers is called a **convex polygon.** Is this true of polygon $RTVWS$ in Fig. 3-8? Check the carrier of $\overline{VT}$. This is called a **concave polygon**. Many people, when speaking of a polygon, mean a convex polygon.

Notice that a convex polygon is one in which a segment joining any two points of its interior contains only interior points. This statement could be used as a description of a convex polygon. In fact, such a statement is frequently extended to describe **convex plane regions** in general, such as the one in Fig. 3-10.

Another common method of distinguishing between convex and concave polygons uses a special set of points related to a polygon. Two endpoints of any side of a polygon are called **adjacent vertices** of the polygon. Consider the segments in Figs. 3-8 and 3-9 that are determined by two nonadjacent vertices. Such segments are called **diagonals** of the polygon. Thus $\overline{AD}$ is a diagonal of polygon *ABCDE*. Now consider what a diagonal is, and recall what we said in the previous paragraph about a convex plane region. If no point of any diagonal of the polygon is an exterior point of the polygon, then the polygon is convex. Thus we are saying that if any diagonal of a polygon contains an exterior point of the polygon, then the polygon is concave. Name all the diagonals of the polygons in Figs. 3-7, 3-8, and 3-9.

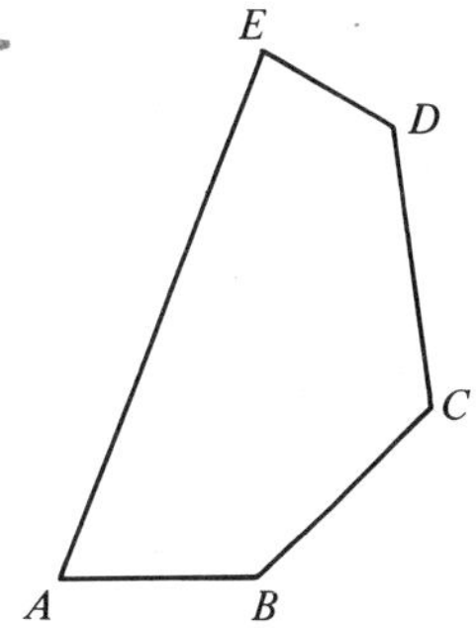

Figure 3-9

There is one more possible issue to decide before we consider further classification of polygons. Is the simple closed broken line in Fig. 3-11 a polygon? Why? We say that this polygon has three vertices, *A*, *B* and *C*, and is really three-sided. Thus in a polygon we will not allow two adjacent sides to have the same carrier. Is this another reason why some people do not allow the sides of an angle to have the same carrier? Notice that our allowing such angles causes us no real trouble.

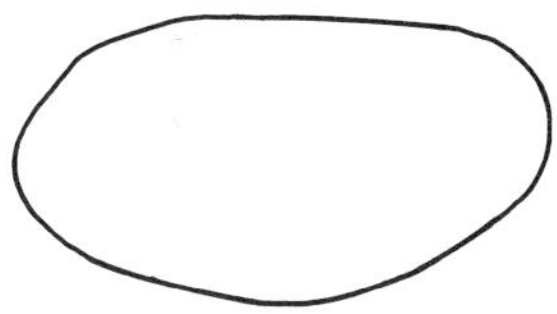

Figure 3-10

Polygons can be classified by their number of sides. Thus we speak of an ***n*-gon** as a polygon with *n* sides. Do you know of another name for a 3-gon? A 4-gon? How about a 7-gon? Here are common names of some *n*-gons:

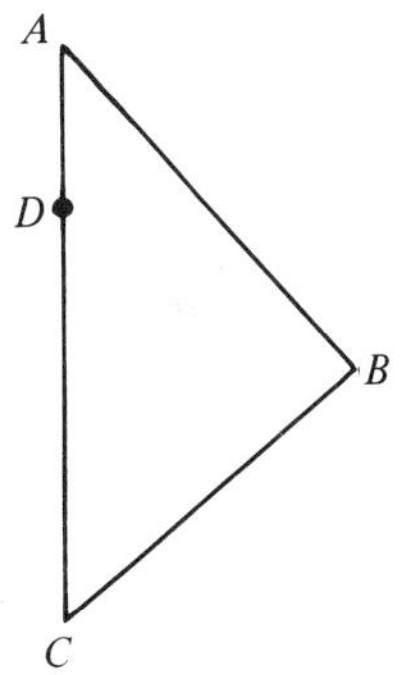

Figure 3-11

n-gon	*Name*	*n-gon*	*Name*
3-gon	triangle	8-gon	octagon
4-gon	quadrilateral	9-gon	nonagon
5-gon	pentagon	10-gon	decagon
6-gon	hexagon	12-gon	dodecagon
7-gon	heptagon (septagon)	20-gon	duodecagon

Names for others exist (for example, an 11-gon is an unagon), but these are the most common ones. Notice that for some the name emphasizes the sides, for others the angles: triangle, three angles; quadrilateral, four sides.

Some of these polygons have other common descriptions. Thus you may find a triangle described as the union of three noncollinear points and the segments joining them. Try making several geometric figures that have only these characteristics to see how they compare with our triangle. Try to find one that satisfies these conditions and is not a triangle by our standards. We mention this to reemphasize that different approaches may produce the same concept. If union of sets is being emphasized, then perhaps this description should be employed. If emphasis is being directed to the classification of subsets of a set, where n distinguishes among the subsets, then surely the idea of a set of n-gons might be preferred.

EXERCISE 3-4

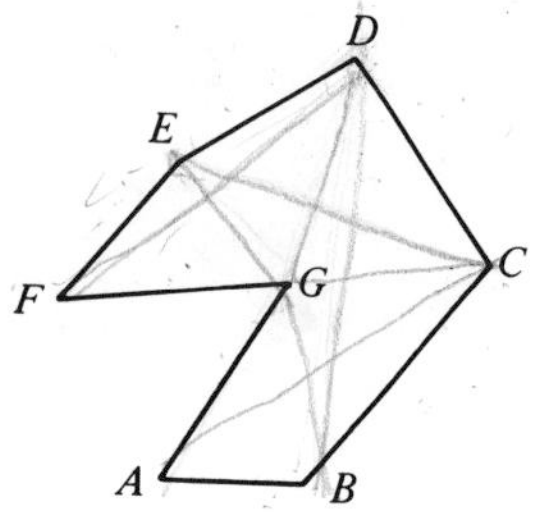

Figure 3-12

1. Given Fig. 3-12:
   (a) How many sides has it? Name three.
   (b) How many angles has it? Name three.
   (c) Name two diagonals.
   (d) Give two word names and one symbolic name for it.
   (e) How many vertices has it?
   (f) Is it convex or concave?
   (g) How many diagonals can be drawn from one vertex?
   (h) How many distinct diagonals can be drawn for it?

2. Assume the following polygons to be convex.
   (a) How many diagonals can be drawn in a triangle?
   (b) How many diagonals can be drawn from one vertex of a quadrilateral? A pentagon? A hexagon? An n-gon? A 30-gon?
   (c) How many distinct diagonals can be drawn in a quadrilateral? A pentagon? A hexagon? An n-gon? A 30-gon?

3. (a) Explain why a 2-gon is not listed in the n-gon table.
   (b) Why is "$\triangle ABC$" an appropriate name for any polygon ABC?

4. In listing the diagonal description of a convex polygon we said, "No point of any diagonal of the polygon is an exterior point." Explain why this is not equivalent to saying, "If every point of every diagonal of the polygon is an interior point, then it is convex."

5. If a polygon is not convex, does the rule for the total number of diagonals determined, developed in problem 2 above, give the correct answer?

6. In Ref. 4, pp. 89–92, there is an interesting and not too difficult discussion about the existence of an inside and an outside of a simple closed planar curve. In general this is called the *Jordan curve theorem*, and it is quite difficult to prove at this level. However, the discussion centers around a polygon. Compare this development with our treatment of separation for a closed curve. Also see Ref. 34, pp. 154–156.

3-5 QUADRILATERALS

Just as we used the concepts of simple, plane, closed, and so on to build the concept of a polygon, we may use combinations of other concepts to bring out other relationships between polygons.

Consider the polygons that are quadrilaterals. Several are illustrated in Fig. 3-13. Which of our other concepts could be used to distinguish among these? Can you see any real difference between (a) and (b)? Between (a) and (c)? Between (b) and (c)? Recall from Sec. 1-7 that two lines are parallel if their intersection is empty and they are coplanar. Do the sides of these quadrilaterals lie in the same plane? How can you be sure? Does it appear that any of the sides of any of these figures are parallel? Suppose we say that each of (a) and (e) is a trapezium (plural: trapezia or trapeziums), (b) is a trapezoid, and the others are parallelograms. Can you state distinctions among these in a way that enables you to decide which quadrilaterals are trapezia, which are trapezoids, and which are parallelograms? Try it.

Is the parallelism concept sufficient to distinguish among (c), (d), (f), and (g)? Would it distinguish between (a) and (e)? Clearly, the answer to these last two questions is no. To make distinctions here we need the concepts of congruence and measurement. (But notice how much material we have developed without such concepts.)

Incidentally, (d) is a rectangle, (e) is a kite, (f) is a rhombus, and (g) is a square, a rhombus, a kite, a rectangle, a parallelogram, a quadrilateral, and so on. Indeed, if we say a trapezium is a least special quadrilateral, then a square is surely a most special quadrilateral. (See Sec. 6-6.)

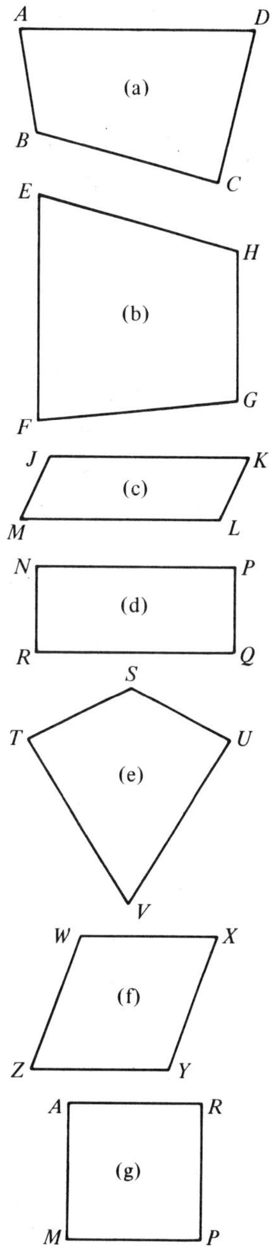

Figure 3-13

EXERCISE 3-5

1. See Ref. 104, p. 646.
2. See Ref. 99, pp. 727–728.

3-6 TRANSVERSAL

The introduction of another term will help us point out some interrelationships between the concepts of parallelism and angle.

Consider two lines in a plane. Their intersection is either empty or nonempty; thus they are either parallel or they intersect. Suppose a third line contains one point of each of these lines such that the three lines are not concurrent? Make sketches to illustrate the possible cases. This third line is called a **transversal** *of* the first and second lines.

In each of the pictures in Fig. 3-14, t is a transversal of the other two lines. In part (a), t is a transversal of m and n. Are there any other lines that might be considered as transversals in Fig. 3-14? Name each transversal and the lines it intersects. In these figures suppose you are asked to select a point P that is on t and between the other two lines? Could you

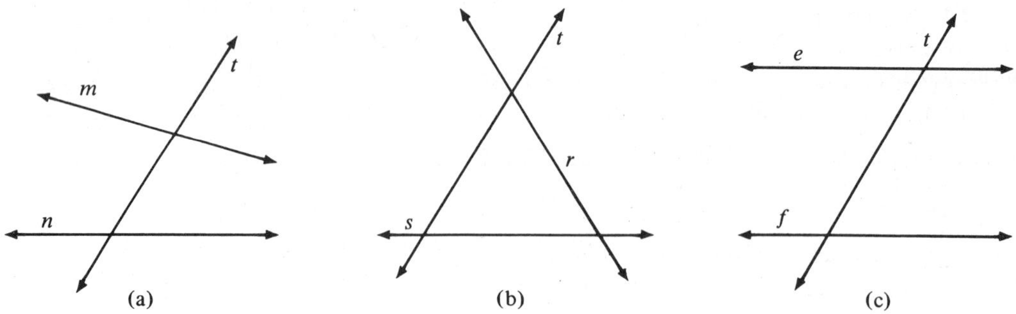

Figure 3-14

decide where to place a dot to correspond to this point in each figure? How much leeway do you have? Notice that these conditions lead us to our previous notion about betweenness.

Would you be willing to say that in Fig. 3-14(c) the plane has been separated into three disjoint subsets: one set consisting of the parallel lines, one set a region containing points between the parallel lines, and the third set a region containing points that are neither on the parallels nor between the parallels?

Now in Fig. 3-15(a) think of transversal *t* in all possible positions that are parallel to the position indicated. In any of these positions which points of *t* are between *m* and *n*? These points we shall refer to as interior points in this type of situation. Lightly shade the interior points of all three parts of Fig. 3-15. Although there might be some question as to how much of part (b) to shade, as long as we are concerned only with transversal *t*, you might shade only those interior points "close to" *t* that is, to the left of *B*. Actually, our description of interior points also includes points to the right of *B*. Notice that we have used betweenness of points on a line to develop interior of lines in a plane. How would Fig. 3-15(b) appear if *t* in its original position contained points *C* and *E*? Thus, in the case of two intersecting lines the region referred to as the interior depends upon the transversal used. Are there two choices if the given lines are parallel?

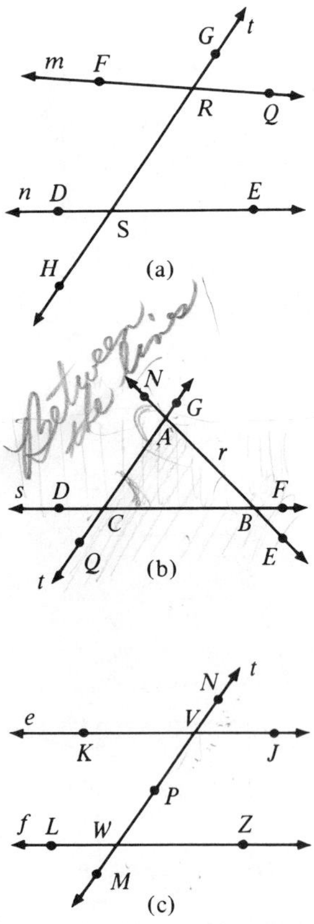

Figure 3-15

3-7 TRANSVERSAL ANGLES

Now let us consider some angles formed by two lines and their transversal, as in Fig. 3-15(a). Name four angles whose interiors would contain interior points of lines *m* and *n*. Do the same with part (c) and the lines *e* and *f*. As you might expect, these are called **interior angles**. Name four angles in each diagram that you would expect to be called **exterior angles**.

Notice that the interiors of $\angle BCA$ and $\angle CAB$ contain interior points of the lines *r* and *s*. Notice also that the interiors of these angles lie in the half-plane that is the *B* side of transversal *t*. Name a pair of angles in Fig. 3-15(c) that you would call interior angles on the same side of the transversal. Now in the same figure name a pair of nonadjacent angles that are interior and have their interior points on opposite sides of the transversal. These are called **alternate interior angles.** Could not the previous pair of angles be referred to as **nonalternate interior angles**?

Angles such as $\angle GRQ$ and $\angle RSE$ in Fig. 3-15(a) are called **corresponding angles.** How might we describe them? Notice that their interiors contain points on the same side of the transversal and are in corresponding positions with respect to the two lines; in this case the interior of $\angle GRQ$ has points "above" *m* and the interior of $\angle RSE$ has points "above" *n*. Likewise $\angle FRS$ and $\angle DSH$ are a pair of corresponding angles. Name two other pairs of corresponding angles in the same figure.

But what has all of this to do with parallelism? After the introduction of some basic measurement concepts in Chapter 5, certain pairs of these angles will be found "equal" or "congruent" when the transversal cuts parallel lines.

EXERCISE 3-7

1. In Fig. 3-15(b) and using *r* as a transversal:
- (a) Name two pairs of corresponding angles.
- (b) Name two pairs of alternate interior angles.
- (c) Name a pair of nonalternate interior angles.
- (d) Name a pair of alternate exterior angles.
- (e) Lightly shade the interior of lines *t* and *s*.
- (f) Make a statement as to which points are considered as being between *t* and *s*.
- (g) What is the boundary of those points not in the interior of *t* and *s*?

2. In Fig. 3-16, which pictures suggest a:
- (a) Polygon.
- (b) Quadrilateral.
- (c) Trapezoid.
- (d) Trapezium.
- (e) Parallelogram.

3. Using $\overleftrightarrow{BC}$ [Fig. 3-16(e)] as a transversal, name one pair of nonalternate interior angles.

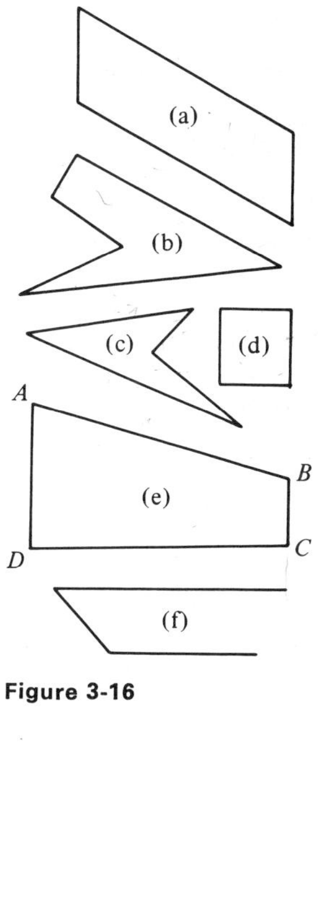

Figure 3-16

3-8 THE PARALLEL POSTULATE

This section illustrates a situation in which mere intuition may lead to incorrect conclusions or provide insufficient information with which to settle an issue. To do this we shall examine one of the fundamental assumptions about parallel lines.

Remember that we are working in a plane. Given a line and a point not in the line, how many lines are there that contain the given point and a parallel to the given line? The answer is not as clear-cut as you might think!

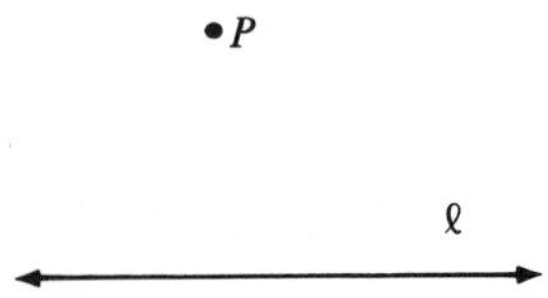

Figure 3-17

In Fig. 3-17, suppose line ℓ and point P are the point and line under discussion. How many lines can you draw through P parallel to ℓ?

To help us along, suppose we consider a different question that bears a close resemblance to our basic question. Two lines that contain P are shown in Fig. 3-18. If each of $\overleftrightarrow{AB}$ and $\overleftrightarrow{RS}$ is parallel to ℓ, then what is their relationship to each other?

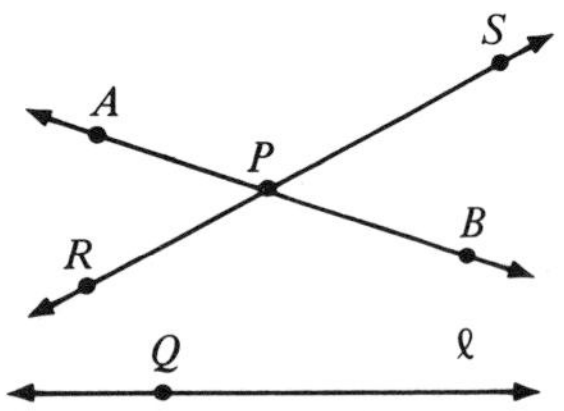

Figure 3-18

As a model of the last question, consider the task of hanging a picture. In Fig. 3-19 the picture is to be hung with its top edge "level" with the floor and the hook slipped over the nail. The husband hangs it so that the top edge has a position suggested by $\overleftrightarrow{AB}$ in Fig. 3-18. His wife comes into the room and "straightens" it so that the top edge takes a position suggested by $\overleftrightarrow{RS}$. The peacemaker, mother-in-law, points out that neither is correct: "Clearly, there is but one position with the picture 'level' with the floor and hooked over the nail, and this position is between the two positions indicated."

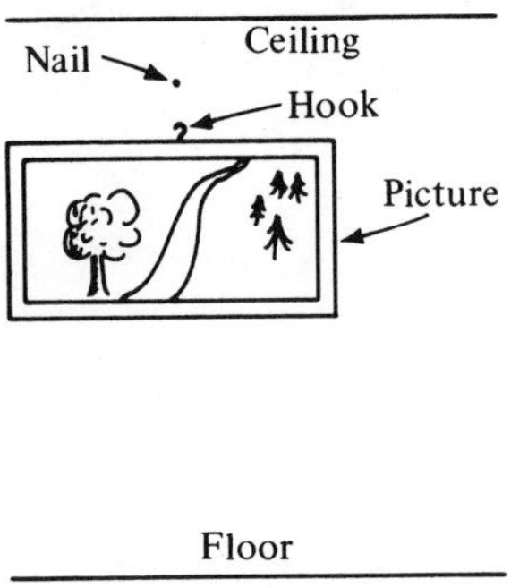

Figure 3-19

This little situation might lead us to believe that there would be only one such line in Fig. 3-17. To reinforce this thought, think of $\overleftrightarrow{MN}$ through P, which is allowed to rotate in the plane in Fig. 3-17. For how many positions of $\overleftrightarrow{MN}$ would *you* think $\overleftrightarrow{MN} \parallel \ell$? We think most readers would say, "Clearly, one."

Consider, however, the following discussion of Fig. 3-18. Think of a point of ℓ such as Q in the diagram. Thinking statically of points in ℓ to the left and right of Q, or thinking dynamically of Q as moving back and forth along ℓ, consider $\overleftrightarrow{PQ}$. As Q moves to the extreme left, well beyond the extent pictured, $\overleftrightarrow{PQ}$ would seem to approach $\overleftrightarrow{RS}$. Intuitively, it seems evident that eventually we would reach a "last" position of $\overleftrightarrow{PQ}$ such that if $\overleftrightarrow{PQ}$ rotated any farther, Q would no longer be a point of ℓ. Could this, then, be $\overleftrightarrow{RS}$? Perhaps. Suppose we move Q on ℓ to the right of the position indicated in Fig. 3-18. We would expect to find a corresponding "last" position of $\overleftrightarrow{PQ}$ such that any further rotation of $\overleftrightarrow{PQ}$ to the right would cause Q to leave ℓ. Might this line then really be $\overleftrightarrow{AB}$? Then what of all of the lines between $\overleftrightarrow{AB}$ and $\overleftrightarrow{RS}$? Will they, or will they not, contain a point of ℓ? Surely not!

Perhaps all of this seems rather strange to you; it should! Some readers may have the impression that there is but one geometry, especially since we have seen that different approaches lead to the same concept. Here we see that we can have two situations in which the same vocabulary is used to arrive at different, even contradictory, concepts.

There is a geometry in which lines such as $\overleftrightarrow{AB}$ and $\overleftrightarrow{RS}$ in the previous discussion are called **parallels**. The lines through P between $\overleftrightarrow{AB}$ and $\overleftrightarrow{RS}$ and not containing a point of ℓ are called **nonintersectors**. Thus, we are led to conclude that in a plane through a point not in the line there are many lines whose intersection with the given line is empty.

Our experimentation has led us to two possible conclusions about the number of parallels to a given line through a given point:

(1) there is one such line,

(2) there are many such lines.

A slightly different approach, which we do not investigate, could lead us to conclude:

(3) there is no such line (see Exercise 1-7, 6).

Among other things, this discussion should (a) show you that intuition has its limitations, and (b) further illuminate the age-old question: What is mathematics? Perhaps you begin to see why some people say mathematics is exactly what you make it.

We are at a point in the development of our geometry where we must select one of the following:

(1) if there is one (such) line, then . . . must follow,

(2) if there are many (such) lines, then . . . must follow,

(3) if there is no (such) line, then . . . must be concluded.

In the Exercises you will be asked to accept first (1), then (2), to use each in reaching further conclusions, and to compare the results.

Historically, discussions such as the foregoing arose when early geometers questioned Euclid's *fifth postulate*. A postulate, like an axiom, is merely a basic agreement. (See Sec. 13-2.) Euclid wrote the first comprehensive treatment of geometry sometime about 300 B.C. In this he assumed a statement (quoted in the summary for this chapter) equivalent to our first conclusion that there is but one such parallel. Our first conclusion is credited to Playfair, who tried, as did many others, to prove or disprove Euclid's fifth postulate. (Playfair's postulate is also stated in the summary.) If we accept either of these statements, then any decisions we make that are based on that statement are a part of the geometry called *Euclidean geometry*. This is the basic form of the geometry studied in high school.

Geometries that deny the Euclidean parallel postulate are called *non-Euclidean* geometries. Geometries using the assumption that through a given point not on a given line there are many parallels to a given line are called Lobachevskian or hyperbolic geometries. Geometries that use the assumption of no parallels are called Riemannian or elliptic geometries. (See Ref. 55 for a discussion of the development of these geometries, from the initial attempts to disprove Euclid's fifth postulate to the formulation of the non-Euclidean geometries.)

At one time, Euclidean geometry provided an acceptable model of "small" physical situations, such as the building of the pyramids or the

John Hancock Building in Chicago. However, the planning of interplanetary travel, and other developments in science, have required a more extensive use of the non-Euclidean geometries.

SUMMARY OF PRINCIPLES

A polygon separates a plane.

Euclid's fifth postulate (sometimes referred to as the eleventh or twelfth): "If a straight line meets two other straight lines so as to make the two interior angles on one side of it together less than two right angles, the other straight lines, if extended indefinitely, will meet on that side on which the angles are less than two right angles."[1]

The postulate of parallels, Playfair's axiom: "Given a line *a* and a point *A* not lying on *a*, then there exists, in the plane determined by *a* and *A*, one and only one line which contains *A* but not any point of *a*."[2]

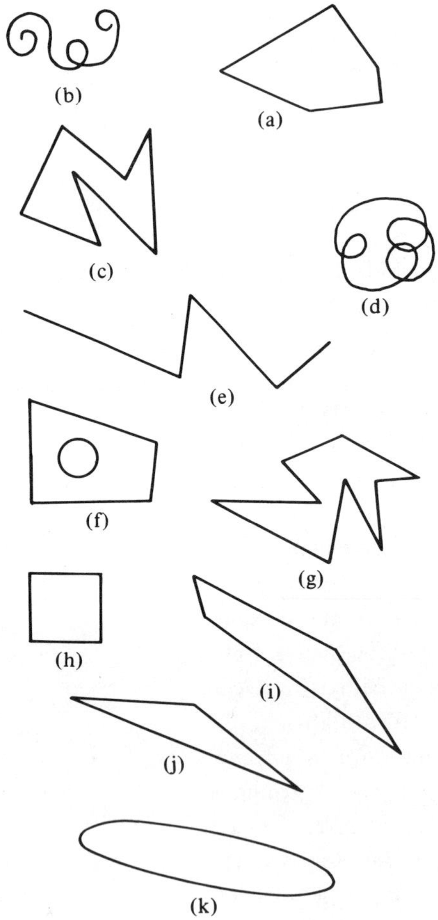

Figure 3-20

EXERCISES

1. On your paper write the letters (a) through (k). Then, referring to Fig. 3-20, copy after each letter any and all boldface terms in this chapter that apply to the figure corresponding to that letter.

2. In Fig. 3-21:
(a) What term best describes this figure?
(b) Name three sides.
(c) Name two angles of the polygon, writing the name for each in two different notations.
(d) Name two pairs of adjacent vertices.
(e) Name the diagonals in the picture.
(f) Using $\overleftrightarrow{AE}$ as a transversal:
(1) Which lines are cut by this transversal?
(2) Using $\overleftrightarrow{AB}$ and $\overleftrightarrow{EF}$, name a pair of:
(a) alternate interior angles,
(b) corresponding angles,
(c) nonalternate interior angles.

[1] See Ref. 11, p. 4.
[2] See Ref. 55, pp. 4, 16.

(g) Shade the interior of $\overleftrightarrow{GD}$ and $\overleftrightarrow{AE}$, considering $\overleftrightarrow{DE}$ as a transversal.

3. Listed in the Summary of Principles are Euclid's fifth postulate and Playfair's axiom. Make sketches that illustrate the principles involved in each. Then write symbolic statements related to each figure that express these same principles.

4. From magazines or catalogs clip pictures that illustrate concepts discussed in this chapter. Consider how they could be arranged as a poster display for your classes.

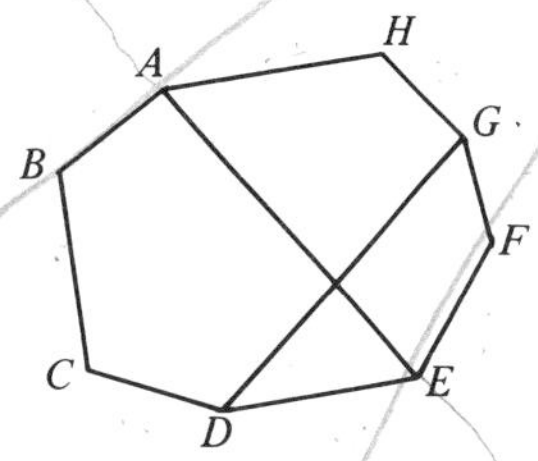

Figure 3-21

5. In Ref. 9, pp. 187–193 and 200–203, a discussion covers the same material as at the end of Sec. 3-8. This is done in a very interesting way, and the material in pp. 194–199 goes into more detail. Outline the first two suggested sections in preparation for a report to the class.

6. (a) In Sec. 3-8 we said: "... some people say mathematics is exactly what you make it." Read Ref. 50. The first five paragraphs of the Preface, pp. 272–273, 276–277, give a little more insight into this comment.

(b) For further insight into the fifth postulate see Ref. 22, pp. 190–198.

7. This exercise is an example of proof by contradiction, sometimes called an indirect proof or proof by *reductio ad absurdum*. Methods of proof are discussed in Chapter 13. Also Ref. 12, pp. 565–571, discusses the nature of mathematics, mentions this type of proof, and comments on Euclidean geometry. This reading might prove valuable before and/or after this exercise.

(a) Below you see an outline of a discussion to prove that if, in a plane, a line intersects one of two parallel lines, then it must intersect the other. You are to supply statements to complete the explanation.

GIVEN: In plane m, $\ell_1 \parallel \ell_2$, $k \cap \ell_1 = \{P\}$.

TO PROVE: $k \cap \ell_2 \neq \varnothing$.

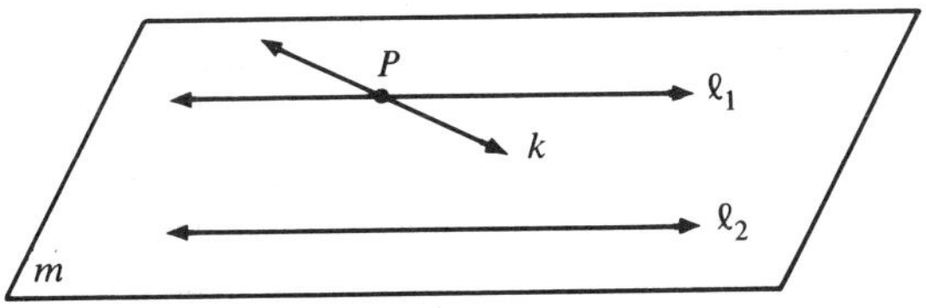

STEPS	EXPLANATION
(1) Assume $k \cap \ell_2 = \varnothing$.	(1) One of two contradictory statements.
(2) Then $k \parallel \ell_2$.	(2) ?
(3) But, this is impossible.	(3) ?
(4) Thus $k \cap \ell_2 \neq \varnothing$.	(4) The contradictory statement has been proved false.

(b) (1) At what step in the proof has the parallel postulate entered the demonstration? Recall that this postulate is equivalent to conclusion (1) in Sec. 3-8.

(2) If we accept conclusion (2) in Sec. 3-8, what would be the result of this proof?

(3) Why does conclusion (3) not apply to this problem?

8. (a) "A *polygonal region* is the union of a finite number of triangular regions, in a plane, such that if two of these intersect, their intersection is either a point or a segment." This quotation is from Ref. 31, p. 291. Make several sketches of polygonal regions and indicate how this quotation applies. Also illustrate a region that is not polygonal because it fails to meet these requirements. Then read Ref. 31, pp. 290–292.

(b) "Let $P_1, P_2, \ldots, P_n$ be a sequence of n distinct points in a plane, with $n \geq 3$. Suppose that the n segments $\overline{P_1P_2}, \overline{P_2P_3}, \ldots, \overline{P_{n-1}P_n}$ have the following properties: (1) No two of the segments intersect except at their endpoints. (2) No two segments with a common endpoint are collinear. Then the union of the n segments is called a *polygon*." This quotation is from Ref. 31, p. 513. As directed in part (a), draw several examples and counterexamples of this statement. Then read pp. 512–514.

(c) How do parts (a) and (b) compare with our discussion of these same items?

9. See problem 7(a), (b) above. Follow the same procedure for the proof below. Statement: If in a plane two distinct lines are parallel to a third line, then these two lines are parallel to each other.

GIVEN: In plane m, $\ell_1 \parallel \ell_3$, $\ell_2 \parallel \ell_3$.

TO PROVE: $\ell_1 \parallel \ell_2$.

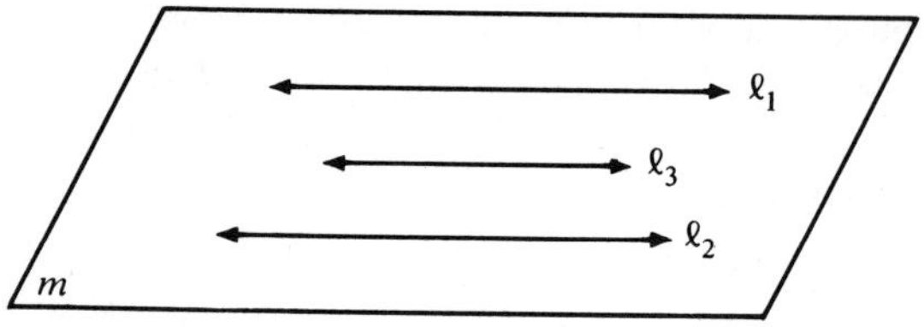

STEPS	EXPLANATION
(1) Assume $\ell_1 \cap \ell_2 \neq \varnothing$.	(1) ?
(2) Then $\ell_2 \cap \ell_1 = \{P\}$.	(2) Definition of parallelism.
(3) Then $\ell_2 \cap \ell_3 \neq \varnothing$.	(3) Result of problem 7(a).
(4) But $\ell_2 \parallel \ell_3$.	(4) ?
(5) Thus step (3) is not possible.	(5) ?
(6) Hence, $\ell_1 \cap \ell_2 = \varnothing$.	(6) ?
(7) And $\ell_1 \parallel \ell_2$.	(7) ?

10. For a model of the third possibility discussed at the end of Sec. 3-8, see Ref. 24, pp. 15-17.

CHAPTER 4

Building in Space: Surfaces, Polyhedrons

4-1 INTRODUCTION

This chapter offers another perspective on many of the ideas and relationships explored earlier. You will learn how these concepts can be used and extended to describe relationships between positions that are not restricted to a line or to a plane. Chapter 4 discusses positions and shapes that are found in ordinary three-dimensional space.

Perhaps you have been wondering why we have not discussed more fully the shapes that correspond to solid objects and positions in three-dimensional space—things for which we have great use in ordinary situations. There are two main reasons for our approach. First, one- and two-dimensional shapes and positional relationships are usually much easier for people to visualize. Second, many of the terms used to describe figures in space are given in terms of concepts developed for one- and two-dimensional figures.

Perhaps you have also been wondering why we have not yet talked about measurement as a topic in geometry, since many of the concepts in geometry seem to be related to ordinary measurement. To have introduced metric ideas before nonmetric concepts would have tended to obscure some of the essential and basic relationships independent of measure but existing between geometric figures.

4-2 FIGURES IN SPACE

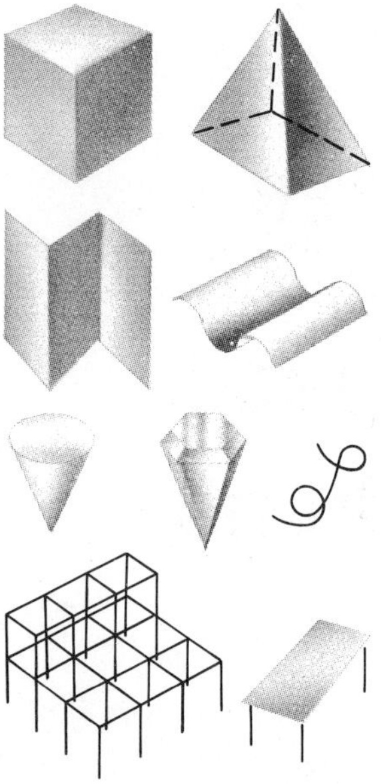
Figure 4-1

The geometric figures we investigate in this chapter are sets of points that are abstractions for common objects, such as a box, a pyramid, a folding door, a corrugated roof, an ice cream cone, a cut diamond, a jungle gym, or a twisted wire. The sets of points that correspond to these real objects, as sketched in Fig. 4-1, we refer to as **figures in space.**

The examples just given happen to correspond to sets of points that are not contained in a single plane. If a figure is comprised of coplanar points, we say that it is **planar.** (See Ref. 4, p. 39, for another interpretation of planar.) If some points of the figure are not coplanar, we shall refer to the figure as **nonplanar**. It is important to note that figures in space include both planar and nonplanar figures. Do planar figures include both figures all of whose points are collinear and figures all of whose points are not collinear?

Do you recall our describing some planar figures by using figures in a line? For example, how did we describe the planar figure called the polygon? In order to describe figures in space, we shall be using both figures in a plane and figures in a line.

EXERCISE 4-2

1. For each of the sketches of figures in space shown in Fig. 4-2 identify some figures in a line and the planar figures contained in the figure in space.
2. Encircle A, S, or N for each statement below.
   (a) A point is a figure in space. A S N

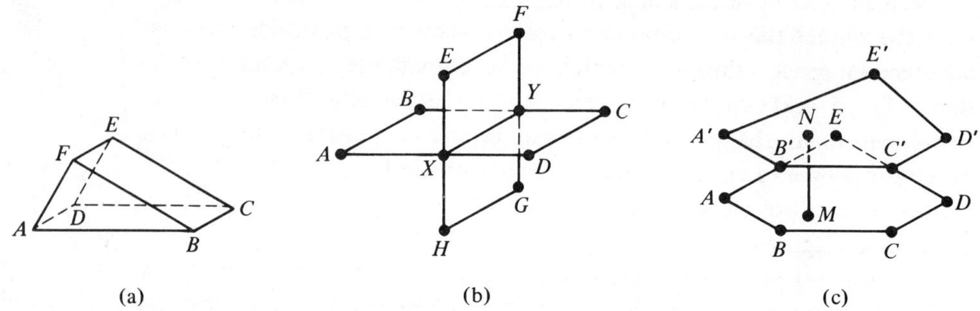

Figure 4-2

(b) An angle is a planar figure. A S N
(c) A dihedral angle is a nonplanar figure. A S N
(d) A line segment is not a figure in space. A S N

4-3 NONPLANAR FIGURES

The nonplanar figures mentioned in Sec. 4-2 can be separated into groups. The first group is represented by the jungle gym. (See Fig. 4-3.)

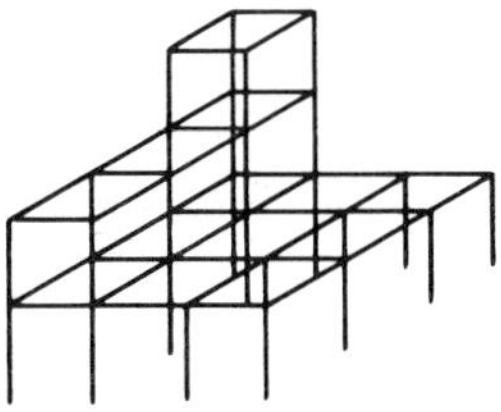

Figure 4-3

Figure 4-4

If you consider only the tubing, do you see that the entire figure could be described as a union of sets of points called line segments? Another model of this group is the ribbing of an umbrella, as shown in Fig. 4-4. These figures are examples of what some people call networks of curves. (See Sec. 12-3.)

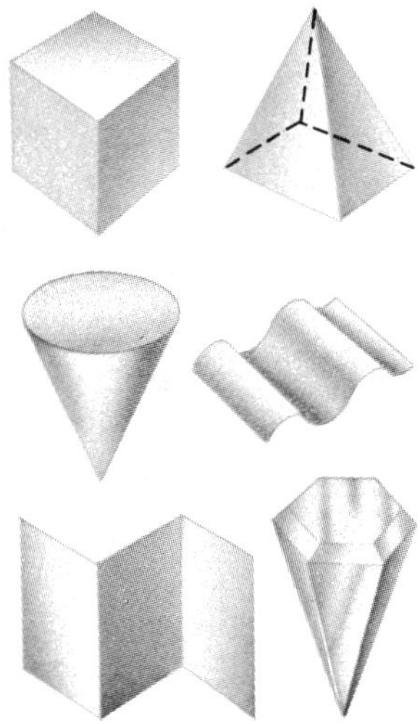

A second group is represented by the empty box, the shell of the pyramid, the cone, the folding door, the corrugated roof, and the surface of the diamond. (See Fig. 4-5.) The figures in this group appear to be made up of regions similar to plane regions, although none of these figures is actually planar. In fact, in some of these figures—for example, the cone—part of the surface is not even flat. These figures are examples of what some people call surfaces.

A third group is represented by the table pictured in Fig. 4-6. It seems to be describable as a combination of figures from the first two groups. Another example of this group is an open umbrella, as sketched in Fig. 4-7.

A fourth group is represented by a filled box, a pyramidal block, a

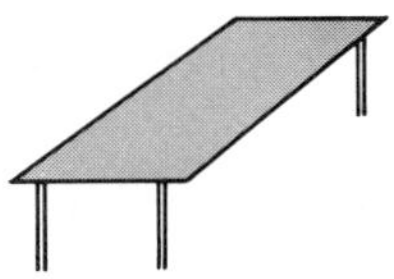

Figure 4-6

Figure 4-7

filled cone, and the cut diamond. These figures are examples of what some people call solids.

Can you think of a figure which is a combination of figures from the first, third, and fourth groups?

EXERCISE 4-3

1. Classify each of the figures sketched in Fig. 4-8 as a network, surface, or combination.

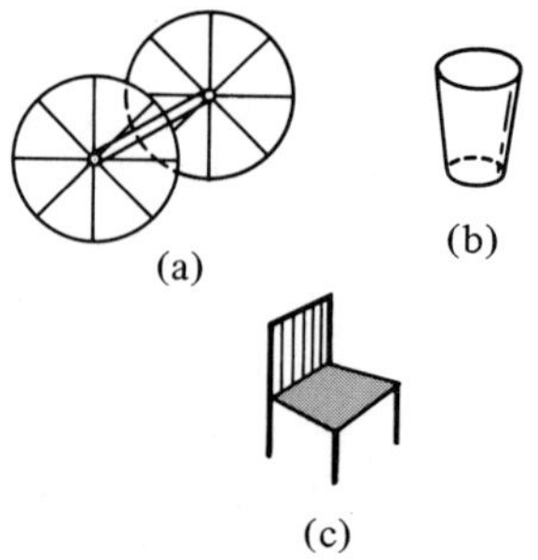

Figure 4-8

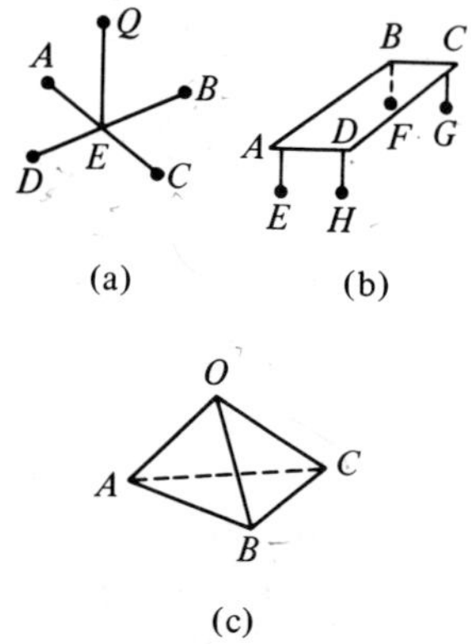

Figure 4-9

2. Match the mathematical phrase with the corresponding diagram in Fig. 4-9:

(1) Polygonal region $ABCD \cup \overline{AE} \cup \overline{BF} \cup \overline{CG} \cup \overline{DH}$.

(2) Triangular region AOB ∪ triangular region BOC ∪ triangular region AOC ∪ triangular region ABC.

(3) $\overline{AC} \cup \overline{BD} \cup \overline{QE}$.

4-4 INTERSECTION AND UNION OF PLANES AND LINES

We have already explored some of the relationships between planes and lines in space. What are the possible intersections of a plane and a line?

If a plane and a line are parallel, then what is the intersection of the plane and the line? [See Fig. 4-10(a).]

If a line is contained in a plane, then what is the intersection of the line and the plane? [See Fig. 4-10(b).]

If a line is not contained in the plane and is not parallel to it, then what is the intersection of the line and the plane? [See Fig. 4-10(c).] In this case, where the intersection of the line and the plane is a point, this point is called the **foot** of the line.

We have also looked at some situations in which we considered the intersection or union of two or more planes.

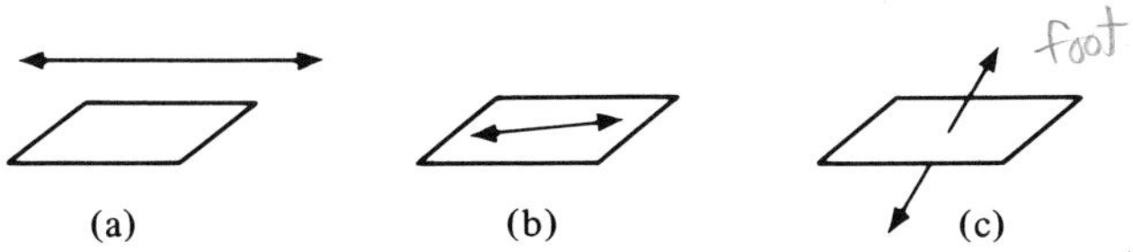

Figure 4-10

If two or more planes are parallel, then what is the intersection of the planes? [See Fig. 4-11(a).] There are no points common to any pair of parallel planes. But notice that the union of the two planes is the set of all points in the two planes.

If two planes are not parallel, then what geometric figure is the intersection of the two planes? [See Fig. 4-11(b).]

What did we call the union of two half-planes and their common separator? (See Fig. 4-12.)

In Sec. 1-7 we noted that if two lines are coplanar and their intersection is empty, then the lines are said to be parallel. However, it is not necessary that two lines whose intersection is empty lie in one plane. For example, using the box pictured in Fig. 4-13, it is apparent that $\overleftrightarrow{AB} \cap \overleftrightarrow{CD} = \varnothing$.

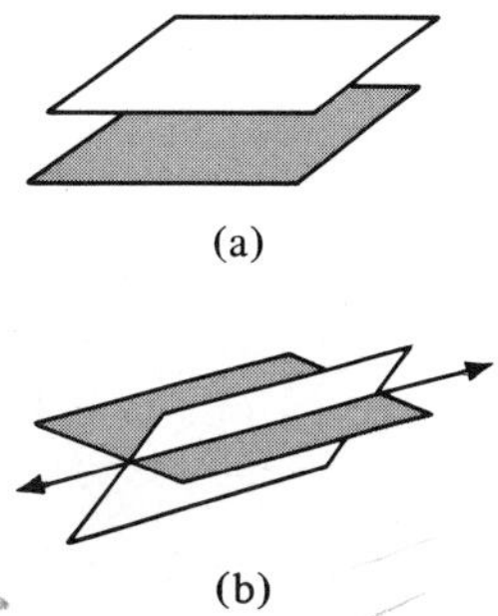

Figure 4-11

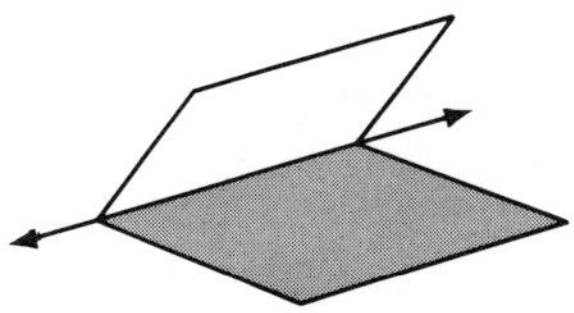

Figure 4-12

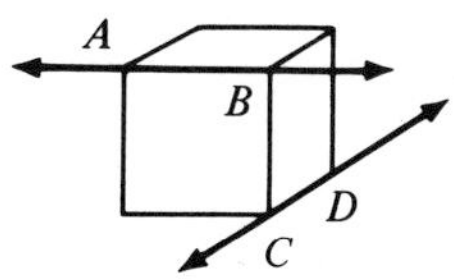

Figure 4-13

Since the lines are not coplanar, they are not parallel. Such lines are called **skew lines.**

EXERCISE 4-4

1. When will the union of a line and a plane be a nonplanar figure? When will the union be a planar figure?

2. In order for two lines to be parallel they must be coplanar. Should this be the case for three lines? For more than three? Use Fig. 4-14 as a basis for your discussion.

3. How might you describe the fence in Fig. 4-15 using geometric figures and set operations?

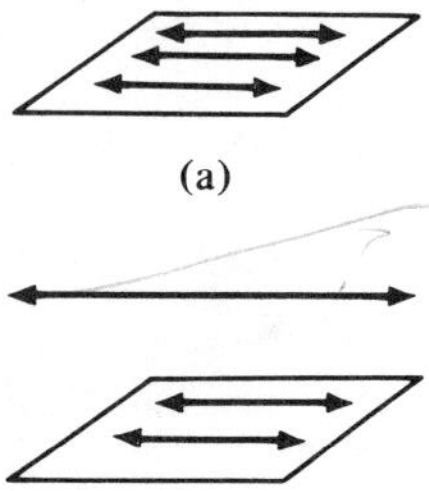

(a)

(b)

Figure 4-14

Figure 4-15

4-5 INTERIOR AND EXTERIOR SPACE REGIONS

Let us return now to consider the second group of figures discussed in Sec. 4-3. You recall that these figures or sets of points have some similarity to regions in a plane and are called **surfaces**. Unlike the plane, not all surfaces have the property of flatness. In fact, most of the surfaces of shapes encountered in ordinary situations are not flat—for example, the shell of an egg, the roof of a house, the surface of a lake, the outside of an ice cream cone, or a corrugated roof. The sets of points suggested by these objects are called surfaces.

Let us explore the relationship of these surfaces to the space in which they are contained. Here we shall investigate three relationships by considering the following examples: the surface suggested by the sides and bottom of a glass, the surface suggested by the shell of an egg, and the surface suggested by the extended sides of an imaginary bottomless pyramid. See Fig. 4-16 for sketches of these objects.

These three situations are analogous to certain relationships between a plane curve and the plane containing the curve. As you recall from Chapter 3, the curves in parts (b) and (c) of Fig. 4-17 separate the plane into three disjoint sets: the figure itself (the boundary) and two regions, one on either side of the boundary. To get from one region to the other it is necessary to cross the boundary. That is, a line segment connecting two points, one in each region, must contain a point of the boundary.

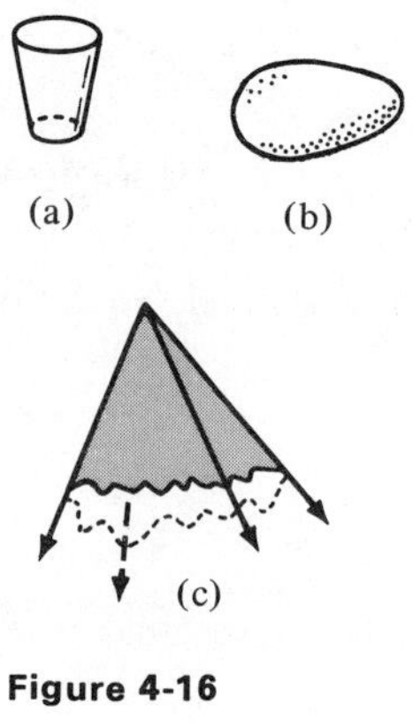

(a) (b) (c)

Figure 4-16

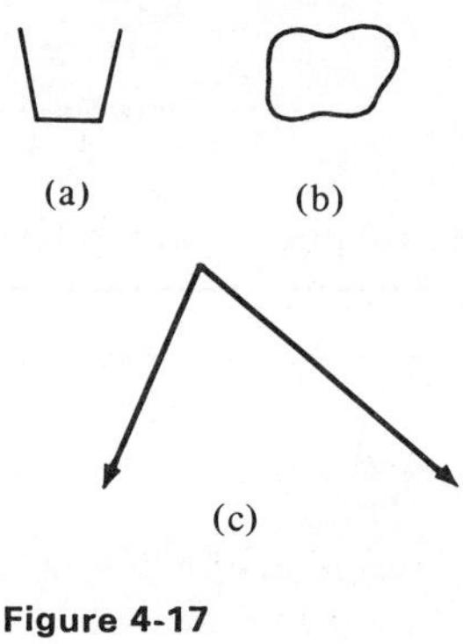

(a) (b) (c)

Figure 4-17

The egg surface and the pyramidal surface are analogous to the curves in parts (b) and (c) of Fig. 4-17. Each surface separates space into three sets: the surface (the boundary) and two sets of points called regions, one on either side of the boundary. To get from one region to the other it is necessary to cross the boundary. The glass and the polygonal path in part (a) of Fig. 4-17 are analogous in that they do not form such regions.

Further, we called the curve in 4-17(b) a simple closed curve. One region of the plane is called the interior and the other region is called the exterior. Similarly, the eggshell surface is called a simple **closed surface**, and its two associated space regions are called the interior region and the exterior region. Notice that in this case the surface encloses a portion of space and the interior region is bounded.

Although neither the angle nor the pyramidal surface is referred to as closed in the same sense as the simple closed curve or the surface of the eggshell, we do use the terms "interior" and "exterior," in a manner analogous to Sec. 2-8, to distinguish the two regions determined by the angle and the pyramidal surface. Furthermore, in Sec. 4-6 we shall investigate some surfaces that are closed in a different sense.

EXERCISE 4-5

1. Do the surfaces suggested by the following objects separate space?
(a) Closed hat box. (b) Tin can open at both ends.
(c) Cup. (d) Ball.

2. All surfaces are nonplanar. True or false?

3. Try to describe what might be meant by each of the following terms:
(a) Simple closed surface. (b) Convex surface.
(c) Concave surface.

4-6 CYLINDRICAL AND CONICAL SURFACES

We shall now take some familiar surfaces and show how to describe them, using the set operations of union and intersection on sets of points with which you are already familiar. (We remind you that this is only one of several possible ways to describe these surfaces.)

You are probably aware that the closed surface suggested by a closed tin can, as in Fig. 4-18, is called a cylinder. If you were to remove the top and bottom and then think of extending the curved part indefinitely, you would have a good idea of the nature of a closed *cylindrical surface*. (See Fig. 4-19.) Notice that this interpretation of closure differs from that used in Sec. 4-5. How?

The closed cylindrical surface is a special case of a cylindrical surface. A cylindrical surface can be described using the concepts of curve, line, parallelism, and union of sets. Consider a curve in a plane *m*, as in Fig. 4-20(a). Now think of a line *e* that intersects that plane in a point of the curve [part (b) of the figure]. Now think of all the lines parallel to line *e* that also intersect this curve [part (c)]. The surface formed by the union of all such lines is called a **cylindrical surface**. Each of the lines is called an **element** of the surface. The surface obtained using the curve in Fig. 4-20(a) should remind you of the corrugated roof sketched earlier. If the plane curve used is a broken line, then the surface obtained will look somewhat like a folding door. Such cylindrical surfaces are called **prismatic surfaces**.

In order to get a closed cylindrical surface, what kind of curve should

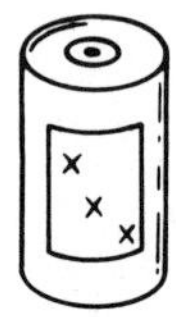

Figure 4-18

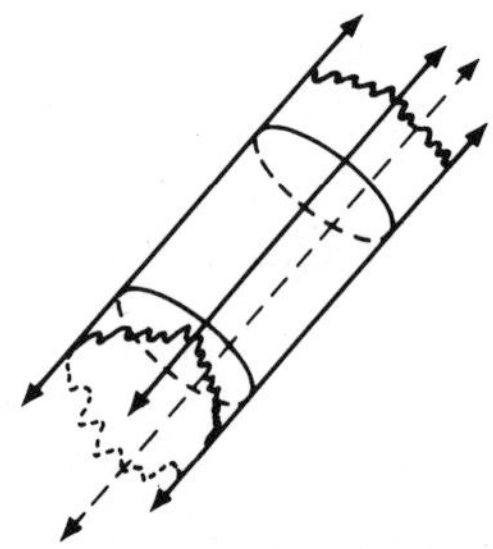

Figure 4-19

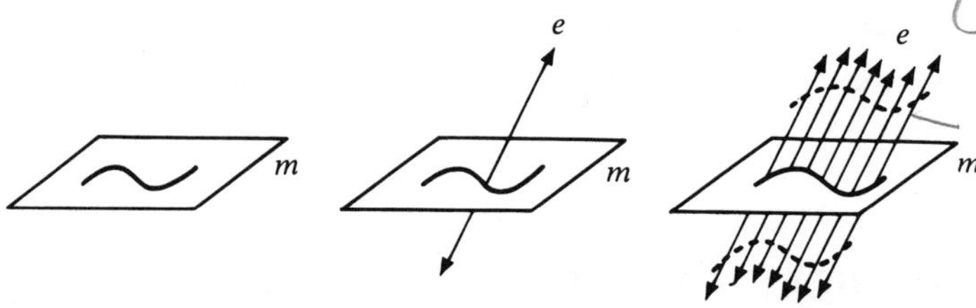

(a) (b) (c)

Figure 4-20

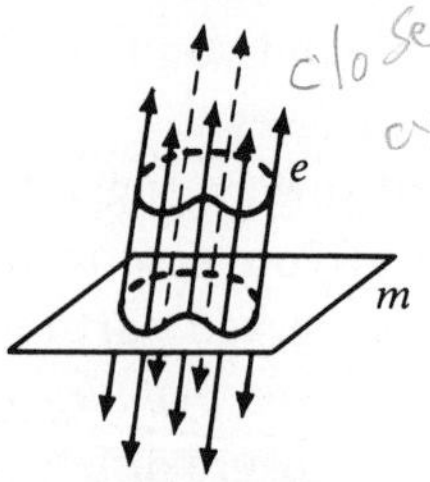

Figure 4-21

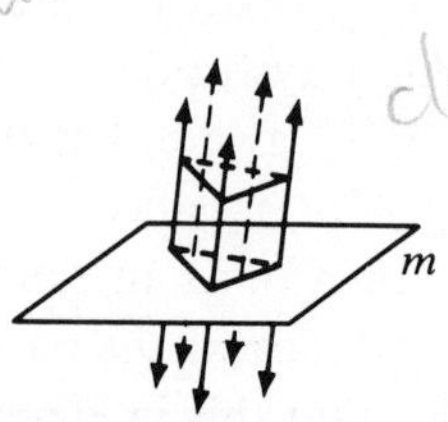

Figure 4-22

you start with? (See Fig. 4-21.) If the curve with which you start is a polygon, then the closed cylindrical surface obtained is called a closed prismatic surface. (See Fig. 4-22.)

Think now of an ice cream cone whose surface extends indefinitely. Figure 4-23 may help. This gives you a good idea of the nature of a *closed conical surface*. We can describe the general conical surface in much the same way as we did the general cylindrical surface. Consider a curve in a plane *r*, as in Fig. 4-24(a). Consider a point *X* not in plane *r*, as in part (b). Now think of a line *e* that contains *X* and also contains a point of the curve in plane *r* [part (c)]. Now try to think of all such lines that contain the given point *X* and also contain a point of the curve. The union of all such lines is called as **conical surface** [part (d)].

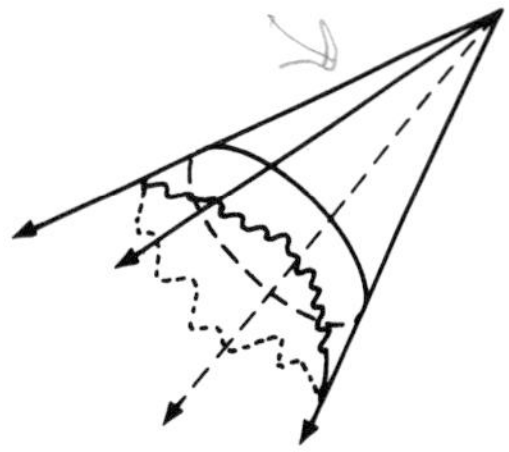

Figure 4-23

In order to obtain a closed conical surface, what kind of curve should you start with? Figure 4-25 may help. If the curve with which you start is a polygonal path, the conical surface is called a **pyramidal surface**. (See Fig. 4-26.)

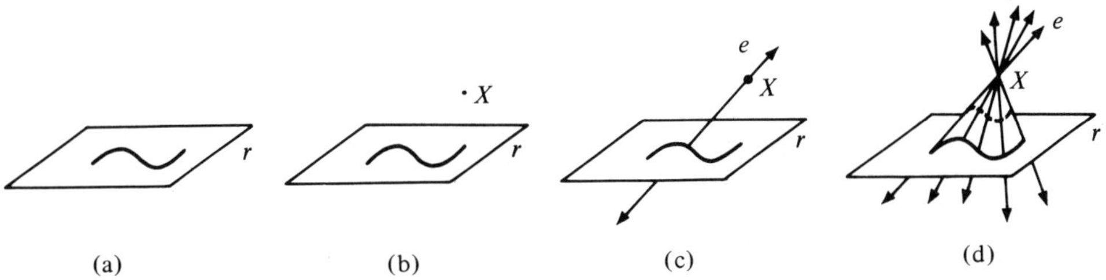

Figure 4-24

You have probably noticed from Fig. 4-25 that since the surface is a union of lines (the elements of the surface), it appears to be the union of two conelike surfaces, each separated from the other by the point *X*. The point *X*, which is common to all the elements of the surface, is called the **vertex** of the conical surface. Each part of the surface on either side of and including the vertex is called a **nappe** of the conical surface.

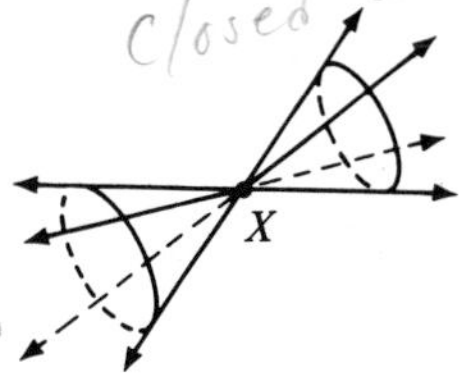

Figure 4-25

Notice that one nappe of a pyramidal surface is a polyhedral angle. Notice also that pyramidal surfaces might be described as unions of angles and the interior regions of these angles.

To return once again to the question of closure, notice that whether a cylindrical or conical surface is closed or not depends entirely on the nature of the plane curve that determines the surface.

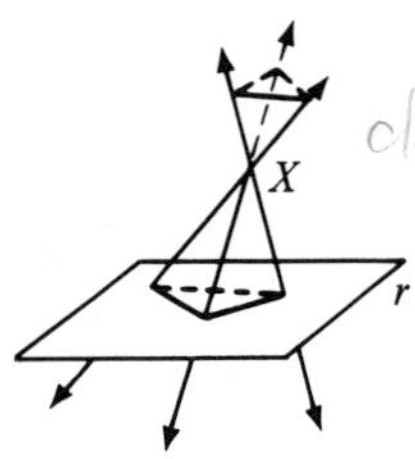

Figure 4-26

Finally, a note on our use of certain terms. When we use the terms "cylindrical surface" and "conical surface," we refer to closed surfaces. If we want to refer to cases in which the surfaces are not closed, we shall use the term "open" as a modifier.

EXERCISE 4-6

1. How many elements are contained in a cylindrical or conical surface?
2. With what planar curve should you start to obtain the open cylindrical surface called a dihedral angle?
3. With what planar curve should you start to obtain one nappe of the pyramidal surface called an octahedral angle?
4. Encircle A, S, or N (always, sometimes, or never true) for each statement below.
   (a) A prismatic surface is a cylindrical surface. A S N
   (b) A conical surface is a pyramidal surface. A S N
   (c) A cylindrical surface has one vertex. A S N
5. Draw a sketch to show how one nappe of a conical surface can be described using the concept of ray.
6. Cylindrical and conical surfaces may also be described from a dynamic viewpoint. In this case the planar curve is called the **directrix** of the surface. The lines used in the static interpretation are called elements. The surface is described in terms of a moving element called the **generator**. Write a description of a cylindrical and a conical surface using the dynamic viewpoint.

4-7 CYLINDERS AND CONES

Notice that the cylindrical and conical surfaces, which are closed surfaces, do not enclose space in the same sense as the eggshell surface. However, these surfaces are related to certain surfaces that do enclose space—namely, cylinders and cones, prisms and pyramids.

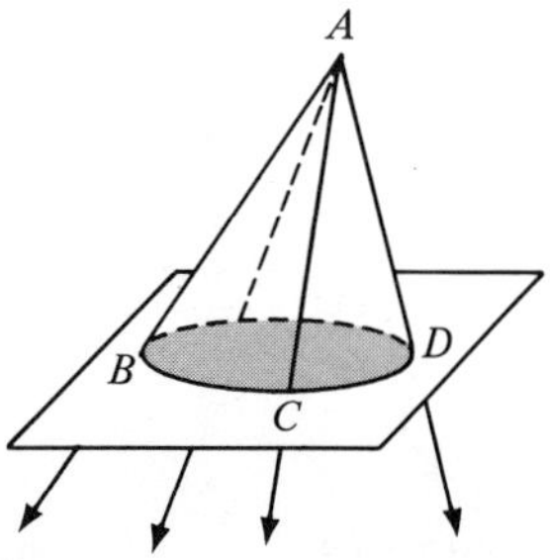

Figure 4-27

A cone is formed when a plane intersects one nappe of a conical surface such that the intersection is a closed curve. (See Fig. 4-27.) The **cone** is the set of points that is the union of the closed planar region and the portion of the conical surface between and including both the vertex and the closed curve. That portion of the conical surface between and including the vertex and the closed curve is called the **lateral surface** of the cone. The closed planar region is called the **base** of the cone.

Cones are often named by associating the name of the vertex with a name of the base. For example, you might call the cone of Fig. 4-27 cone *A-BCD*.

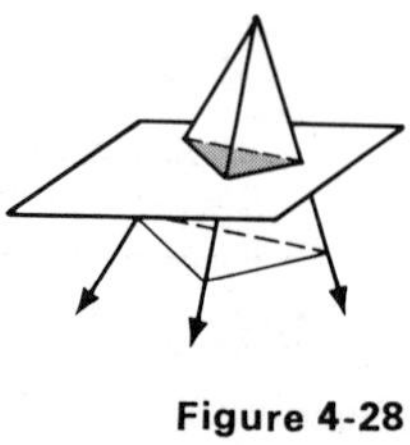
Figure 4-28

If the intersection of the conical surface and the plane is a polygon, the union of the closed planar region and the portion of the surface between and including the vertex and polygon is called a **pyramid.** (See Fig. 4-28.) Again the closed planar region is called the base of the pyramid, and the portion of the pyramidal surface between and including the vertex and closed curve is called the lateral surface of the pyramid.

If we want to consider the surface of the cone and its interior points, we speak of this set as a **conical solid.** Note, however, that this convention is not uniform. You often hear people using the word "cone" to mean either the surface or the conical solid. We make the distinction between a cone and a conical solid. Likewise, when considering a pyramid and its interior points we refer to such a set as a **pyramidal solid.**

In general, we shall call any region in space that is the union of a closed surface and its interior points a **solid.** What parts of an egg correspond to a geometric solid?

A cylinder is formed when two parallel planes intersect a cylindrical surface so that the intersection is a pair of closed curves. The **cylinder** is the union of the two closed planar regions and the portion of the cylindrical surface between the two parallel planes. (See Fig. 4-29.) The two closed planar regions are called the bases of the cylinder, and the portion of the cylindrical surface between and including the two planar curves is called the lateral surface of the cylinder.

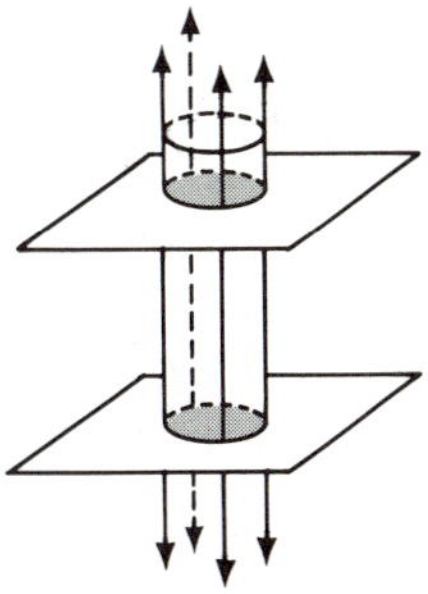

Figure 4-29

If the intersection of the parallel planes with the cylindrical surface is a pair of polygons, the associated closed surface is a **prism.** (See Fig. 4-30.) The polygonal regions in the parallel planes are called the bases of the prism. Both bases will always be the same kind of polygonal region. Why must this be so? If the bases are octagonal regions, as in Fig. 4-31, then the prism is called an octagonal prism. What would you call the prism shown in Fig. 4-32?

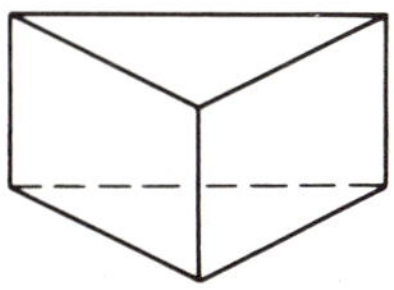

Figure 4-30

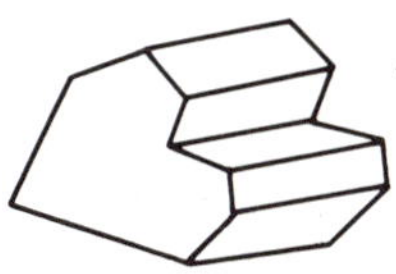

Figure 4-31

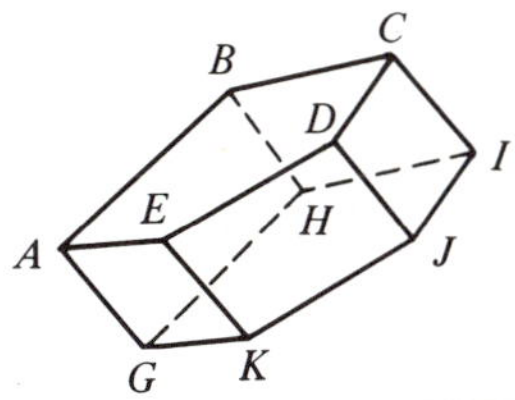

Figure 4-32

What kind of regions make up the lateral surface of a prism? Did you notice that they were always quadrilateral regions? Why is this? Note, moreover, that these quadrilaterals are always parallelograms.

If the bases of the prism are also parallelograms, then the prism is known as a **parallelepiped.** (See Fig. 4-33.)

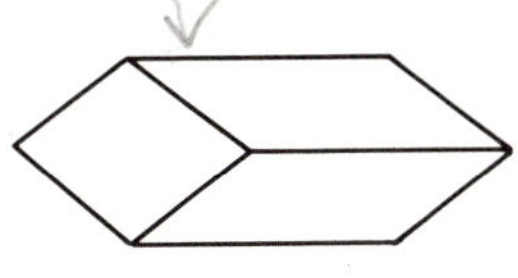

Figure 4-33

Prisms are often named by associating the names of the two bases of the prism so that the order of the vertices of the base polygons is maintained. The prism in Fig. 4-32 may be called pentagonal prism *ABCDE-GHIJK*. Notice that *A* and *G* are endpoints of the same segment.

EXERCISE 4-7

1. Draw sketches of each of the following figures:
   (a) Cone. (b) Pyramid.
   (c) Cylinder. (d) Prism.
2. Write a symbolic name for each of the geometric figures pictured in Fig. 4-34.
3. Name the polygonal regions that comprise the lateral surface of the surface in Fig. 4-34(d).
4. What geometric figure might be considered an element of the lateral surface of a cone or cylinder? Why can a line not be an element of the lateral surface of either of these figures?
5. How might a solid prism be defined in terms of segments?
6. Devise a classification system for prisms, using the fact that the base of a prism is a polygonal region.

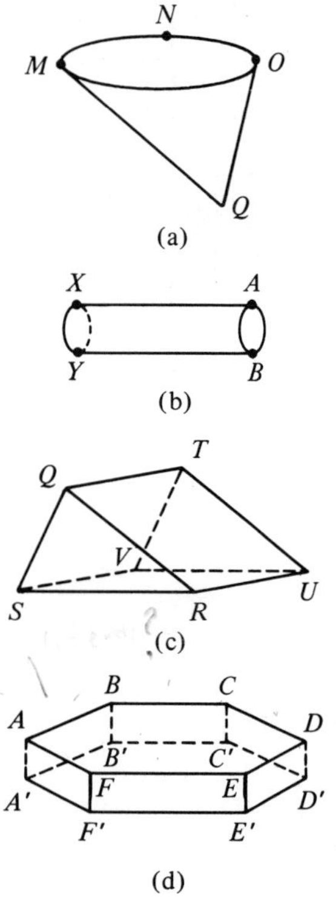

Figure 4-34

4-8 POLYHEDRONS

You have probably noticed by now that a prism has the property of being a union of polygonal regions. For example, the pentagonal prism in Fig. 4-35 is the union of two parallel pentagonal regions and five parallelogram regions. Thus, the pentagonal prism is a union of seven polygonal regions.

Are there closed surfaces that are describable as unions of polygonal regions but that are not prisms? Perhaps the most common example is the pyramid mentioned in Sec. 4-7. Are there closed surfaces that are unions of polygonal regions and that have parallel bases, as do prisms, but that are not prisms? (See Fig. 4-36.)

All closed surfaces that enclose a portion of space in the manner of the eggshell and that can be described as the union of polygonal regions are called **polyhedrons** (sometimes written "polyhedra"). Each of the closed polygonal regions that make up the polyhedron is called a **face** of the polyhedron. Polyhedrons are often classified according to the number of faces they have.

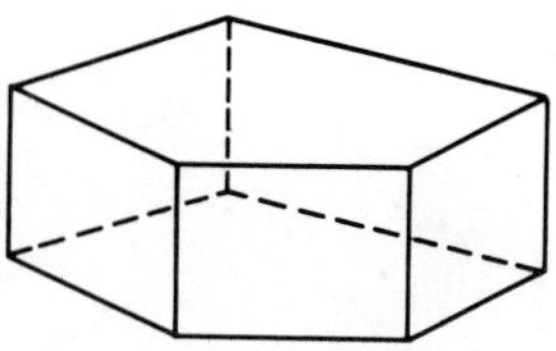

Figure 4-35

Number of Faces	*Name*
4	tetrahedron
5	pentahedron
6	hexahedron
7	heptahedron
8	octahedron
9	nonahedron
10	decahedron
20	icosahedron

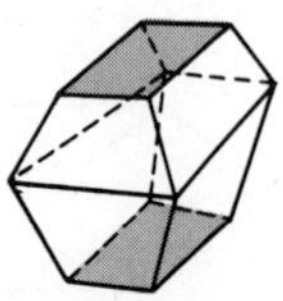

Figure 4-36

Figure 4-37 shows two hexahedrons. Note that one is a prism and the other is not.

Each of the segments forming the figure that is the boundary of the faces is called an **edge** of the polyhedron. The points that are the intersections of the edges are called **vertices** of the polyhedron. In Fig. 4-37(a) $\overline{AB}$ is an edge; C is a vertex.

Did you notice that each pair of faces that intersect in a line segment determine a dihedral angle? Did you notice that each vertex of the polyhedron is also the vertex of a polyhedral angle? Does the polyhedron actually contain any polyhedral angles?

A **polyhedral solid** is the union of a polyhedron and its interior points.

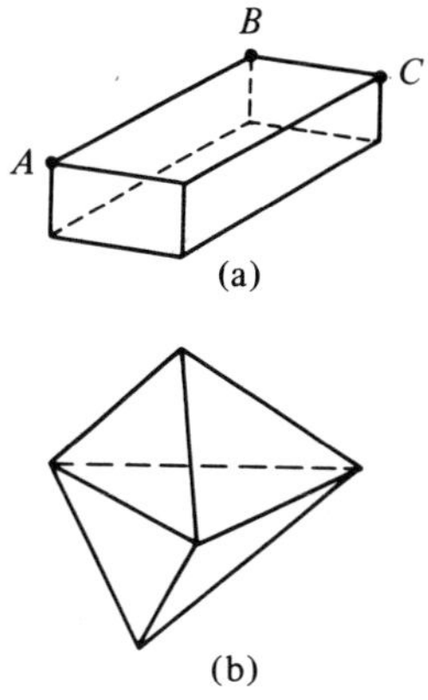

Figure 4-37

EXERCISE 4-8

1. Which of the sketches in Fig. 4-38 represent:
   (a) Polyhedrons. (b) Prisms.
2. Name the edges and vertices of Fig. 4-38(a).
3. How many faces does a decahedron have?
4. How many faces does a decagonal prism have?

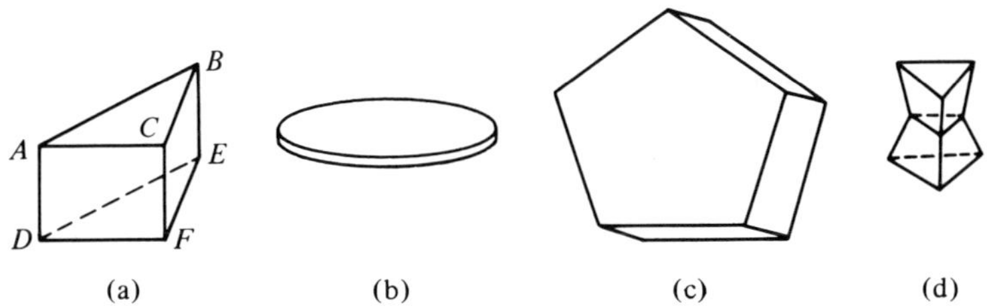

Figure 4-38

4-9 SECTIONS

In Sec. 1-7, when talking about the concept of plane, we mentioned that planes were especially important in describing sections of figures. We are now ready to consider sections, using more precise geometric terminology.

A **section** of a surface or solid is simply the intersection of the surface or solid and a plane. It may help you to visualize the intersection if you think of slicing through the surface with a knife. Try to visualize a plane that intersects a triangular prism. What does the intersection look like?

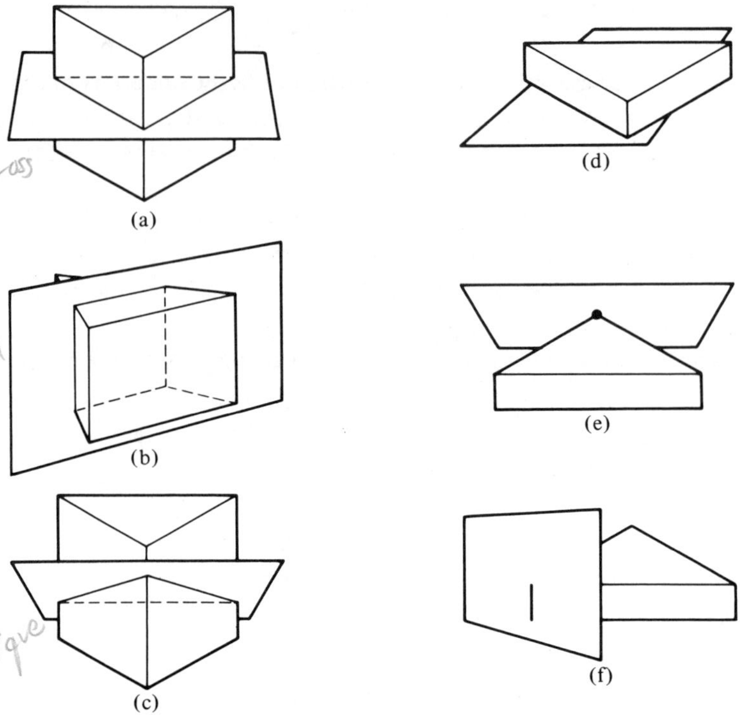

Figure 4-39

It depends on how you slice it. Several possibilities are illustrated in Fig. 4-39.

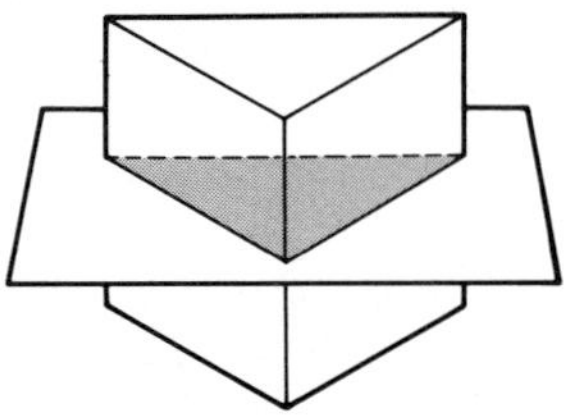

Figure 4-40

Figure 4-39(a) shows a cross section of the prism. A **cross section** of a prism (cylinder, pyramid, or cone) is the nonempty intersection of the prism (cylinder, pyramid, or cone) and a plane parallel to the base of the figure. Part (b) shows a "vertical" section, and part (c) shows an oblique section. [Note that for cases (a) and (c) the plane intersects every element of the lateral surface of the prism.] The closed curves that are the intersections are called sections of the prism. Other intersections of a plane with a prism are possible, though we seldom think of them as sections. For part (d) we might say that the knife missed! The intersection is empty. In part (e) the intersection is a vertex of the prism, and in part (f) the intersection is an edge. Are there other possible intersections?

Figure 4-40 shows a section of a solid. Note that the section is a closed planar region.

EXERCISE 4-9

1. Draw sketches to show several sections of a prism whose bases are quadrilaterals.

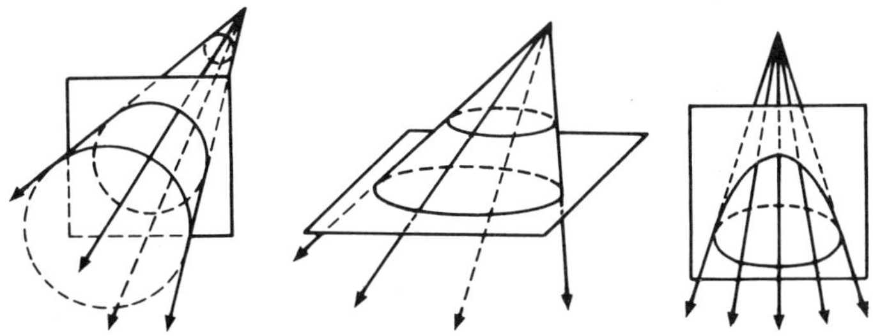

Figure 4-41

2. The cross sections of an octagonal pyramid are octagonal regions. A S N
3. The sections of a conical surface are simple closed curves. (See Fig. 4-41.) A S N
4. Read Ref. 95, pp. 415–421.

4-10 OTHER SURFACES

Obviously, we have not talked about all the various figures that exist in space. In order to describe some space figures, such as the sphere, we need concepts from metric geometry (we shall begin to investigate these in the next chapter). Other surfaces are sometimes adequately described as unions of simpler space figures. For example, how might you describe the surface in Fig. 4-42?

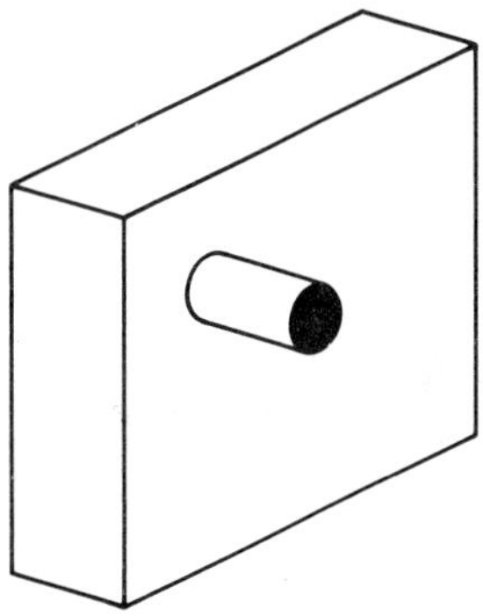

Figure 4-42

Perhaps you have noticed that (1) many figures can be described without considering any metric properties, (2) space figures are closely related to figures in a plane, and (3) planar figures are closely related to figures in a line. All of these are related to the concept of point. Point, line, plane, and space! Is there something beyond the kind of space we have been considering?

EXERCISES

1. Encircle A, S, or N (always, sometimes, or never true) for each statement below.
   (a) A section of a closed conical surface is a closed curve. A S N
   (b) A section of a triangular pyramid is a triangle. A S N
   (c) Two elements of a cylindrical surface may have a nonempty intersection. A S N
   (d) Three planes intersect in a point. A S N
   (e) A prism whose base has ten sides is called a decahedron. A S N
   (f) Two lines that are skew lie in one plane. A S N
   (g) The intersection of a plane and a pair of parallel lines is a pair of points. A S N

(h) The intersection of a cylindrical solid and a plane is a closed parallelogram region. A S N

(i) A cylindrical surface is the same as a pyramidal surface. A S N

(j) A prism is a polyhedron. A S N

(k) A polyhedron is a prism. A S N

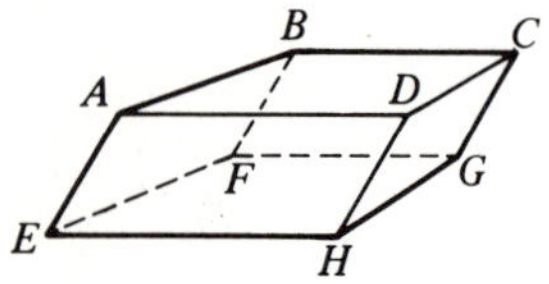

Figure 4-43

2. In Fig. 4-43, *ABCD-EFGH* is a parallelepiped.

(a) Will $\overleftrightarrow{CD}$ and $\overleftrightarrow{EF}$ ever intersect?

(b) Are $\overleftrightarrow{CD}$ and $\overleftrightarrow{EF}$ skew lines?

(c) Think of a plane containing $\overleftrightarrow{CD}$ and $\overleftrightarrow{EF}$. Describe the two closed space figures determined by this plane.

3. In Fig. 4-44, $\overline{AX}$ might be called a diagonal of the prism. How many such diagonals of the prism contain interior points of the prism?

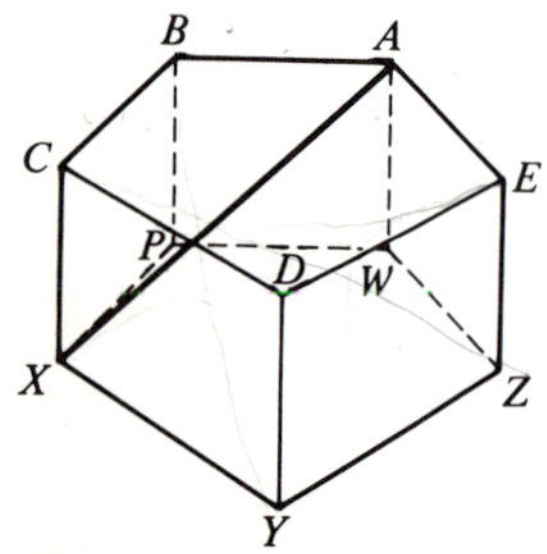

Figure 4-44

4. In Fig. 4-45, how many segments connecting vertices of the polyhedron contain exterior points of the polyhedron?

5. Describe the intersection of a plane and a sphere.

6. In the text we have noted that surfaces may be closed in several different senses. Figure 4-46 shows three different closed surfaces. Which of these surfaces can contain in its interior:

(a) A line segment. (b) A ray.

(c) A line.

7. Find a set of planar curves for which the relations will be the same as for (a), (b), and (c) in problem 6.

8. Draw sketches of the possible intersections of a cylinder and a quadrilateral prism.

9. Describe each object pictured in Fig. 4-47 in terms of space figures.

10. Make a list of names for pyramids, using the scheme that the name depends on the number of sides of the base polygon.

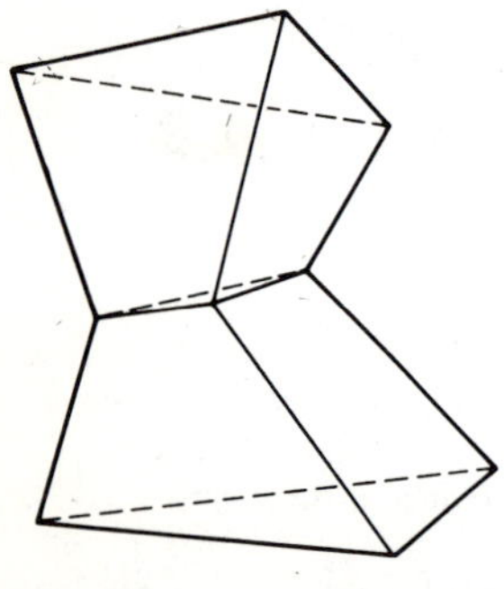

Figure 4-45

11. In Appendix C are patterns for making models of polyhedrons. Make the models indicated by your instructor.

12. Look up the etymology of "polyhedron" and "lateral."

13. We noted in Sec. 4-4 that if the intersection of a line and a plane is empty, then the line is parallel to the plane. We might call this the definition of "line parallel to a plane." Suppose, now, that a given line is parallel to a second line that lies in a particular plane. If the given line is not in the particular plane, what is the relationship of this line to the plane? Below you will find an outline of a discussion leading to a proof of the statement: "If a line, not in a given plane, is parallel to a line in that plane, then it is parallel to the plane." You are to complete the explanation. You may wish to refer to problem 7 in the Exercises following Chapter 3.

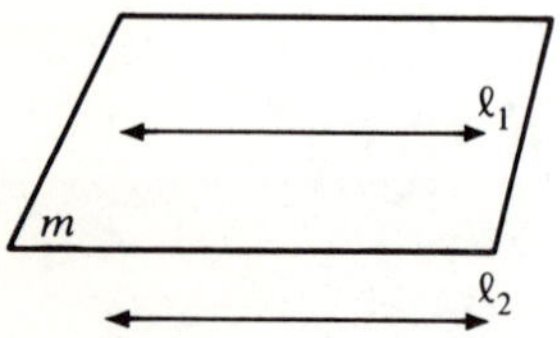

GIVEN: ℓ_1, ℓ_2, plane m, $\ell_1 \neq \ell_2$, $\ell_1 \parallel \ell_2$, $\ell_1 \in$ plane m,
$\ell_2 \notin$ plane m.

TO PROVE: $\ell_2 \parallel$ plane m.

PROOF: The proof is indirect. We shall assume the contradiction of $\ell_2 \parallel m$; that is, $\ell_2 \cap m \neq \varnothing$.

STEPS	EXPLANATION
(1) ℓ_1, ℓ_2 determine a plane p.	(1) See problem 14 of the Exercises following Chapter 1.
(2) $p \cap m = \ell_1$.	(2) What is the intersection of two planes? What set of points is common to the two planes?
(3) Assume that $m \cap \ell_2 \neq \varnothing$.	(3) One of two contradictory statements.
(4) If $\ell_2 \cap m$, it must do so in a point of ℓ_1.	(4) Why?
(5) But this is impossible.	(5) Why?
(6) Thus, $\ell_2 \cap m = \varnothing$, which implies that $\ell_2 /\!/ m$.	(6) The contradictory statement has been proved false.

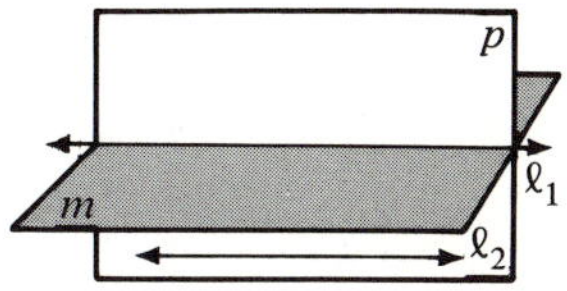

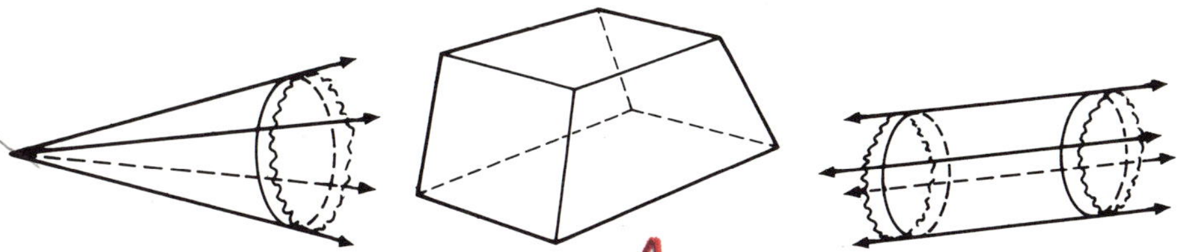

(a) (b) (c)

Figure 4-46

(a) (b) (c) (d)

Figure 4-47

14. (a) Complete the following table and try to determine the relationship among the three variables. V refers to the number of vertices of a prism, F to the number of faces of a prism, E to the number of edges of a prism.

Name	*V*	*F*	*E*
Triangular prism		5	
		6	
Pentagonal prism		7	
Hexagonal prism	12		
			21
Octagonal prism			24
		11	
Decagonal prism			

(b) For the polyhedrons you made for problem 11, make a table like that in part (a). Which relationships for prisms are also valid for polyhedrons?

15. For a possible answer to the question that ends the text of this chapter, read Ref. 88, pp. 11–13; Ref. 1, and Ref. 8. Ref. 1 should give you a better understanding of the importance of line, plane, and space in three-dimensional geometry.

16. Reread problem 9 of the Exercises following Chapter 3. Then prove that if three noncoplanar lines are related so that the first and second lines are coplanar and each is parallel to the third, then the first and second lines are parallel to each other. [*Hint*: Use the results of problem 14 above.]

CHAPTER 5

Measurement

5-1 INTRODUCTION

None of the properties that we have considered thus far has required attention to the relative sizes of objects. For this reason these properties are called **nonmetric**. However, as must be fairly obvious, size, and the shapes that depend on size considerations, are among the most important attributes of many objects. Many of the ideas in the remaining chapters are concerned with the "size" of geometric figures. These geometric ideas are a part of what is often called **metric geometry**.

The process of measurement is not strictly a geometric consideration, since we can also speak of weight measurement, time measurement, psychological measurement, and so on. These situations are not usually classified as geometric. What we shall be looking at in the following chapters is the process of measurement applied to geometric situations. In the physical situation we might be measuring lengths of rope, surfaces of playgrounds, capacities of containers, or degrees of steepness of inclined planes. You probably recognize that these four examples correspond to the geometric concepts of length, area, volume, and angle measure.

In this chapter we shall examine a mathematical idealization of the measuring process in the physical world and the relationship of this abstraction to practical problems. In particular, we shall focus on mathematical models for measuring line segments and for measuring angles.

5-2 MEASUREMENT AS SIMPLE COMPARISON

Unfortunately, man has not recorded the many steps that culminated in present-day systems of measurement. We can only speculate on some of the earliest approaches. There is not much doubt, however, that the process of measuring is based on the notion of comparing.

Suppose you have a straight piece of rope with two knots in it, as in Fig. 5-1. We say that knot B is farther from end E than is knot A. Thinking of A and E as endpoints of a straight piece of rope, we see that this situation might be represented using line segments. Figure 5-2 illustrates two segments that can be associated with this comparison situation. Notice that $\overline{BE}$ and $\overline{AE}$ have a common endpoint. In saying that B is farther from E than is A (or that $\overline{BE}$ is longer than $\overline{AE}$), we seem to be taking account of the fact that A is between B and E. We could also have noted that $\overline{AE}$ is a proper subset of $\overline{BE}$. Using this kind of analysis, why is $\overline{CD}$ longer than $\overline{MD}$? (See Fig. 5-3.) When would the two segments have the same length?

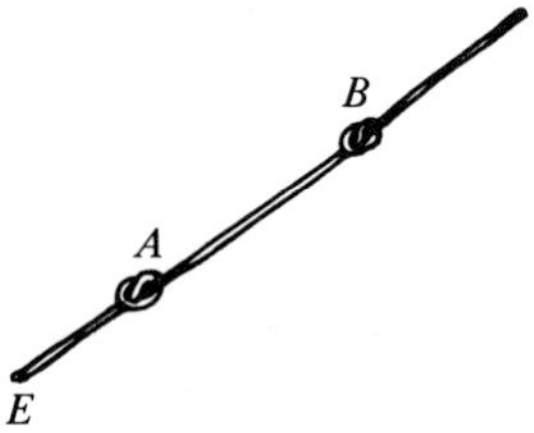

Figure 5-1

Suppose now that the two segments do not have a common endpoint, or that if they do, one segment is not a subset of the other. To create a fanciful situation, one can conjure up the image of the caveman father looking at his sons and noting that Em and On are the same height, that Ut is taller than On, and that Ag is shorter than Em.

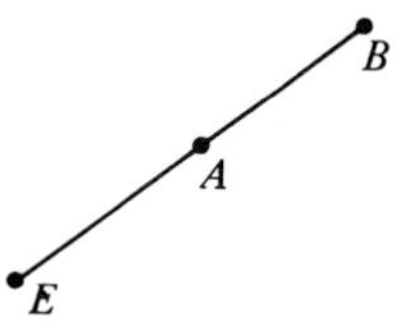

Figure 5-2

Let us see how this situation is related to certain properties of geometric figures. Since we are concerned only with the height of each son, we might replace each son's picture in Fig. 5-4 with a figure that eliminates all the extraneous properties, such as arms and legs, retaining only the relevant property of **distance** between two points, the tip of the head and the bottom of the feet. Line segments, in the geometry we have been considering, fit this criterion very nicely. In Fig. 5-5, $\overline{ON}$ corresponds to son On and $\overline{UT}$ corresponds to son Ut.

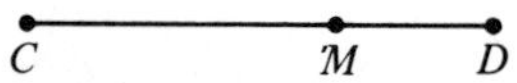

Figure 5-3

In order to determine who was taller, the caveman father probably had On stand beside Ut. He noticed that a position on Ut corresponded to the tip of On's head. (See Fig. 5-6.)

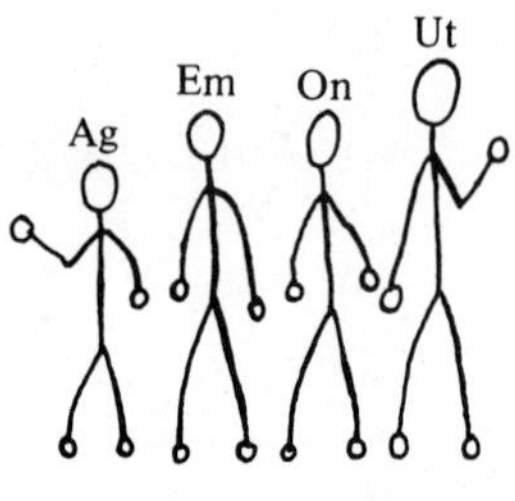

Figure 5-4

A similar analysis could be made in terms of the line segments that represent the heights, $\overline{ON}$ and $\overline{UT}$. Using Fig. 5-7 as a guide, we see that U on $\overline{UT}$ corresponds to O on $\overline{ON}$. There is a point on $\overline{UT}$ that corresponds to N on $\overline{ON}$, so that the distance from this point to U is the same as the distance from O to N. We shall call this point N'. Do you see that U and

N' determine a line segment that is a subset of segment $\overline{UT}$? What is the name of this segment? Notice that this segment, $\overline{UN'}$, is not equal to $\overline{ON}$, since $\overline{UN'}$ is one set and $\overline{ON}$ is a different set. But notice that the distance from O to N is related to the distance from U to N'. What is the relationship between these two distances?

When two segments are related in such a way that the distances between their endpoints are equal, we say that the segments are **congruent**.

How do we determine that $\overline{UT}$ is longer than $\overline{ON}$? Intuitively, one might say that there is more to $\overline{UT}$ than there is to $\overline{ON}$. How might we express this, using relations between geometric figures? Assuming that we can find the point N' that corresponds to N, we have $\overline{UT} = \overline{UN'} \cup \overline{N'T}$. (The reader will note that we have not yet discussed how to locate the point N' that corresponds to N such that the corresponding segments are congruent. We simply assume that we can do it. This is simple enough to do with the two sons—they can be moved into position. But can line segments be moved? This question will be investigated further in Chapter 7.)

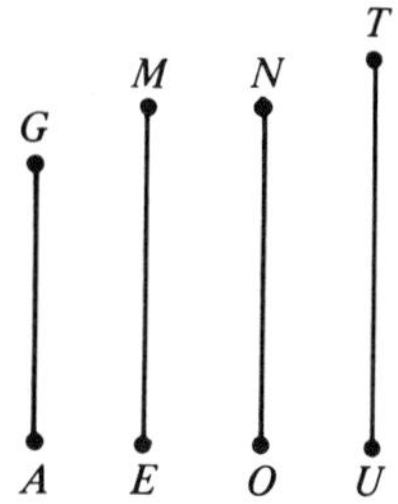

Figure 5-5

Note that N' partitions $\overline{UT}$ into three sets: the point N' and the two sets on either side of N'. We shall say that the two segments UN' and $N'T$ do not **overlap.** By **nonoverlapping segments** we shall mean two segments that have the same carrier and whose intersection is no more than one point. Since $\overline{UN'}$ and $\overline{N'T}$ have the same carrier and since $\overline{UN'} \cap \overline{N'T} = \{N'\}$, the two segments are nonoverlapping. Recalling that a point has no size, it seems right to say that the length of $\overline{UT}$ is the same as the sum of the lengths of $\overline{UN'}$ and $\overline{N'T}$. Remember that $\overline{ON}$ and $\overline{UN'}$ are congruent.

Figure 5-6

By saying that $\overline{UT}$ is longer than $\overline{ON}$, then, we seem to be saying that $\overline{UT}$ can be thought of as the union of at least two segments whose intersection is a single point, where one of the segments is congruent to $\overline{ON}$.

EXERCISE 5-2

1. Using Fig. 5-8:
   (a) Name two pairs of nonoverlapping segments.
   (b) Name two pairs of overlapping segments.
2. Construct a discussion similar to the one in the text that explains why $\overline{AB}$ is longer than $\overline{XY}$ in Fig. 5-8.

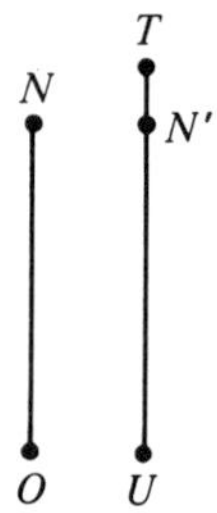

Figure 5-7

5-3 MEASUREMENT AND THE TRANSITIVE PROPERTY

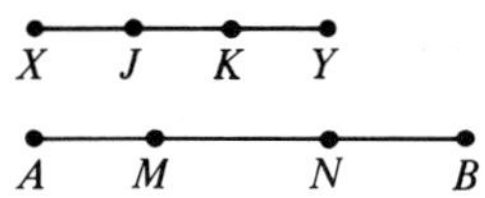

Figure 5-8

To return to the prehistoric situation, the caveman father can determine that son Ag is shorter than son Ut in at least two different ways. He can compare the two sons directly by having one stand next to the other. But he can also reason that since Ut is taller than On, and since On is taller than Ag, then it must be that Ut is taller than Ag. This is precisely the relational property known as transitivity.

Much the same information is conveyed if we have available the height measurements of each of the four boys: Ut, 56 in.; On, 48 in.; Em, 48 in.; and Ag, 40 in. We say that Ut is taller than On since $56 > 48$; On is as tall as Em since $48 = 48$; and Ag is shorter than On since $40 < 48$. Using the transitive property of the order relation for whole numbers, we can show that Ut is taller than Ag. That is, $56 > 48$ and $48 > 40$ implies that $56 > 40$.

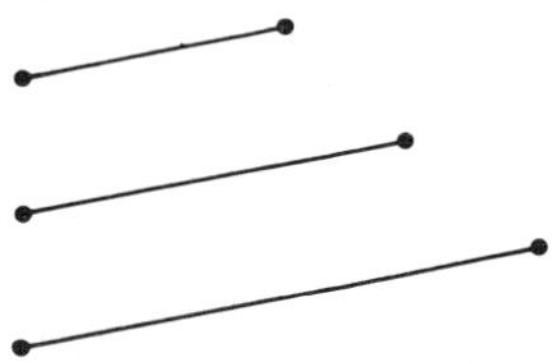

Figure 5-9

Of course, the caveman did not have at his disposal our system of measurement, which uses a combination of numbers and units. The caveman could put his sons in order according to their heights, but he could not state how much taller one son was than another. What is there about our present system of measurement that allows us to do just this?

This new question applies to practically all measurement situations in geometry. It is usually easy enough to order three segments according to their lengths, as in Fig. 5-9, but what allows us to say that one segment is so many feet longer than another? Often we can visually order three rectangular regions according to the size of their interiors, as in Fig. 5-10, but what allows us to say that one region is so many square feet larger than another? Similar questions might be posed for the size of closed regions in space and the size of angles.

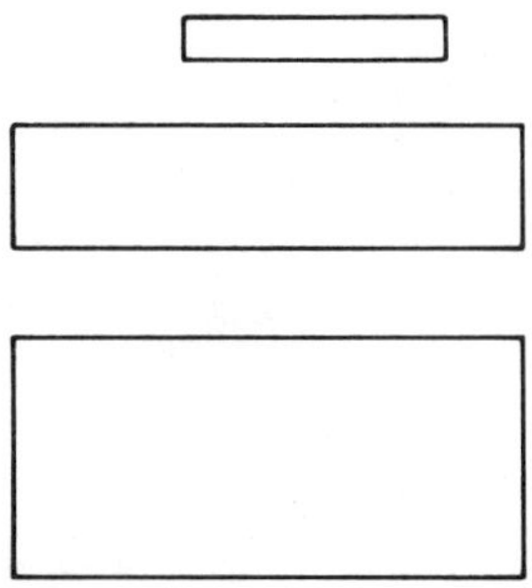

Figure 5-10

5-4 MEASUREMENT AND COUNTING

No one really knows how the transition from "taller than" to "so many inches taller than" occurred. It probably did not happen until man developed the ability to count objects (and defined the unit inch). In fact, as we shall soon see, one way we can interpret the process of measuring is by converting the measuring act into a counting act.

Two questions seem relevant to any explanation of the modern concept of measurement. First, why does man want to know how much taller (longer, greater in area) one thing is than another? This question is answerable by the psychologist or the social historian, and we shall not attempt to answer it here. Second, how does man go about deciding how much taller one thing is than another?

The brief review that follows will help clarify the role of counting in the measurement process. The end result of counting a set of objects is the association of a counting number with the set. Given two sets of objects, one can find how many more objects are in one set than in the other by subtracting one counting number from another.

Now, for sets of physical objects, each object in the set is distinguishable from every other object in the set. Even if the objects all look alike, as in a set of red pencils, we are able to refer to this pencil or that pencil and know which one we are talking about. We usually refer to this situation by saying that the set consists of a collection of **discrete** objects. Examples are familiar to everyone: How many fingers on a hand? How many books in a library? How many articles in a house?

This notion of discrete objects is not immediately apparent when we are trying to determine the length of a segment or the area of a closed region. (See Fig. 5-11.) There seems to be no collection of discrete objects that can be counted (except that there is one segment or one region). In these two cases the sets involved are said to be **continuous**. There is no readily apparent, natural subdivision of the set into a number of discrete objects. The process of measurement is essentially a way of associating a unique number with a continuous set. (It should be noted here that simple counting is not measurement.) A first approximation to measurement then would be a reduction of the problem to one of counting discrete objects. But what discrete objects are to be counted?

Figure 5-11

Let us go back to the caveman and his sons. The father takes a big step forward in the measurement process when he notices that not only is Ut taller than On, but Ut is one head taller than On. The father also notices that On is one head taller than Ag. (What assumption do you think he is making about the comparative size of Ut's head and On's head?) With some imagination and some skill at counting, father can figure that Ut is two heads taller than Ag. Since father is also quite smart, he notices that Ag is five heads tall. (See Fig. 5-12.) How tall are On and Ut? How tall is Em? Thus is born, at least in our fanciful historical description, the primitive idea of measurement as a counting process. In the primitive measurement situation, the object that is counted is called a **unit**. What unit is father using?

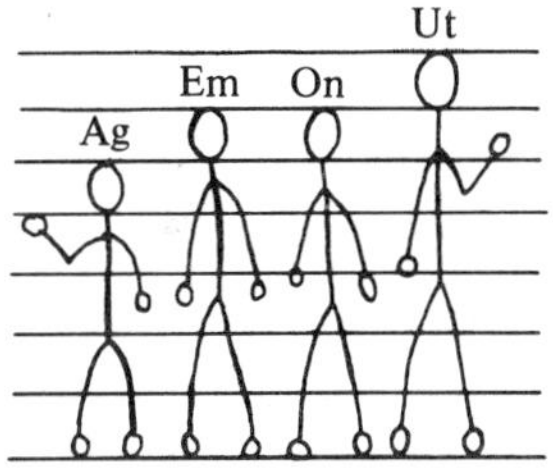

Figure 5-12

Referring to Fig. 5-13, we can examine this situation using the geometric ideas of line segments and congruence. The height of Ut's head corresponds to $\overline{N'T}$. As in Sec. 5-2, N' corresponds to N and $\overline{ON}$ and $\overline{UN'}$ are congruent. What segment shown in the figure is associated with the height of On's head? Now, what must be the relationship between $\overline{G'N}$ and $\overline{N'T}$? Why did you say congruent?

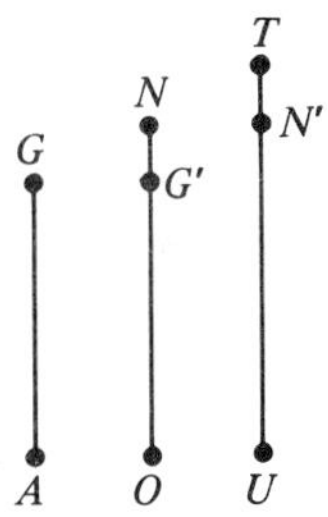

Figure 5-13

The segment TN' is called a unit, since it establishes the basis for the counting process associated with primitive measurement. We shall further agree that its **measure**, the number to be associated with the length of the unit segment, is one. Since $\overline{G'N}$ is congruent to $\overline{N'T}$ what must its measure be?

Earlier, the caveman said that Ag was five heads tall. Notice that $\overline{AG}$ may be thought of as the union of five nonoverlapping segments, each congruent to the unit segment, $\overline{N'T}$. That is, $\overline{AG} = \overline{AB} \cup \overline{BC} \cup \overline{CD} \cup \overline{DE} \cup \overline{EG}$. (See Fig. 5-14.) If we use the unit $\overline{N'T}$, the number associated with the length of $\overline{AG}$ is five. That is, the measurement of $\overline{AG}$ is five units.

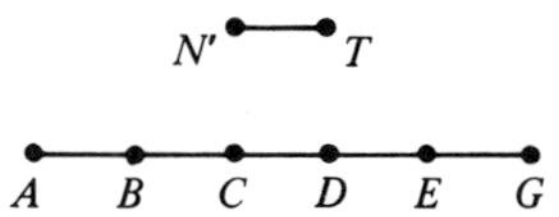

Figure 5-14

Will the measurement of $\overline{AG}$ still be five units if we use a different unit segment longer than $\overline{N'T}$? If the new unit segment is shorter than $\overline{N'T}$?

Your next question might well be: What happens if the segment to be measured cannot be thought of as the union of segments all of which are congruent to the unit segment? If $\overline{XY}$ is the unit, $\overline{YZ}$ in Fig. 5-15 is an example of this condition. Segments XY, YA, and AB are congruent. Segment BZ is congruent to none of these and is, in fact, shorter than the unit segment. Your knowledge of the ruler gives a clue as to the probable next steps—subdivision of the unit, using either fractional numbers or smaller units.

Figure 5-15

Let us digress to consider some other aspects of measurement in geometry. First, we want to show that the ideas about area, volume, and angle measurement are analogous to the ideas about linear (segment) measurement. Second, we shall discuss some of the terminology used to refer to measures and measurement. Third, we shall consider some geometric shapes whose descriptions depend upon certain metric properties of the figure.

EXERCISE 5-4

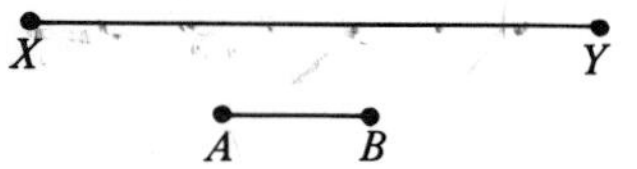

Figure 5-16

1. Using Fig. 5-16, make copies of the unit segment AB on $\overline{XY}$ to find the length of $\overline{XY}$.
2. Use $\overline{CD}$ in Fig. 5-17 as the unit and find the length of $\overline{XY}$.
3. In saying that the lengths of two congruent segments are always equal, what assumption are we making about the unit?

Figure 5-17

5-5 FOUR BASIC MEASUREMENT SITUATIONS

As may be obvious by now, measurement in geometry is a procedure for assigning unique numbers to properties of geometric figures, using as a basis a preestablished unit of measurement. The number associated with the unit of measurement is always the number one.

We consider four basic measurement situations in the geometry we are discussing:

(1) How long is a segment? (How tall, how long, how far?) The unit of measurement is a segment, and the number associated with the measure of the segment is called **length**. The length of a segment is

equal to the **distance** between its endpoints. The length reported with the unit of measurement is called the measurement of the segment. This is **linear measurement.**

(2) How large is a closed region in a plane? (How much surface? What expanse?) The unit of measurement is a closed plane region, and the number associated with the measured region is called its **area.** This is **area measurement.**

(3) How large is a closed region in space? (What capacity?) The unit of measurement is a closed region in space, and the number associated with the measured closed region is called its **volume.** This is **volume measurement.**

(4) How large is an angle? The unit of measurement is an angle, and the number associated with the measured angle is called simply the *measure of the angle.* This is **angular measurement.**

Thus, when we measure some property of a figure, we use something with a like property for the unit.

When we are working with physical objects—the real-world situation—the most direct (but not necessarily the easiest) way to find the measure of something is to make copies of models of the unit of measurement and count how many are required to "fill" the object being measured. A model of the unit is called a **standard.** In the abstract situation—the mathematical problem—we can analyze the geometric figure in terms of the union of a certain number of nonoverlapping figures, each congruent to the unit of measurement. To clarify this idea, let us investigate an area problem and a volume problem.

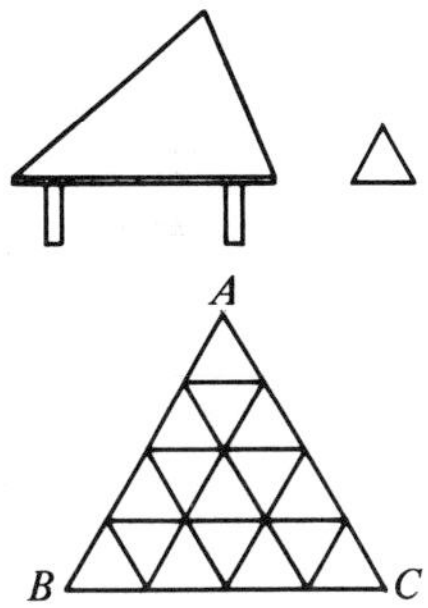

Figure 5-18

AREA: How many triangular regions are needed to cover the surface of the triangular table shown in Fig. 5-18? The direct attack involves cutting out several models of triangular regions, all having the same shape and size as the unit, placing them on the table top, and counting how many are needed. We can represent the same situation by considering the triangular region ABC to be the union of sixteen nonoverlapping congruent triangular regions.

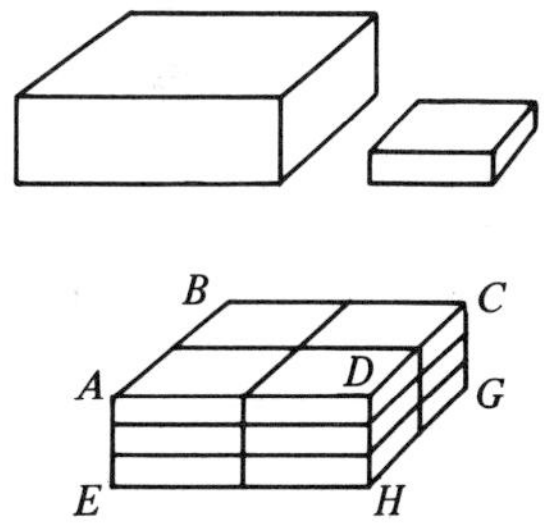

Figure 5-19

VOLUME: How many bricks, all of same size and shape, will it take to fill a certain box? (See Fig. 5-19.) The direct solution is to fill the box with bricks and then count how many were used. The mathematical description involves saying that a polyhedral region $ABCD$-$EFGH$ may be considered as the union of twelve nonoverlapping congruent polyhedral regions.

EXERCISE 5-5

1. Using a pencil as a model of a unit, find the length of a yardstick.
2. Make a copy of the hexagonal region pictured in Fig. 5-20 and use it as a unit to find the area of the cover of this book.

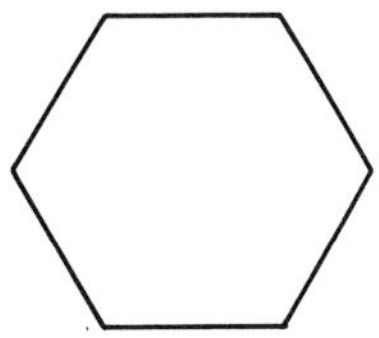

Figure 5-20

3. Using a child's toy block as a model of a unit, find the volume of a half-gallon milk carton.
4. Make a copy of the angle pictured in Fig. 5-21 and use it to find the measure of the angle formed by the hands of a clock at 9:30.

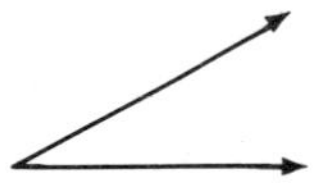

Figure 5-21

5-6 SOME TERMINOLOGY

Since one is apt to encounter different conventions in various books on geometry, we include a brief review of some current terminology for measurement situations. Suppose you have measured a segment, $\overline{AB}$, and have found the result to be 6 in. If you wish to speak only of the number of inches—that is, 6—then you have the **measure** of $\overline{AB}$. A measure is a number. This point is often made to distinguish between a segment and the number associated with some property of it by the measuring process. Clearly, a segment is not a number, but many people make the mistake of writing "$\overline{AB} = 6$." This is technically incorrect, for "6" in this case is not a name for a segment; it is a name for a number. To eliminate this confusion, one of two conventions is usually adopted. If you wish to speak of the measure of $\overline{AB}$, write either "$m(\overline{AB}) = 6$" or "$AB = 6$." Notice that the bar is left off the segment symbol in the second form.

In the Exercises for Chapter 1 we introduced the use of lowercase letters to name segments. In figures this lowercase letter often plays a dual role: naming a segment and being a variable that represents the measure of the segment. The context usually makes clear which interpretation is intended.

Sometimes the measure of a segment is distinguished from its measurement. We have said that the measure of a segment is its length. The measure is a number; the **measurement** is the number recorded with the name of the unit of measurement used. Thus, "6" records a measure, while "6 in." records a measurement. The measurement is 6 inches. There appear to be two reasons for making this distinction. First, a given segment may have different measures, depending upon the unit of measurement used. Recording the unit eliminates confusion on this score. Second, it is often necessary to combine several measures, using the operation of addition (or multiplication, or whatever). Technically speaking, the operations deal only with pure numbers, measures, not with what are sometimes called denominate numbers, measurements. Many mathematicians would argue that it is not correct to write "2 in. + 3 in.," since "+" is used to signify an operation on numbers. The expression "2 in." represents a measurement, not a number. It is suggested that it is better to write "2 + 3," find the sum, and then write the measurement, "5 in." (Notice that you must be careful when finding the combined linear measure of two segments, one of which has measurement 2 *ft* and the other 3 *in*. To find the sum 2 + 3 would not produce the commonly accepted answer.)

The words "length," "area," and "volume" are often used ambiguously. Sometimes they refer to the measure associated with a particular figure; sometimes they refer to the measurement of a particular figure; less frequently they refer to a figure itself. In most instances the context will make the meaning clear.

5-7 GEOMETRIC FIGURES WHOSE DESCRIPTION DEPENDS ON METRIC PROPERTIES

Certain geometric figures and relations between figures are distinguishable only by reference to certain metric properties. Perhaps you have wondered why little mention has been made of such familiar figures as the square and the circle. Simply, it was desirable for us to introduce measurement ideas before characterizing these figures.

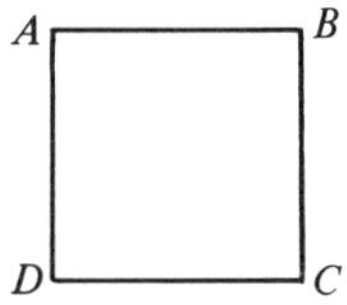

Figure 5-22

For example, in order to completely characterize the square (Fig. 5-22), we must state three conditions (according to one definition), only one of which is a nonmetric property. Both of the other conditions are metric properties. A figure is a square if and only if: (1) the figure is a quadrilateral, (2) each of the four segments has the same measure, and (3) each of the four angles determined by the quadrilateral has the same measure. If the figure is known to satisfy these conditions, then it is a square. If the figure is known to be a square, then it will have the properties stated in these conditions.

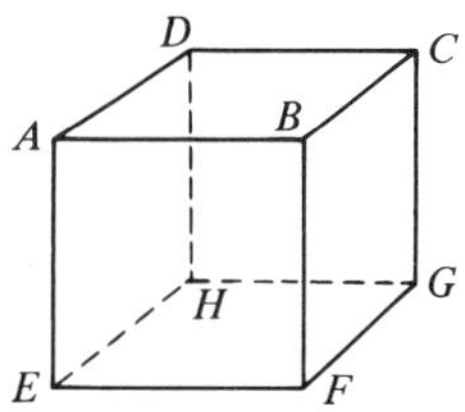

Figure 5-23

Another common figure that requires metric properties in the statement of its defining conditions is the **cube** (Fig. 5-23). A set of necessary and sufficient conditions for a cube is: (1) the figure is a hexahedron, (2) each edge of the hexahedron has the same measure as every other edge, and (3) all the polyhedral angles determined by the faces are the same size.

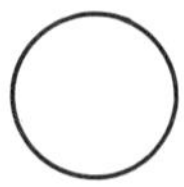

Notice the metric and nonmetric properties stated in this definition of a circle: A **circle** is a simple closed plane curve such that all segments that have one endpoint at a given point (point O in Fig. 5-24) and the other endpoint on the simple closed curve are the same size.

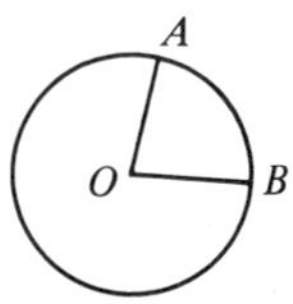

Figure 5-24

The given point, O, is called the **center** of the circle. The segments mentioned in the definition are the radii of the circle. Each segment is called a **radius** of the circle. (The word "radius" also often denotes the measure of the radius segment.)

The space figure corresponding to the plane figure circle is called a sphere. A **sphere** is a closed surface in space such that all segments that have one endpoint at a given point (point A in Fig. 5-25) and the other endpoint on the closed surface have the same measure.

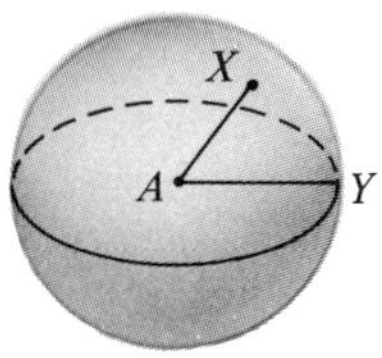

Figure 5-25

Again the given point is called the center and one of the segments is called a radius of the sphere.

EXERCISE 5-7

1. Draw a figure that satisfies conditions (1) and (2) for a square given in the text but does not satisfy condition (3).
2. Sketch a picture of the space figure that satisfies conditions (1) and (2) for a cube given in the text but does not satisfy condition (3).
3. How many radii does a circle have?
4. What is the intersection of a plane with a sphere?
5. What do you think is meant by "congruent circles"?

5-8 LINEAR MEASUREMENT— A MATHEMATICAL ABSTRACTION

Now that we have noted some of the basic similarities among four measurement situations and have seen how metric properties function in characterizations of geometric figures, it is time to finish describing the mathematical abstraction of linear measurement we began in Sec. 5-4.

Linear measurement has been introduced as a simple counting process in which a given segment is expressed as the union of segments congruent to a given unit segment. This interpretation of measurement fails to give satisfactory results in certain situations. In Fig. 5-26, with $\overline{XY}$ the unit, with $\overline{XY}$, $\overline{AB}$, and $\overline{BC}$ congruent, and with $\overline{CD}$ shorter than the unit, a simple count of the segments congruent to the unit would indicate that $m(\overline{AD}) = 2$. Similarly, with $\overline{XY}$ the unit, with $\overline{XY}$, $\overline{EF}$, and $\overline{FG}$ congruent, and with $\overline{GH}$ shorter than the unit but longer than $\overline{CD}$, a simple count of the segments congruent to the unit would indicate that $m(\overline{EH}) = 2$. Under this interpretation of measurement, $\overline{AD}$ and $\overline{EH}$, which are not congruent, would have the same measure associated with them. Ideally, this is not a desirable result. If two segments are congruent, we want them to have the same measure. Conversely, if two segments have the same measure, we want them to be congruent. Clearly, this second condition does not hold in the present case. Is there some way to assign numbers as measures of segments so that both of the conditions stated above will hold?

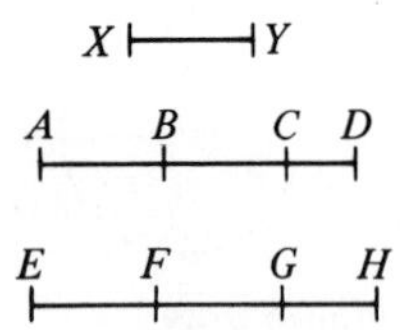

Figure 5-26

One might first try to apply nonnegative rational numbers (sometimes called fractional numbers) to the situation. Let us return for a moment to the problem of counting discrete objects. Sometimes we wish to know more than just the number of objects in a given set; we may want to know the relationship of the number of objects in one set to the number of objects in another set or to the number of objects in a subset of one of the sets. One can ask the question: How many more? But one can also ask: What is the ratio of the number of elements in one set to the number of elements in another set? For example, considering a class as a set of thirty pupils and knowing that there are twelve girls in the class, the ratio of the number of girls in the class to the number of students in the class is $\frac{12}{30}$. This ratio is the quotient of two whole numbers, the denominator indicating the number of elements of the unit, the numerator indicating the number of elements in the subset of the unit. In any case, the set of numbers that associates pairs of whole numbers as quotients is often called the set of nonnegative rational numbers or the set of fractional numbers. (Are there sets of numbers other than the set of counting numbers and the set of rational numbers?)

Look again at $\overline{AD}$ in Fig. 5-26. Segment CD is shorter than the unit $\overline{XY}$. Using Fig. 5-27(a) as a guide, if C corresponds to X, then we have agreed that there is a point D' on $\overline{XY}$ that corresponds to D such that $\overline{XD'}$ has the same measure as $\overline{CD}$. D' separates $\overline{XY}$. We can consider that $\overline{XY} = \overline{XD'} \cup \overline{D'Y}$. Now, if $\overline{XD'}$ has the same measure as $\overline{D'Y}$, then we could say that the unit segment has been subdivided into two congruent parts. We could associate the number $\frac{1}{2}$ with the measure of each of these segments. If the number 0 is associated with endpoint X of the unit segment and the number 1 is associated with endpoint Y, it seems natural to associate the number $\frac{1}{2}$ with the endpoint D'.

Considering $\overline{EH}$ of Fig. 5-26 in a similar manner, suppose that H corresponds to H' on $\overline{XY}$ and that there is another point, P, on $\overline{XY}$ such that $\overline{XY} = \overline{XP} \cup \overline{PH'} \cup \overline{H'Y}$. Suppose further that $\overline{XP}$, $\overline{PH'}$, and $\overline{H'Y}$ are congruent. What number might naturally be associated with one of these segments? [See Fig. 5-27(b).] What number does it seem natural to associate with H' if 0 is associated with X and 1 with Y?

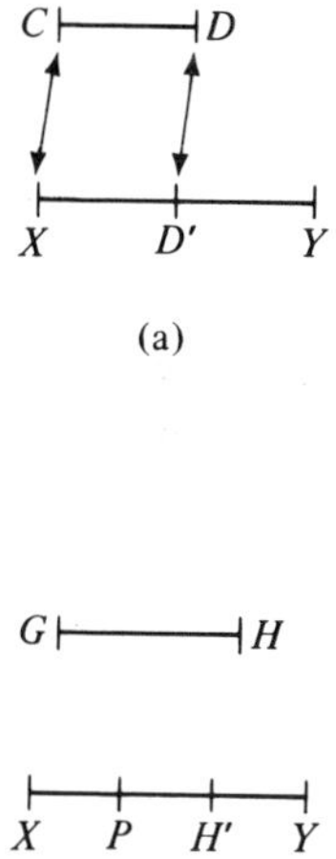

Figure 5-27

These two situations are examples of the process of subdividing the unit segment into a given number of congruent, nonoverlapping segments. It seems evident that it is logically possible to subdivide the unit segment into as many nonoverlapping, congruent segments as is desired. Then, depending upon the number of subdivisions, it is possible to associate certain rational numbers with the endpoints of these congruent segments. These numbers are called the **coordinates** of the endpoints. For example, in Fig. 5-28 there are ten congruent segments so that the endpoints of the segments have coordinates 0, 0.1, 0.2, 0.3, . . . , 0.9, 1.0.

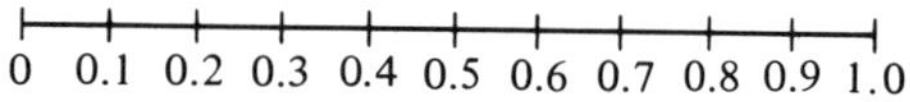

Figure 5-28

To find the measure of any segment [say $\overline{ST}$ of Fig. 5-29(a)] it would seem sufficient to: (1) establish a unit (say $\overline{AB}$), (2) express the segment as the union of nonoverlapping segments congruent to the unit segment and one other segment that is shorter than the unit:

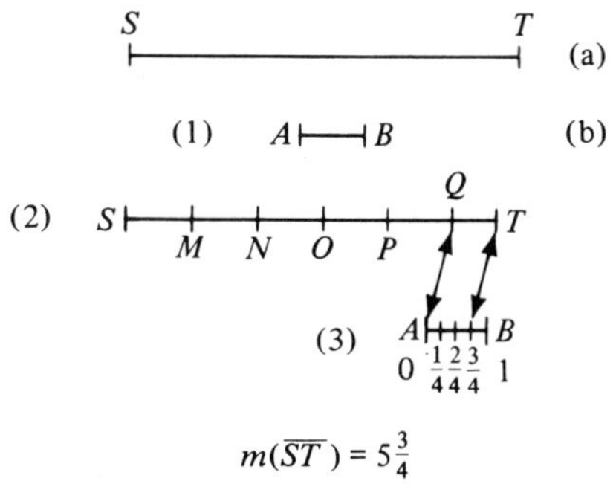

Figure 5-29

$$\overline{ST} = (\overline{SM} \cup \overline{MN} \cup \overline{NO} \cup \overline{OP} \cup \overline{PQ}) \cup \overline{QT},$$

(3) find the subdivision of the unit so that one endpoint of the shorter segment corresponds to an endpoint of the unit and so that the other endpoint of the shorter segment corresponds to one of the points that subdivides the unit, or (3′) at this point, having found the subdivision of the unit, you may wish to redefine the unit as one of the subdivisions.

One then obtains the measure of the segment by counting the number of units (five in the case of $\overline{ST}$) and deciding which rational number corresponds to the endpoint of the shorter segment (in this case $\frac{3}{4}$). [Thus, using the subdivision approach, $m(\overline{ST}) = 5\frac{3}{4}$. Using the unit-redefinition approach, $m(\overline{ST}) = 23$.]

The question now is: Is it always possible to find a subdivision of the unit segment into a set of congruent segments so that the endpoint of the shorter segment corresponds to an endpoint of one of the congruent segments of the subdivision? In other words, is the set of rational numbers sufficient to allow us to assign a unique number as the measure of every possible segment, once a unit has been selected? The answer is, "No!" There are some segments whose measure is neither a counting number nor a nonnegative rational number. (A more complete analysis of this situation will be possible after we have studied the Pythagorean theorem in Sec. 10-3.) If we are to be able to assign a number to every possible segment, the set of rational and irrational numbers is required and is sufficient.

We can summarize this problem, and present an idealization of the linear-measurement situation, by considering an abstraction called the **coordinatized ray**.

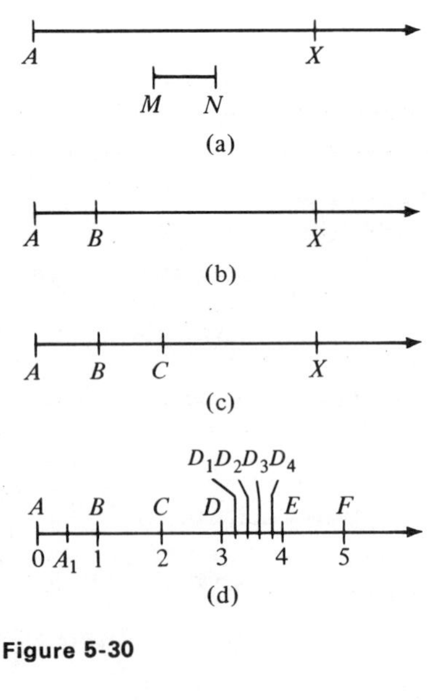

Figure 5-30

Figure 5-30(a) shows a picture of a ray, $\overrightarrow{AX}$, and a unit segment, $\overline{MN}$. If M corresponds to A, we shall agree that there is a point, call it B, on $\overrightarrow{AX}$ such that $\overline{AB}$ has the same measure as $\overline{MN}$. [See part (b) of the figure.] If we now consider $\overrightarrow{BX}$, M corresponds to B, and there is a point, call it C, on $\overrightarrow{BX}$ such that $\overline{BC}$ has the same measure as $\overline{MN}$. [See part (c).] Notice that this point is also on $\overrightarrow{AX}$. Why? This analysis can be continued indefinitely, leading us to believe that $\overrightarrow{AX}$ can be thought of as a union of nonoverlapping segments, each of which is congruent to the unit segment. Thus,

$$\overrightarrow{AX} = \overline{AB} \cup \overline{BC} \cup \overline{CD} \cup \overline{DE} \cup \ldots.$$

If we consider point A to be associated with the number 0, it seems natural to associate the number 1 with point B, since the measure of $\overline{AB}$

is 1. What is the measure of $\overline{BC}$? What is the measure of $\overline{AC}$? Since this measure is 2, what number could be associated with C? Similarly, the number 3 corresponds to D, 4 to E, and so on. All the counting numbers are associated with points on the ray.

Recalling the discussion of the subdivision of the unit segment into congruent segments, what point on $\overrightarrow{AX}$ in Fig. 5-30(d) should correspond to the number $\frac{1}{2}$? What point should correspond to $3\frac{1}{5}$? Does it not seem evident that every rational number can be associated with a unique point on the ray? (Is $\frac{2}{4}$ a different rational number from $\frac{1}{2}$?) Every rational number can be associated with a unique point on the ray.

Once a unit segment has been chosen, there are some points on the ray that cannot be associated with any rational number. These points, together with the endpoint of the ray, determine segments that were known even to the ancient Greeks as incommensurable with the unit. They knew no number to associate with these segments. We would encounter the same kind of situation in trying to find a number for point A_1 in Fig. 5-30(d) if there were no subdivision property.

Today we realize that the points corresponding to the endpoints of the "incommensurables" have coordinates in the subset of the real numbers known as the set of irrational numbers. Thus, to be able to assign a unique number to every point on a ray, we will find it necessary and sufficient to have at our disposal the set of nonnegative real numbers.

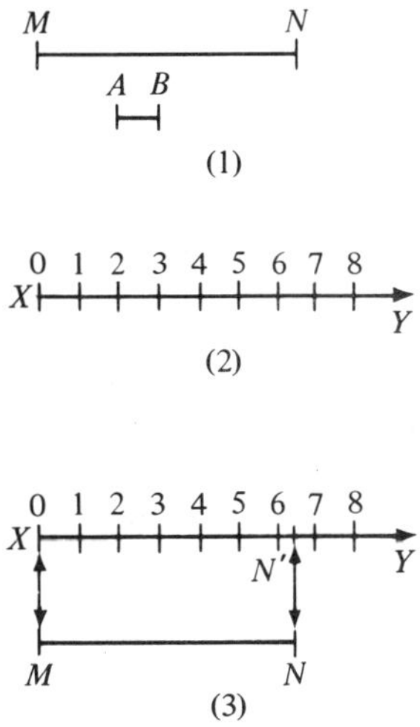

Figure 5-31

This procedure of associating the set of real numbers with points of the ray such that each real number corresponds to a unique point and each point corresponds to a unique real number is called **coordinatizing the ray**. We have established a one-to-one correspondence between the set of points of the ray and the set of nonnegative real numbers. For measurement purposes this correspondence must have certain properties, which we shall discuss later.

What does the coordinatized ray have to do with measurement? You will note that the coordinatized ray is in effect a "ruler." Using a ray coordinatized on the basis of a given unit segment, we can find the measure of any segment in the following manner (use Fig. 5-31 as a guide): Given a segment to measure ($\overline{MN}$):

(1) Establish the unit (say $\overline{AB}$).

(2) Determine the coordinatization of the ray ($\overrightarrow{XY}$).

(3) On this ray one endpoint of the segment to be measured corresponds to the endpoint of the ray. (Thus, M corresponds to X.) The other endpoint of the segment corresponds to some other point on the ray. (Thus, N corresponds to N'.) The segment determined on the ray is congruent to the segment to be measured.

(4) The measure of the segment to be measured ($\overline{MN}$) is the coordinate of the corresponding endpoint (N') on the ray.

In order for the one-to-one correspondence between the set of points on the ray and the set of nonnegative real numbers to be useful for measurement purposes, it must fulfill certain conditions. This is necessary because there is more than one way to establish the one-to-one correspondence. These properties are as follows:

(1) A segment on the coordinatized ray has the same measure as the unit segment if and only if the numerical difference of the coordinates of its endpoints is one.

(1′) Two segments on the coordinatized ray are congruent if and only if the numerical difference between the coordinates of the endpoints of one segment is equal to the numerical difference between the coordinates of the endpoints of the other segment. This assures us that congruent segments will have the same measure.

(2) Given any three points of the ray such that one point is between the other two, then the coordinate of this point will be greater than the coordinate of one of the other points and less than the coordinate of the third point. This maintains the order property.

(3) If $\overline{MN}$ is a segment on the coordinatized ray and O is a point between M and N, then $m(\overline{MO}) + m(\overline{ON}) = m(\overline{MN})$. This guarantees that linear measures are additive.

EXERCISE 5-8

1. Using the unit pictured in Fig. 5-32, find the measure of each of the other segments pictured by using a ray coordinatized on the basis of this unit. Subdivide the unit into five congruent segments.

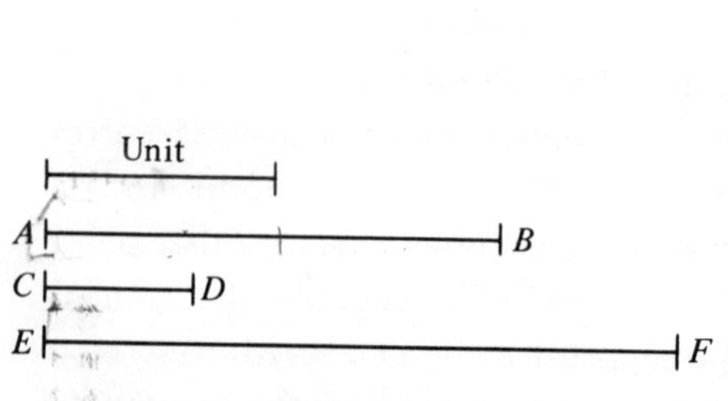

Figure 5-32

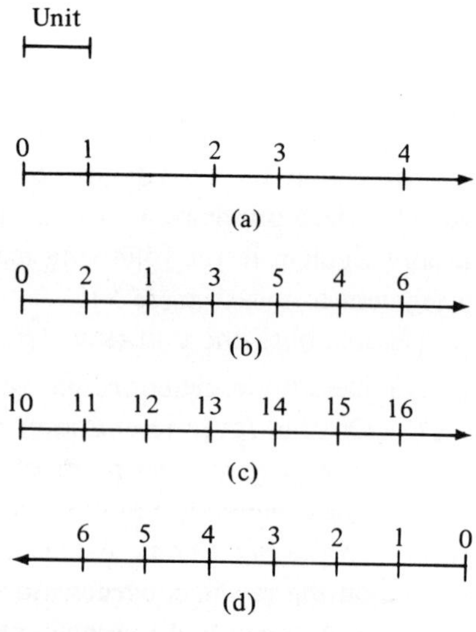

Figure 5-33

2. For each of the one-to-one correspondences between the points of a ray and the set of nonnegative real numbers indicated in Fig. 5-33, tell which of the properties listed above will not hold.

5-9 STANDARD UNITS

More than likely, you have been consciously or subconsciously using your knowledge of the ruler based on inch units to help you follow the discussion in the previous section. We have avoided introducing any common units for linear measurement, because we wish to emphasize that the choice of unit is quite arbitrary. In designing a system for linear measurement, one can choose any segment whatsoever to be the unit for the system.

What then, in the practical measurement situation, determines the choice of unit? Two factors seem to be important: (1) What is the nature of the thing to be measured? (2) What units are available?

Concerning the nature of the thing to be measured, it appears that units smaller than the segment to be measured are, for some reason, preferred. Perhaps this is an outgrowth of early attempts at measuring that simply entailed counting. With the introduction of the real-number system, theoretically it is always possible to express the measure of a segment as a number less than one if the unit segment happens to be longer than the segment to be measured. In any event, it is not usual to measure the length of a piece of notebook paper in terms of yards; it is not typical to measure the height of an ordinary building in terms of miles. Yet it also seems to be the case that units that are "too small" are not preferred. For example, building heights are not usually reported in terms of inches, and distances between cities are not given in terms of feet.

The second factor concerns the availability of units. That is, what standards are available on a **ruler** that can be conveniently brought to the measuring situation as a model of the coordinatized ray? Although we are apt to take the common units of linear measurement (inches, feet, yards, miles) for granted, they have not always been a part of man's history. In the early history of the use of measurement different tribes and communities (and even different persons within communities) had their own units. It was not until people began to trade extensively that certain conventional units came into general use. These units, which are accepted and used by the various peoples of the world, are called **standard units.** At present, they are usually defined by legal statute. The Exercises list some articles containing brief histories of the development of standard units.

Although the most commonly used system for linear measurement in the United States is the English system (inches, feet, yards, miles, and so on), the official standard unit of linear measurement is the *meter*. Legislation enacted in 1866 by the Congress made it legal to use the meter in this country. The Bureau of Standards adopted the meter as the standard unit in 1893 and then defined the yard to be $\frac{3600}{3937}$ of the meter.

The meter was originally defined as one ten-millionth of the distance along the meridian through Dunkirk and Barcelona from one of the poles to the equator. The official standard was kept in Paris. Since then a new standard has been established in terms of the wavelength of the light emitted by krypton 86. This is about one hundred times as precise as the platinum standards kept by bureaus of standards.

For those who have never seen a "meter stick," a meter is approximately 3 in. longer than a yard. The subdivisions of the meter have been deliberately constructed to conform to the decimal base of our system of numeration; that is, each subdivision is further subdivided into ten subunits. The prefixes used in the names of the various subunits indicate the relationship of the subunit to the meter. "Deci-" means one tenth; "centi-" means one hundredth; and "milli-" means one thousandth. (The same set of prefixes is used with units for weight, time, and volume measurement.) Special prefixes are also used to indicate the various powers of ten as they are combined with the root word "meter" to indicate units larger than the basic meter. "Deka-" means ten: "hekto-" means hundred; and "kilo-" means thousand. Table 5-1 lists names for some of the metric linear units and indicates the relationship to the basic unit.

TABLE 5-1
LINEAR UNITS OF THE METRIC SYSTEM

Name and Abbreviation		*Relation to the Meter*
kilometer	km	1000 m
hektometer	hm	100 m
dekameter	dkm	10 m
meter	m	1 m
decimeter	dm	0.1 m
centimeter	cm	0.01 m
millimeter	mm	0.001 m

This system makes it very easy to interpret measurements in terms of the other units of the system. A measurement of 6.234 meters is clearly 6 meters, 2 decimeters, 3 centimeters, and 4 millimeters. One need read only the place values to convert a measurement expressed in terms of the other units. Since there are 100 cm in 1 m, 0.23 m must be the same as 23 cm. Using the relationship that there are 1000 mm in 1 m, 15 mm must be the same as 0.015 m. How many centimeters is 15 mm?

Figure 5-34

Figure 5-34 pictures a segment whose measurement is 1 cm.

Table 5-2 shows that the relationship between the standard units of the English system is not as systematic as in the metric system.

TABLE 5-2
LINEAR UNITS OF THE ENGLISH SYSTEM

Name and Abbreviation		*Equivalent*
mile	mi	5280 ft
nautical mile		6076.1 ft (approx.)
furlong		40 rods
rod		$5\frac{1}{2}$ yd
yard	yd	3 ft
foot	ft	12 in.
inch	in.	

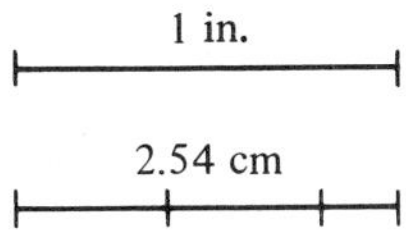

Figure 5-35

To get some idea of the relative sizes of the units of the two systems, note that 1 m is approximately 3.3 ft and 1 cm is approximately 0.4 in. Figure 5-35 shows the relationship between the inch and the centimeter.

EXERCISE 5-9

1. Look up the meaning of the following units:
(a) Cubit. (b) Hand.
(c) Fathom. (d) Myriameter.

2. For brief histories of units of measurement read Ref. 17 and Ref. 20, pp. 1–21.

3. Complete each of the following to make them true statements:
(a) 4 in. = _____ yd. (b) 12 ft = _____ in.
(c) $2\frac{1}{2}$ mi = _____ ft. (d) 3 cm = _____ m.
(e) 1 km = _____ mm. (f) 10 in. = _____ cm (approx.).

4. A standard and a standard unit are the same thing. True or false?

5. "Standard units" are not always as standard as they would seem to be. Look in a standard reference work to find various interpretations of "nautical mile" and "mile."

6. For readings on the use of the metric system in the elementary grades see Ref. 65, pp. 349–353, Ref. 76, p. 179, and Ref. 89, pp. 576–580. For junior high school grades see Ref. 111, pp. 621–623.

5-10 PRECISION AND ACCURACY

In Sec. 5-8 we discussed linear measurement in terms of a mathematical abstraction called the coordinatized ray. For purposes of analysis this conception serves to lay bare the structure of the measurement situation. In this ideal situation it is always possible to uniquely determine the length of any segment by finding the coordinate of a certain point on the ray. In the physical measuring situation it is not possible to determine the exact coordinate of this point on the ruler model of the ray. Physical measuring always gives approximate results. But this should not detract from the mathematical abstraction—for, as we shall see, the abstraction provides a means of explaining some of the questions about these errors in measurement.

In actual practice we determine length either by applying some measuring instrument, such as a ruler, directly to the object or by using some indirect means, such as observations obtained with surveying instruments. (See Chapter 10.) The measurements obtained in the practical situation are rarely, if ever, the "true" measurements of the segment in question. The use of an instrument of any sort introduces error. Any number of things can be the source of the error—for example: (1) the object being measured actually changes its shape and size (for example, the change might be attributed to expansion due to temperature change), (2) the person doing the measuring might be careless, or (3) limitations are built into the measuring instrument.

Errors of type (1) are perhaps best explained by the physicist. Errors of type (2) are best controlled by being careful. Errors of type (3) are beyond our control; we can explain this by referring to the mathematical abstraction for the measuring situation.

Design limitations are related to the familiar concepts of precision and accuracy. These important concepts are often used ambiguously and misunderstood. If a ruler were a perfect measuring instrument, we could determine exactly the two points of the ruler that correspond to the endpoints of the measured segment. Furthermore, for each point of the ruler we would know its real-number coordinate. A real ruler cannot satisfy either of these conditions. One cannot mark every point on the edge of the ruler, nor can one indicate the number that corresponds to each point. Because of the continuous nature of a ray, the corresponding marks would simply run into one another indistinguishably. We must, then, decide how many marks, or subdivisions, are feasible and necessary for a given measuring situation. Notice that for some measuring situations, such as measuring children's heights, it is sufficient if we know the length to the nearest inch. In other situations, such as in machine construction, it may be necessary to obtain measurements to the nearest thousandth of an inch, so that in effect we are using a much smaller unit.

The smaller the unit of measurement, the closer you get to the "true" measurement of the segment being measured. This seems to be the idea people are using when they ask how precise a measurement is. That is, given a particular unit of linear measurement, the more subdivisions of the unit used, the more *precise* will be the measurement obtained. The **preci-**

sion of a measuring instrument is often indicated by listing the smallest subdivision of the unit used on the instrument. For example, given the ruler shown in Fig. 5-36(a), the precision of the instrument—and hence of the measurements made with it—is $\frac{1}{4}$ in. What is the precision of the instrument shown in (b)?

Recently it has been suggested that precision can be indicated in the numerals with which we record measurements. The suggestion is made that the denominator of the fractional number associated with a measurement should always indicate the smallest subdivision of the unit on the instrument used. In Fig. 5-37(a) the measurement would be reported as $\frac{23}{4}$ or $5\frac{3}{4}$ in. In (b) the measurement would be reported as $3\frac{2}{4}$ in. rather than the customary $3\frac{1}{2}$ in. Using this system, how would you report the measurement in (c)?

We can establish similar conventions when using decimal numerals and decimal fractions; the number of decimal places can indicate the precision of the measurement. For "5.241" the precision is one thousandth of the unit. In order to record a measurement of five units and show precision to the thousandth of a unit you would write "5.000 units."

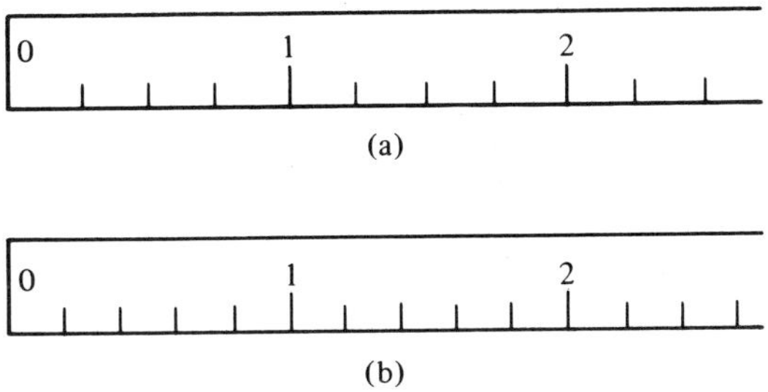

Figure 5-36

Determining the degree of precision from the numeral is not so simple when the measurement involves only numerals for whole numbers. There seems to be no problem with recorded measurements such as "55 units," "1001 units," or "30,302 units." The unit of precision is one unit. Doubts arise, however, when we consider "60 units," "10,000 units," or "30,300 units." Do the zero symbols actually indicate precision to the place value in which they occur, or do they indicate that the measurement has been rounded to the next largest place value? Thus, "60" may indicate precision to the nearest unit; it may also indicate that the measurement has been rounded to the nearest 10 units. Notice that for "10,000" there are five possible interpretations: nearest unit, nearest 10, nearest 100, nearest 1000, and nearest 10,000. Which was actually intended? It would seem that the simplest way to avoid misinterpretation is to state explicitly the actual degree of precision. Otherwise, we are entitled to claim only the least possible precision implied by the numeral.

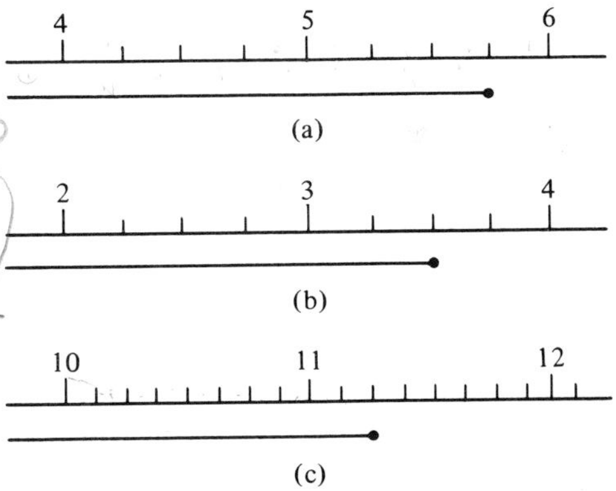

Figure 5-37

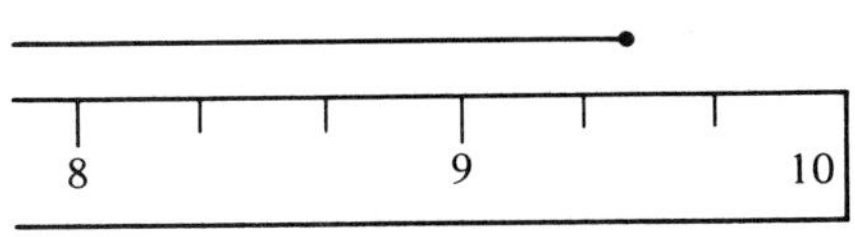

Figure 5-38

To return to the error due to instrument design, it is not difficult to see where error is going to occur when the instrument is used. If you use a ruler to measure a segment, as in Fig. 5-38, and if the endpoint of the measured segment does not happen to fall on one of the marked points of the ruler, you will certainly not get a reading of the true measure of the segment. The question is now: How large an error can occur?

A close examination of the diagrams in Fig. 5-39 will help answer this question. In part (a) there is no question that the measurement is $2\frac{2}{4}$ units, and there seems to be no error. In part (e) there is no question that the measurement is $2\frac{3}{4}$ units, and there seems to be no error. The difference between the two measures is $\frac{1}{4}$. What is the precision for this instrument? Looking at part (b), we realize immediately that any measurement reported will be in error if we use the precision of this instrument. Do you agree that this measurement is best reported as $2\frac{2}{4}$ units? We underestimate the true measure, but the error is certainly less than $\frac{1}{4}$. In part (d) we note again that error will be involved, but we agree to call the measurement $2\frac{3}{4}$ units. This time we overestimate the true measure. Again the error is less than $\frac{1}{4}$. In fact, in both (b) and (d) the error is less than $\frac{1}{8}$, or one half of the smallest subdivision.

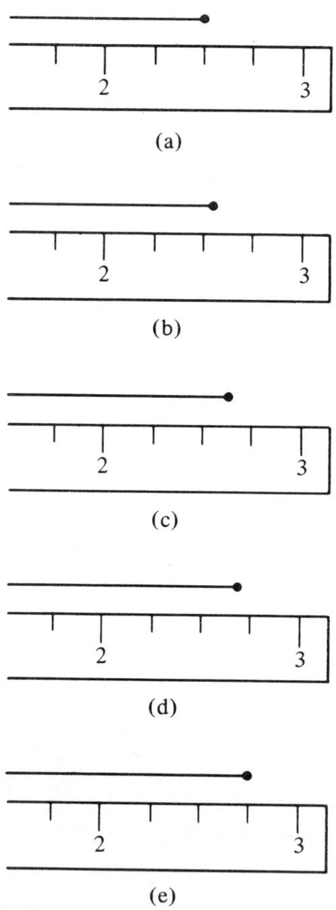

Figure 5-39

Look at part (c) of Fig. 5-39. Here the endpoint seems to be halfway between the two marked points that correspond to $2\frac{2}{4}$ and $2\frac{3}{4}$. If you call the measurements $2\frac{2}{4}$ units, then the error of measurement is one half of the subdivision, or $\frac{1}{8}$. ($\frac{1}{2} \times \frac{1}{4} = \frac{1}{8}$.) But the error is still $\frac{1}{8}$ if you call the measurement $2\frac{3}{4}$ units. Now, notice that if the endpoint of the segment is as in part (b) or (d), the error must be less than one half of the subdivision. The greatest possible error for any measurement with this ruler is $\frac{1}{8}$ unit.

In general, the **greatest possible error**, also known as **absolute error**, for a given ruler will be one half of the smallest subdivision on the ruler. What is the greatest possible error for the ruler shown in Fig. 5-40? Explain why you said "$\frac{1}{10}$ unit." What do we mean when we say one measurement is more precise than another?

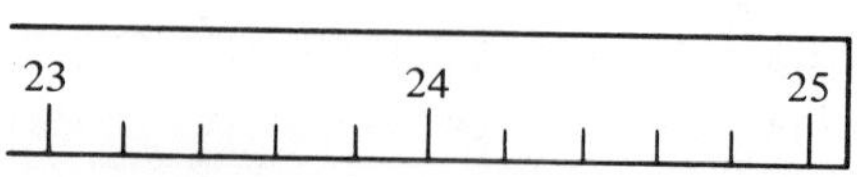

Figure 5-40

Perhaps this discussion reminds you of the concept of accuracy. In fact, many people consider accuracy to be the same as precision. In the technical sense, however, the two are not the same. To appreciate this, consider the following example. Which gives a more accurate result—measuring the length of a pencil to the nearest inch or measuring the length of a room to the nearest inch? In each case the precision, and hence the absolute error, is identical; but surely the room measurement is a more accurate result than the pencil measurement. (Notice that we do not use the word "accurate" here in the sense of being careful, although we do need to be careful when measuring.) *Accurate* here implies a better measurement, relatively speaking. What makes the difference? It seems to be in the relationship of the size of the unit used to the size of the measurement obtained. Technically, in fact, **accuracy** is given by the ratio of the size of the greatest possible error in a measurement to the size of the measurement actually obtained—that is, the ratio:

$$\frac{\text{size of greatest possible error}}{\text{size of obtained measurement}}.$$

This ratio is called **relative error**. Since the greatest possible error is also known as absolute error, the ratio above can be renamed to give the following relationship:

$$\text{relative error} = \frac{\text{size of absolute error}}{\text{size of obtained measurement}}.$$

Returning to the pencil and the room, suppose that the measurement for the pencil is 6 in. and for the room 60 in. The precision is 1 in. for both cases, so that the absolute error is $\frac{1}{2}$ in. For the pencil, the accuracy or relative error is $\frac{1}{2} \div 6$ or $\frac{1}{12}$. For the room, the accuracy is $\frac{1}{2} \div 60$ or $\frac{1}{120}$. Thus, the smaller the relative error, the more accurate is the measurement.

As a final note, let us consider what happens to the amount of error when measurements are combined. Measurements were made to determine the amount of fencing required to enclose the triangular garden shown in Fig. 5-41. Assuming that the measurements are precise to the nearest inch, what is the absolute error of the sum of the three measures?

Our first inclination is to say that 120 in. of fencing will be needed. But the absolute error for each of the three measurements is $\frac{1}{2}$ in. If all the measurements underestimate the true measurements, the error in the sum could be as large as $1\frac{1}{2}$. Thus, the sum of the true measures could be as large as $121\frac{1}{2}$. If the three measurements overestimate the true measurements, the error again could be as much as $1\frac{1}{2}$. Thus, the sum of the true measures could be as small as $118\frac{1}{2}$. What would the sum of the true measures be if two of the measurements were overestimated by $\frac{1}{2}$ unit each and the other was underestimated by $\frac{1}{2}$ unit? For this example, the sum of the true measures can range from $118\frac{1}{2}$ to $121\frac{1}{2}$. This range is sometimes called the **interval of tolerance**. In the future, we shall report

Figure 5-41

sums of measures in terms of the least precise of the measurements associated with the sum.

EXERCISE 5-10

1. For a given measurement the precision is twice the absolute error and the accuracy is given by the relative error. However, precision and absolute error are not synonymous, nor are accuracy and relative error. To see why this is so, complete the following statement. Within a given unit system:
   (a) as precision increases, absolute error (increases, stays the same, decreases);
   (b) as accuracy increases, relative error (increases, stays the same, decreases).
2. Explain the following statement. Relative error is derived from knowledge of absolute error, which is in turn derived from knowledge of precision.
3. Find the absolute error and relative error for each measurement:
   (a) 25 cm (precision: nearest cm).
   (b) 125 mi (precision: nearest mi).
   (c) 2.04 yd (precision: nearest hundredth yd).
   (d) $7\frac{10}{16}$ in.
   (e) 250 mi (precision: nearest 10 mi).
   (f) 0.030 m (precision: nearest mm).
4. No precision was indicated for each of the following measurements. What is the maximum precision we are allowed to assume, using the convention cited in the text?
   (a) 153 in. (b) 150 in.
   (c) 0.235 m. (d) 0.230 m.
   (e) 0.023 m.
5. Report the relative errors obtained in problem 3 as percents.
6. Given measurements of 15 in. and 15 cm:
   (a) Which is the more precise measurement?
   (b) Which is the more accurate measurement?
   (c) Size of unit affects __________ but not __________, for a given number of units.
7. Discuss: The length of the interval of tolerance for the sum of two measures is greater than the precision for either measure.
8. Unfortunately, "tolerance" is another of those words whose use seems ambiguous. Essentially it refers to the interval in which the true measure corresponding to an approximation of that measure lies. At times, the boundary values of the interval are given explicitly. In another fairly standard notation used to report tolerance intervals, the boundary values are expressed in terms of the absolute error or relative error for the measurement. For example, a measurement of 5 in. may have the tolerance interval of $5 \pm \frac{1}{2}$. The upper boundary is found by adding $\frac{1}{2}$ to 5. The lower boundary is found by subtracting $\frac{1}{2}$ from 5. The same interval can be reported as

$5 \pm 10\%$, where 10% equals the ratio of $\frac{1}{2}$ to 5. Find the upper and lower boundary values for each of the following intervals.

(a) $15 \pm \frac{1}{2}$.
(b) $15 \pm 10\%$.
(c) 200 ± 50.
(d) 0.007 ± 0.0005.
(e) $25 \pm 2\%$.
(f) $0.05 \pm 10\%$.

9. Complete the following table.

Measurement	*Absolute Error*	*Relative Error*
2.5 m		
24.8 m		
123.4 m		
2345.1 m		

If the *absolute error* remains constant, the *relative error* (increases, stays the same, decreases) as the numerical value of the measurement increases.

10. Complete the following table, in which precision is indicated by a caret (^) over the digit in the appropriate place-value position.)

Measurement	*Absolute Error*	*Relative Error*
$0.020\hat{6}$ in.		
$0.20\hat{6}$ in.		
$20.\hat{6}$ in.		
$20\hat{6}$ in.		
$20\hat{6}0$ in.		

If the *relative error* remains constant, the *absolute error* (increases, stays the same, decreases) as the numerical value of the measurement increases.

11. A quick comparison of the accuracies for two measurements can be made by simple inspection of the digits in a decimal numeral that reports an approximation to a measurement. Before we can do this, however, we must make some agreements as to which digits in the numeral are accuracy-relevant. We first agree to abide by the convention on reporting precision cited in the text; that is, the smallest place value in a decimal fraction indicates the precision, and only the minimal unquestionable precision is permitted (unless otherwise indicated) when the measure is reported as a whole number.

The following examples will help clarify the agreements to be made.

Consider two measurements: 9 units 6 units
Their accuracies: $\frac{1}{2} \div 9$ $\frac{1}{2} \div 6$
The larger measurement has the greater accuracy.

Consider: 85 units 7 units
Accuracy: $\frac{1}{2} \div 85$ $\frac{1}{2} \div 7$
The measurement with more digits is more accurate.

But more digits do not mean more accuracy when some of the digits are zero.

Consider: 8500 units .850 unit

Accuracy: $\frac{1}{2}(100) \div 8500$ or $\frac{\frac{1}{2}}{85}$ $\frac{1}{2}(.001) \div .850$ or $\frac{\frac{1}{2}}{850}$

In this case the measurement with fewer digits is the more accurate.

The problem lies in the fact that the several zero digits in "8500" serve only to place the decimal point. The first "significant" digit occurs in the place-value position occupied by "5." The place-value positions occupied by the zero digits are "not significant."

Let us consider the effect of zero digits on accuracy. Recall the precision reporting convention.

Consider: 85 units 8500 units

Accuracy: $\frac{1}{2} \div 85$ $\frac{1}{2}(100) \div 8500$ or $\frac{\frac{1}{2}}{85}$

The zero digits are not significant.

Consider: 85 units .0085 unit

Accuracy: $\frac{1}{2} \div 85$ $\frac{1}{2}(.0001) \div .0085$ or $\frac{\frac{1}{2}}{85}$

The zero digits are not significant.

Thus far, it seems that zero digits used to place the decimal point are not significant; that is, they do not affect the accuracy of the measurement. (Perhaps you are confused by what is meant by "used to place the decimal point." Notice that ".0085" is read "85 ten-thousandths" and "8500" is read "85 hundreds." To write the decimal numerals corresponding to these readings, we must introduce zero digits that do not appear in the verbal readings of these symbols.)

Consider: .85 unit .8500 unit

Accuracy: $\frac{1}{2}(.01) \div .85$ or $\frac{\frac{1}{2}}{85}$ $\frac{1}{2}(.0001) \div .8500$ or $\frac{\frac{1}{2}}{8500}$

Here the zero digits *do* affect the accuracy.

Finally, consider: 85 units 805 units

Accuracy: $\frac{\frac{1}{2}}{85}$ $\frac{\frac{1}{2}}{805}$

Again, the zero digit does make a difference.

Then, under our agreement about precision, those digits significant to the question of determining accuracy are (1) all nonzero digits, (2) all zero digits that occur between two nonzero digits, (3) all zero digits that occur to the right of the nonzero digits in a decimal fraction. These we call **significant digits**.

The quick test to determine the relative accuracies of measurements expressed as decimal numerals, then, is:

(1) the greater the number of significant digits in the numeral, the more accurate is the measurement;
(2) if two measurements have the same number of significant digits, then the sequence of significant digits with the greater numerical value corresponds to the more accurate measurement.

Find the number of significant digits in each of the following, and then rank-order the measures according to their accuracies:

(a) 0.027. (b) 35.

(c) 3006. (d) 0.270.
(e) 0.0300. (f) 2.
(g) 300. (h) 405.
(i) 315. (j) 3600.

12. For further reading on accuracy, precision, and significant digits see Ref. 20.

13. The diagram depicts a measure being made that is between 2 and 3. Find the following:

(a) $m(\overline{AB}) =$ _____ to the nearest unit.
(b) $m(\overline{AB}) =$ _____ to the nearest 0.1 unit.
(c) $m(\overline{AB}) =$ _____ to the nearest 0.01 unit.
(d) $m(\overline{AB}) =$ _____ to the nearest 0.001 unit.
(e) What is the square root of 5?

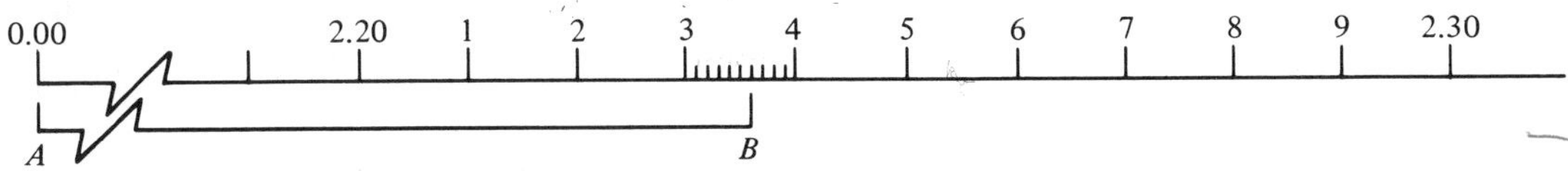

5-11 COMPARISON—ANGULAR MEASURE

Turning our attention to the problem of determining the size of an angle, we shall find that much of the discussion about angular measurement parallels the discussion for segment measurement. We shall look at the nature of angular measurement, a mathematical abstraction for the process, standard units, and the instrument (the protractor) commonly used to determine the size of an angle in the physical measuring situation. We look first at what might be meant by the statement that one angle is larger than another.

Man's preoccupation with the heavens may well have led him to consider relative sizes of angles. If we select two stars in the night sky, one higher above the horizon than the other, and select a point on the ground from which to observe, we have a situation that we can represent by using the angle concept. Taking the observation point, O, as endpoint, there is a ray that contains the position of star B. There is also a ray that contains the position of star A. Now if we use a point, H, on the horizon for land reference, we see that there is also a ray from O through H. Star A occupies a position on $\overrightarrow{OA}$ of $\angle AOH$, and star B occupies a position on $\overrightarrow{OB}$ of $\angle BOH$.

Intuitively, we would say that $\angle BOH$ is larger than $\angle AOH$, since star B is higher in the sky than is star A. But notice that the notion "higher" in this case more or less assumes that we know how to compare angles, since the reference for higher is the horizon. Let us look at Fig. 5-42 to see how we might determine that one angle is larger than another without referring to stars, "higher," or the like. Angles *BOH* and *AOH* have a common side. What is it? Where does side $\overrightarrow{OA}$ lie in relation to $\angle BOH$? With the exception of the endpoint *O* this ray lies in the interior of $\angle BOH$. That is, $\overset{\circ\rightarrow}{OA}$ is in the interior of $\angle BOH$. This observation suggests a way of describing what we mean when we say that one angle is larger than another.

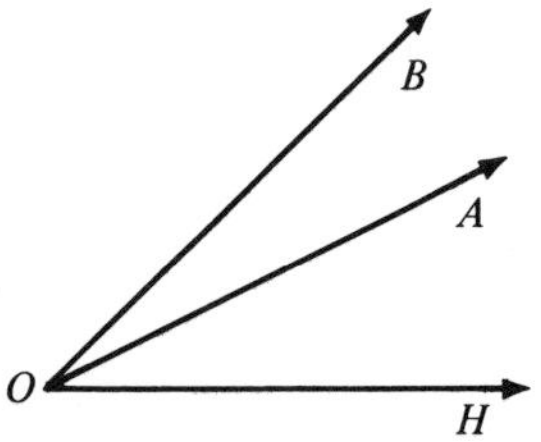

Figure 5-42

We begin by saying that one angle is larger than another angle if and only if the two angles have a common side and the half-line of the other side of the second angle is contained in the interior of the first angle. When would two angles be equal in size? How did we first describe two segments that have the same measure?

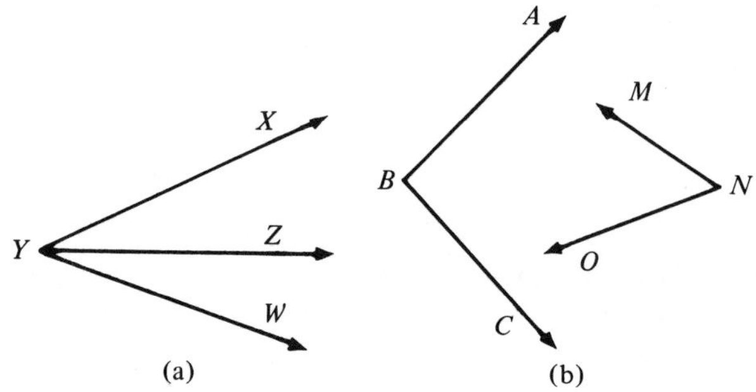

Figure 5-43

This criterion seems appropriate only for those situations in which the two angles have a common side, and a half-line of a ray of one is contained in the interior of the other. What analysis can be made for situations such as those in Fig. 5-43(a) and (b)? Which is larger, $\angle XYZ$ or $\angle ZYW$? Which is larger, $\angle ABC$ or $\angle MNO$? In both cases one of the angles seems to be larger than the other. But in situation (a), even though the two named angles have a common side, the half-line of the second side of neither angle is contained in the interior of the other angle. In (b) there is no common side.

Is it possible to change the description of "larger than" given above to include these situations? Yes—if we are willing to make an assumption, much as we did in the case of segment measurement. Given any angle, say $\angle FGH$ in Fig. 5-44(a), and any ray, say $\overrightarrow{ST}$, if one side of the angle, $\angle FGH$, corresponds to $\overrightarrow{ST}$, then there is another ray with *S* as endpoint, say $\overrightarrow{SQ}$ in (b), such that $\angle TSQ$ is the same size as $\angle FGH$. We shall make this assumption and say that the two angles are **congruent**.

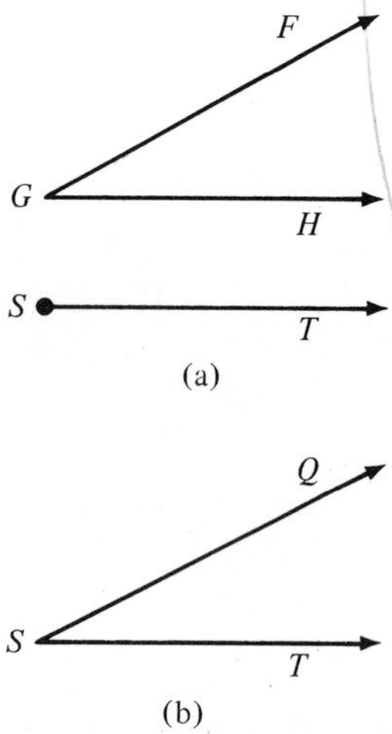

Figure 5-44

Is there only one such angle congruent to a given angle? Figure 5-45(b) shows that there may be two angles that satisfy the conditions—one on either side of the carrier of the ray. Are there more than two such angles? Intuition would say no. We shall strengthen our assumption and say: Given any angle, say $\angle FGH$, and any ray, say $\overrightarrow{ST}$, there is one and only one ray, $\overrightarrow{SQ}$, such that $\angle TSQ$ is congruent to $\angle FGH$ and Q is on the A side of $\overleftrightarrow{ST}$.

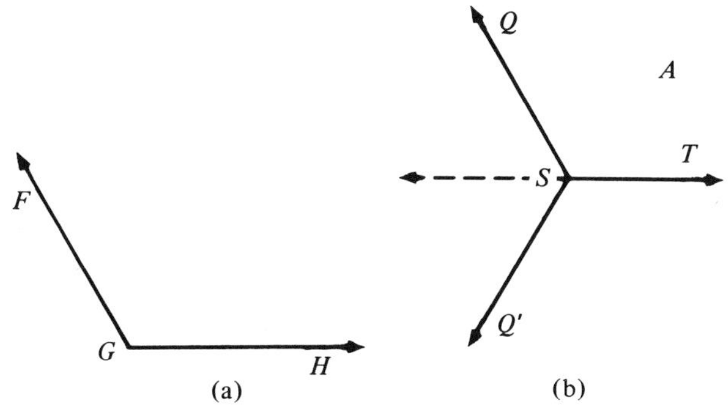

Figure 5-45

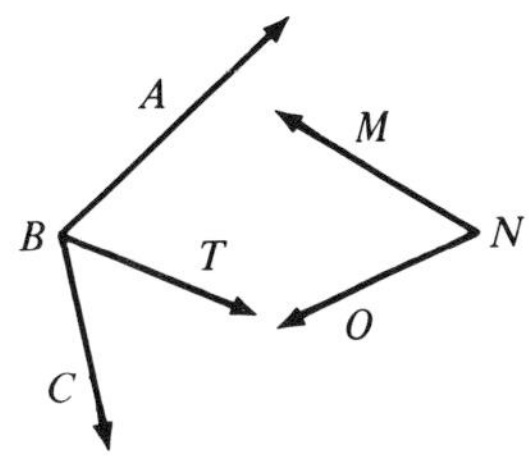

Figure 5-46

Thus, for situation (b) of Fig. 5-43, we can say that $\overrightarrow{NO}$ corresponds to $\overrightarrow{BC}$ and that under our assumption about congruent angles there is another ray through B with at least one point on the A side of $\overleftrightarrow{BC}$, say $\overrightarrow{BT}$, which makes $\angle TBC$ congruent to $\angle MNO$. If $\overset{\circ\!\!\longrightarrow}{BT}$ is in the interior of $\angle ABC$, as in Fig. 5-46, then $\angle ABC$ is larger than $\angle MNO$. What would we conclude if $\overrightarrow{BT} = \overrightarrow{BA}$? What would we conclude if $\overset{\circ\!\!\longrightarrow}{BT}$ is in the exterior of $\angle ABC$?

5-12 ANGULAR MEASUREMENT

The preceding section clarifies what we mean by saying "one angle is larger than another," but we have not yet analyzed what is meant by "one angle is a certain amount larger than another angle." As in the case of linear measurement, this requires the association of a set of numbers with the set of angles. Further, since angular measurement involves the problem of continuous quantities, a set of numbers with the properties of the real numbers will be required. Since the development is so similar to that for line segments, we simply give the angular analogue to the set of steps listed in Sec. 5-8. In Sec. 5-13 we shall note an essential difference between linear and angular measurement.

Given any angle to measure (say $\angle ABC$ of Fig. 5-47):

(1) Establish a unit angle (say $\angle XYZ$).

(2) Starting with a given ray (say $\overrightarrow{QN_0}$), numbering counterclockwise,

establish a coordinatization of the set of angles in a plane all of which have Q as endpoint.

(3) Using this set of rays, one ray of the angle to be measured (say $\overrightarrow{BC}$) corresponds to the ray associated with 0 ($\overrightarrow{QN_0}$), and the other ray of the angle ($\overrightarrow{BA}$) corresponds to some other ray in the set ($\overrightarrow{QT}$).

(4) The measure of the given angle ($\angle ABC$) is the number associated with this second ray.

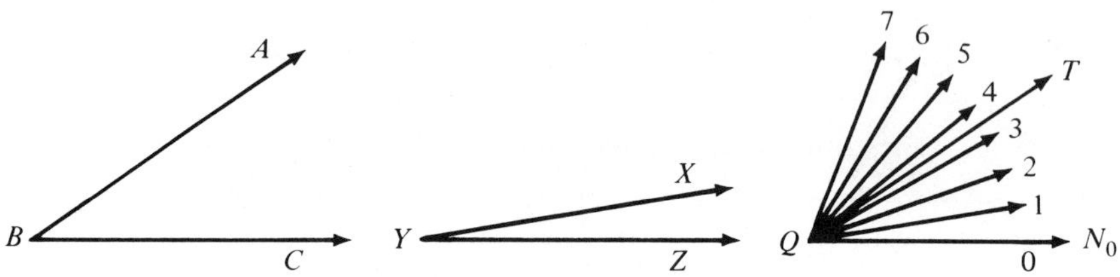

Figure 5-47

Note the importance of each of the following remarks. First, the angle selected as the unit is strictly arbitrary. Any angle will do. Second, Fig. 5-48 will aid in clarifying how the numbers are associated with the rays. Ray YZ corresponds to $\overrightarrow{QN_0}$, and $\overrightarrow{QN_0}$ is associated with the number 0. Using the assumption about the existence of congruent angles, $\overrightarrow{QN_1}$ corresponds to $\overrightarrow{YX}$ in such a way that $\angle XYZ$ and $\angle N_1QN_0$ are congruent. Ray QN_1 is associated with 1. Now, using $\overrightarrow{QN_1}$ to correspond to $\overrightarrow{YZ}$, $\overrightarrow{QN_2}$ corresponds to $\overrightarrow{YX}$ such that $\angle N_1QN_2$ is congruent to the unit angle.

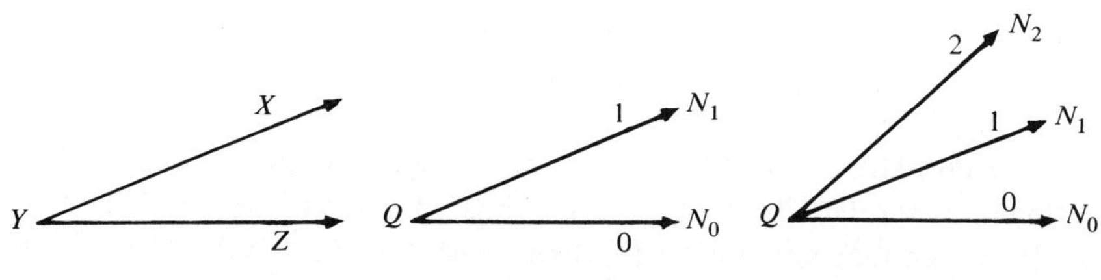

Figure 5-48

Ray QN_2 is associated with 2. This process is continued. Can you continue it indefinitely?

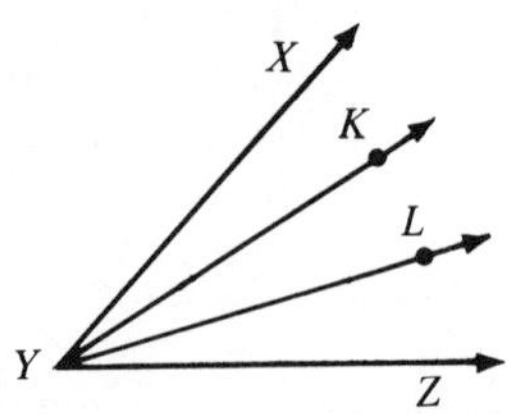

Figure 5-49

Third, the subdivisions of the unit follow a pattern illustrated in Fig. 5-49. Points K and L are in the interior of $\angle XYZ$ and determine $\overrightarrow{YK}$ and $\overrightarrow{YL}$. If these two points are so situated that the three angles XYK, KYL, and LYZ are congruent, and if 0 is associated with $\overrightarrow{YZ}$ and 1 with $\overrightarrow{YX}$, what numbers might be associated with $\overrightarrow{YL}$ and $\overrightarrow{YK}$? Subdividing in this fashion requires the use of the rational numbers as coordinates of the rays.

Just as for segments, there are angles whose measures are neither rational numbers nor whole numbers. We require, and find sufficient, the set of real numbers (in fact, a subset of the real numbers) in order to have a coordinate for every ray in the set.

This correspondence of numbers with the set of rays has certain properties similar to those noted for the coordinatized ray used in linear measurement.

(1) In the **coordinatized set of rays**, if the numbers associated with the rays of an angle have a difference of one, then the angle is congruent to the unit angle and they have the same measure.

(1′) More generally, given two angles formed by rays in the set such that the differences between the numbers associated with the rays of the angles are equal, then the two angles are congruent.

(2) In one of the half-planes determined by the carrier of the ray associated with 0, if the half-line of one ray lies in the interior of the angle formed by two other rays of the set, then the number associated with this ray must be larger than that associated with one of the two rays and smaller than the other.

(3) If point C of $\overrightarrow{AC}$ is in the interior of $\angle BAD$, then $m(\angle BAD) = m(\angle BAC) + m(\angle CAD)$. [Note: "$m($ $)$" is read "the measure of."]

EXERCISE 5-12

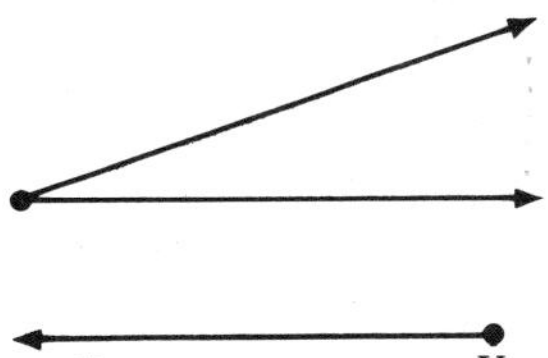

Figure 5-50

Figure 5-51

1. Using the angle pictured in Fig. 5-50 as the unit and a copy of the ray drawn below it, make a coordinatized set of rays, numbering in a counterclockwise direction. (You may find it convenient to trace the picture of the angle to make a cutout copy of it.)
2. How many rays could you number in problem 1 above before you started "covering old territory"?
3. Make a coordinatized set of rays but number clockwise this time. Use the same unit as in problem 1 with a copy of the ray shown in Fig. 5-51.
4. Try to subdivide one of the angles congruent to the unit angle.

5-13 STANDARD UNITS

One of the most common standard units for angular measurement is the degree. There are several ways to approach the definition of an angular degree. Some of these might help us select a suitable definition of angle. (See Sec. 2-8.) The derivation of the degree as a unit angle also illustrates the basic difference between linear measurement and angular measurement.

A most ancient approach to defining the degree is undoubtedly related to the approximate yearly cycle of 360 days. In any event, it is related to the Ancients' use of 60 as a base for a numeration system. Using Fig. 5-52 as a guide, we see that we can conceive of a plane surface as representing a year's time. Selecting a point Q as endpoint, we can choose 360 points in the plane, which with Q determine a set of rays. These points are so selected that each successive pair of rays determines an angle that is congruent to all other angles formed in the same manner. An angle whose measurement is 1 *degree* is then defined to be an angle that is congruent to any one of these 360 angles.

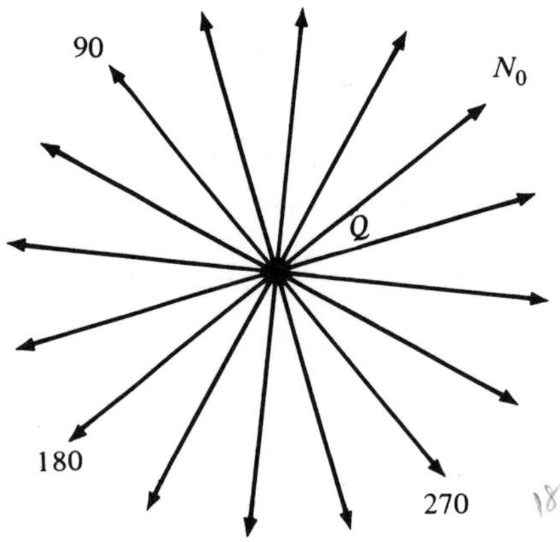

Figure 5-52

We could associate 0 with one ray in this set. Then, numbering successively so that the properties listed above are satisfied, we find that there is a limit to the process unless we want more than one number associated with each ray. (Sometimes this is desirable.) With what number is $\overrightarrow{QN}$ associated in Fig. 5-52? With what ray will 360 be associated? 361? This phenomenon corresponds to the overlap situation you encountered in problems 1 and 2 of Exercise 5-12. Thus, in order to associate a unique real number with every ray in this set of rays with common endpoints, we need use only the real numbers between 0 and 360, including 0. Herein lies an essential difference between linear measurement and angular measurement. Linear measurement requires the complete set of nonnegative real numbers; static angular measurement requires only a special subset of the nonnegative real numbers.

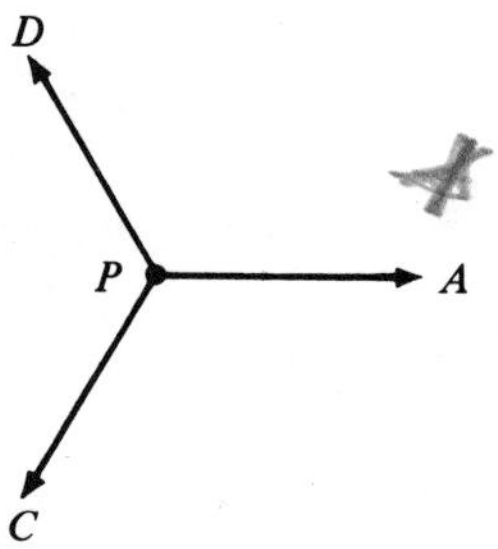

Figure 5-53

An angle of one degree can be separated into any number of subdivisions. The usual number of subdivisions is 60. One of these subdivisions has a measurement of 1 *angular minute*. The angular minute is further subdivided into 60 subunits called *angular seconds*.

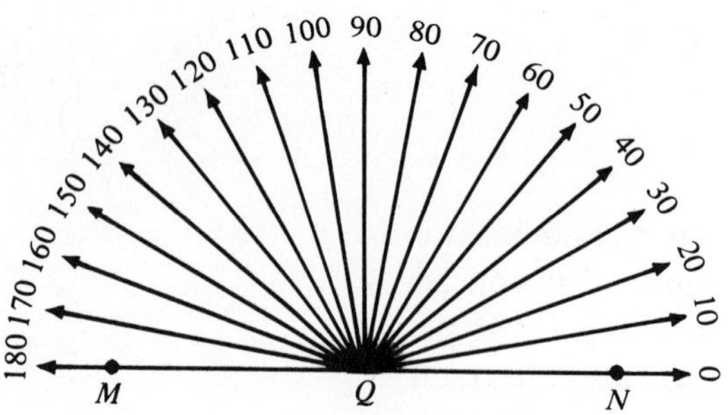

Figure 5-54

If we use the dynamic interpretation of angle, the size of an angle can have a measure anywhere between 0 and 360 in this system. If $\overrightarrow{PA}$ in Fig. 5-53 represents the initial side of the angle (associated with 0) and the rotation is counterclockwise, then $\angle APD$ has a measure less than 180, while $\angle APC$ has a measure between 180 and 360. This second angle corresponds to the reflex angle mentioned in problem 6(g) of Exercise 2-10.

Using a static interpretation of angle, a more recent formulation does not allow "reflex angles."

The unit of measure is still the degree, but the measure of an angle is restricted to values between 0 and 180 (not inclusive). Consider a line MN as in Fig. 5-54. The line determines two half-planes. With point Q on $\overleftrightarrow{MN}$ as vertex, a set of points in one half-plane can be found such that, with M and N, 180 nonoverlapping congruent angles are determined. Any angle congruent to one of these has measurement of 1 angular degree. The number 0 is associated with $\overrightarrow{QN}$, 180 with $\overrightarrow{QM}$, and 1 through 179 with the other rays such that the properties listed in Sec. 5-12 are satisfied. Subdivisions are the same as previously considered.

EXERCISE 5-13

1. How many angular seconds in 1 angular degree?
2. Do you know of any other standard units used to measure angles?

5-14 THE PROTRACTOR

Just as the ruler is simply a device for bringing a model of a coordinatized ray to places where linear measurement is required, the **protractor**, pictured in Fig. 5-55, is simply a device that brings a set of numbers approximately associated with a given set of rays to the angle-measuring situation. You should notice immediately the relationship between the protractor and the coordinatized set of rays shown in Fig. 5-54. V on the protractor corresponds to Q, the vertex of all the rays. The heavy "line" on the protractor corresponds to $\overleftrightarrow{MN}$ in the diagram. The marks at the edge of the protractor are points that correspond to points on the rays. The pictures of the rays have been omitted on the protractor. The numerals at the edge of the protractor are those associated with the coordinates of each of the "missing" rays.

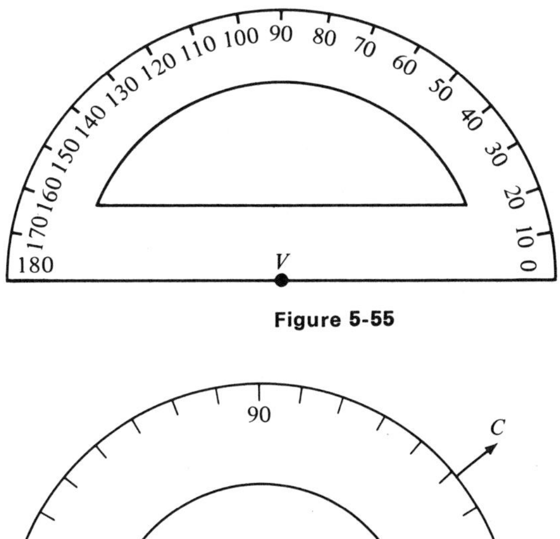

Figure 5-55

Once the mathematical abstraction for measuring angles is understood, using a protractor to measure angles in the physical situation is easy. To find the measure of a given angle, $\angle CAB$, using a protractor, place the dot at V on the protractor on top of the vertex A in the picture of the angle. (See Fig. 5-56.) Align the dark "line" of the protractor with $\overrightarrow{AB}$ of the angle. Ray AC will pass

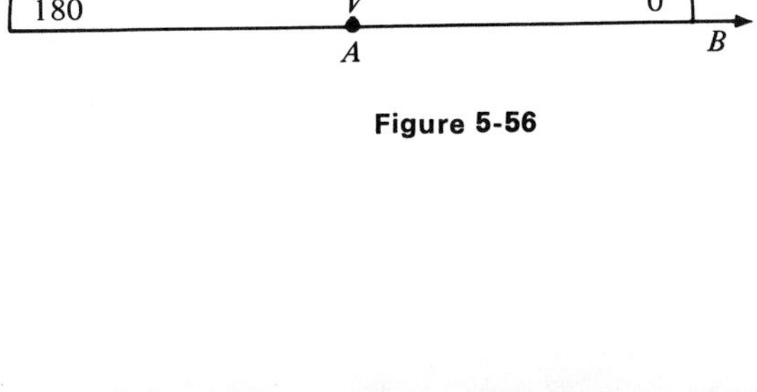

Figure 5-56

under the curved edge of the protractor. If the picture of $\overrightarrow{AC}$ is not long enough to reach the edge of the protractor, simply extend the drawing. (How long is a ray anyway?) The measure of $\angle CAB$ is obtained by reading the scale value corresponding to $\overrightarrow{AC}$. Why is this measure only an approximation to the true measure of the angle?

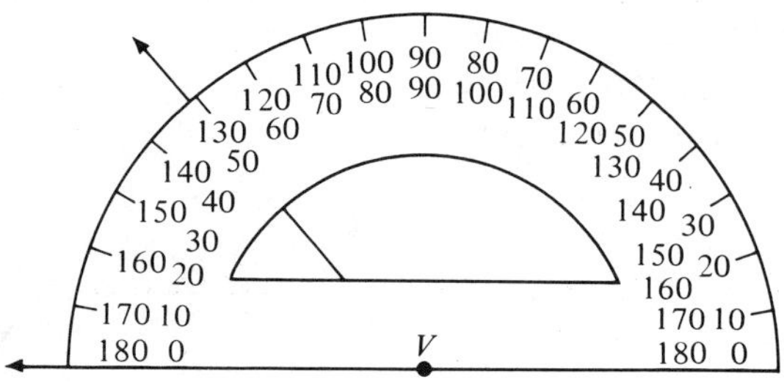

Figure 5-57

Note in Fig. 5-57 that some protractors combine two scales on one instrument. One scale reads clockwise, the other counterclockwise. One ray of the angle to be measured is usually made to correspond to one of the zero marks on the protractor. One then must be careful to use the scale corresponding to this zero when reading the measure of the angle. For the angle pictured in Fig. 5-57, the measure is approximately 50; the measurement is approximately 50 angular degrees. What is the precision of the instrument pictured?

EXERCISE 5-14

1. Make a protractor as patterned in Appendix C.
2. Using this protractor, find the approximate measures of each angle pictured in Fig. 5-58.
3. What is the precision of this measuring instrument?

SUMMARY OF PRINCIPLES

Given any segment, $\overline{AB}$, and any ray, $\overrightarrow{MN}$, there is one and only one point, X, on the ray such that $\overline{MX}$ is congruent to $\overline{AB}$.

Given any segment there exists a set of $n - 1$ points in the interior of the segment that subdivide the segment into n congruent, nonoverlapping segments.

There exists a one-to-one correspondence between the set of points on a ray and the set of nonnegative real numbers.

Given any angle, $\angle XYZ$, and any ray, $\overrightarrow{AB}$, then there is one and only one ray, $\overrightarrow{AC}$, such that $\angle BAC$ is congruent to $\angle XYZ$ and C is on the M side of $\overleftrightarrow{AB}$.

Given any angle there exists a set of $n - 1$ points in the interior of the angle which with the vertex and sides of the given angle determine a set of n congruent angles, no two of whose interiors intersect.

As precision of measurement increases, absolute error decreases.

As accuracy of measurement increases, relative error decreases.

If absolute error remains constant, relative error decreases as the numerical value of the measurement increases.

If relative error remains constant, absolute error increases as the numerical value of the measurement increases.

A digit in a numeral used to report a measurement is significant if it indicates the number of units involved in the measure.

EXERCISES

1. Encircle A, S, or N (always, sometimes, or never true) for each of the following statements.
- (a) The measuring process is one of counting discrete objects. A S N
- (b) The units for linear measurement are segments. A S N
- (c) Nonoverlapping segments are nonintersecting segments. A S N
- (d) The units for volume measurement are closed regions. A S N
- (e) $AB = \overline{AB}$. A S N
- (f) Increasing the precision used in measurement increases the accuracy. A S N

2. On a given line, how many segments are there:
- (a) Congruent to a given segment.
- (b) Congruent to a given segment so that the segment has a given point of the line as endpoint.

3. In measurement notation there is a set of symbols commonly used to abbreviate "angular degree," "inch," and "foot." The abbreviation of "n angular degrees" is "$n°$." A single prime is used to abbreviate "foot" and a double prime is used to abbreviate "inch." Complete the following to make them true sentences.
- (a) 25 angular degrees = _____°. (b) 36″ = _____′.
- (c) 2.5′ = _____″.

4. Draw segments corresponding to each of the following measurements.
- (a) 0.15 m. (b) 30 cm.
- (c) $4\frac{6}{16}''$. (d) 0.5′.

5. Draw angles corresponding to each of the following measurements.
- (a) 48°. (b) 90°.
- (c) 125°.

6. Find the measure of each side of the quadrilaterals shown in Fig. 5-59. Find the measurement to the nearest millimeter and record the measure in the table below.

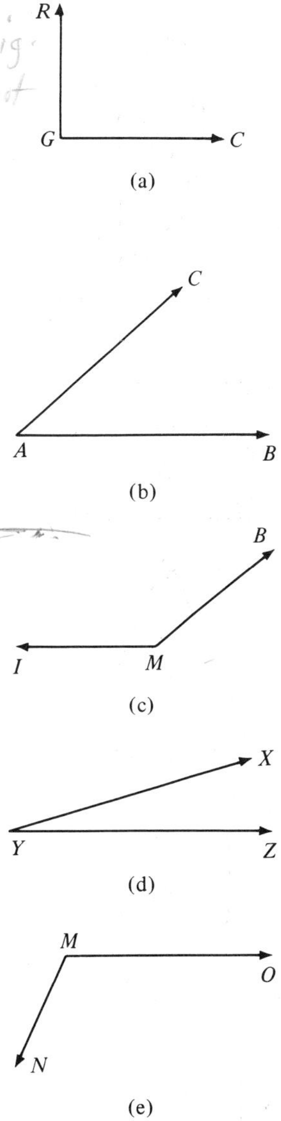

Figure 5-58

Quadrilateral	AB	CD	BC	AD
(a)				
(b)				
(c)				
(d)				
(e)				
(f)				
(g)				

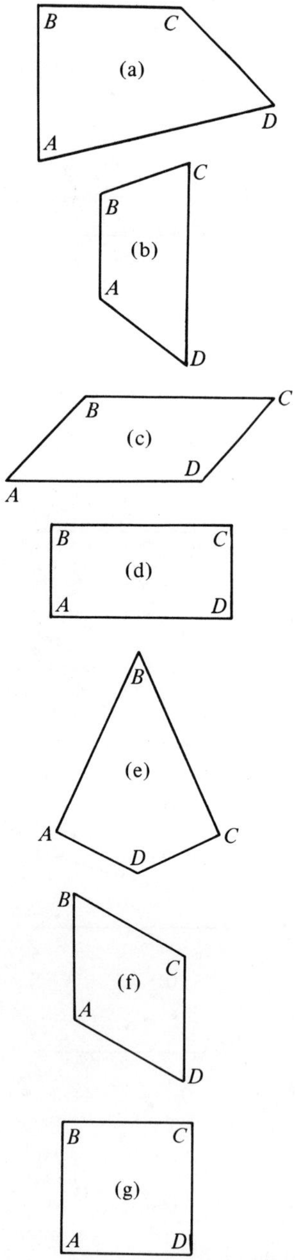

Figure 5-59

7. Find the measure of each angle of the quadrilaterals shown in Fig. 5-59. Find the measurement to the nearest degree and record the measure in the table below.

Quadrilateral	A	B	C	D
(a)				
(b)				
(c)				
(d)				
(e)				
(f)				
(g)				

8. Find the sum of the measures of the angles for each quadrilateral in problem 7.

9. Find the measure of each side of the triangles shown in Fig. 5-60. Find the measurement to the nearest $\frac{1}{16}$ in. and record the measure in the table below.

Triangle	AB	BC	AC
(a)			
(b)			
(c)			
(d)			

10. Find the measure of each angle of the triangles shown in Fig. 5-60. Find the measurement to the nearest degree and record the measure in the table below.

Triangle	A	B	C
(a)			
(b)			
(c)			
(d)			

11. Find the sum of the measures of the interior angles of each of the polygons shown in Fig. 5-61.

12. Make a drawing of a pair of adjacent angles whose noncommon sides have the same carrier. Measure each angle and find the sum of the measures.

13. Make a drawing of a pair of vertical angles. Name the pairs of vertical angles. Find the measure of each angle in each pair.

14. Find the measure of the angles formed by the hands of a clock at each of the following times.
    (a) 3:00. (b) 3:30.
    (c) 3:45. (d) 4:00.

15. Find the measurement to the nearest $\frac{1}{4}$ in. of each edge of the polyhedron models you constructed in problem 11 of the Exercises following Chapter 4.

16. Find the measurement to the nearest degree of each of the angles determined

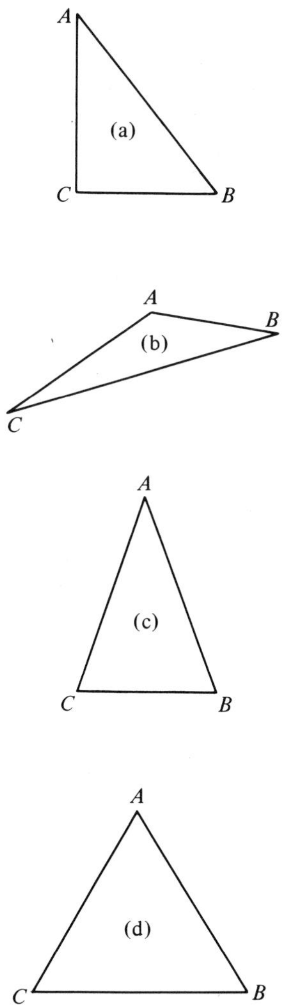

Figure 5-60

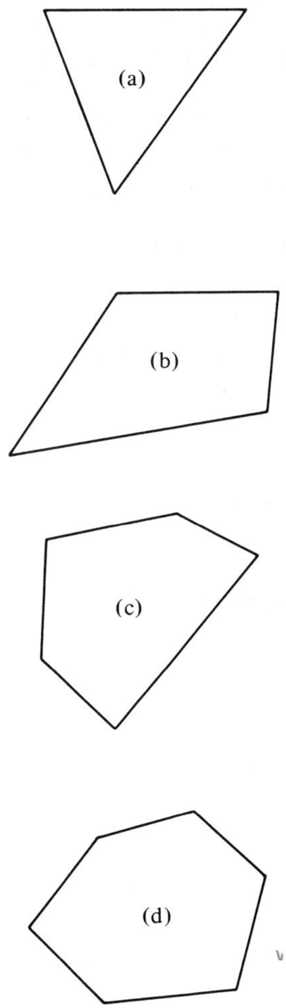

Figure 5-61

by the edges of the polyhedron models you constructed in problem 11 of the Exercises following Chapter 4.

17. A circular cylinder is one whose cross sections are circles. What is a circular cone?

18. Draw a pair of nonparallel lines. Draw a transversal and find the measures of:

(a) A pair of alternate interior angles.
(b) A pair of nonalternate interior angles.
(c) A pair of corresponding angles.

19. Draw a pair of parallel lines. Draw a transversal and find the measures of:

(a) A pair of alternate interior angles.
(b) A pair of nonalternate interior angles.
(c) A pair of corresponding angles.

20. **Scientific notation** is a way of expressing a given number as a product of two numbers: one of these is a number between 1 and 10 (including 1), the other is a power of 10. For measurement notation purposes, the number between 1 and 10 is indicated by a numeral that contains the same number of significant digits as the given number. The following examples will show you how to write a number in scientific notation.

$2350 = 2.35 \times 1000$	$1000 = 10^3$	$2350 = 2.35 \times 10^3$
$23.50 = 2.350 \times$ ____	$10 = 10^?$	$23.50 = 2.350 \times 10^1$
$2.350 = 2.350 \times$ ____	$1 = 10^?$	$2.350 = 2.350 \times 10^0$
$0.2350 = 2.350 \times$ ____	$0.1 = 10^?$	$0.2350 = 2.350 \times 10^{-1}$
$0.02350 = 2.530 \times$ ____	$0.01 = 10^?$	$0.02350 = 2.350 \times 10^{-2}$

Write each of the following in scientific notation.

(a) 403. (b) 4030.
(c) 0.275. (d) 0.0275.
(e) 0.00020.

21. Discuss: A digit in a numeral used to report a measurement is significant if it indicates the number of units involved in the measure.

22. For readings on the concept of measurement see Ref. 34, pp. 108–126, and Ref. 81, pp. 283–287.

CHAPTER 6

Metric Properties of Plane Curves

6-1 INTRODUCTION

The preceding chapter introduced you to basic properties of a concept of measurement, thereby opening for us a new field of investigation. We shall now see how metric properties can be used to bring out more relationships among the many concepts at our disposal.

Henceforth, for the most part, metric and nonmetric ideas will be used without our stating whether we are in a metric or nonmetric situation. However, perhaps on occasion you might ask yourself, "How does a metric property play a role in this particular discussion?" By the end of this chapter the point should have been made: a great deal of geometry can be developed without recourse to metric properties, but the introduction of a metric allows other relationships to be emphasized.

6-2 ANGLE MEASURE

As mentioned in the previous chapter, "things" are measured by like "things"; that is, a property is measured by something else exhibiting this same property. In preparation for a discussion about polygons, let us make some useful observations of a more basic concept. We are interested in obtaining the measure of an angle by comparing it with other angles.

Suppose we have a clock with only the hour and minute hands. In this treatment we are not thinking of the clock really as a clock, with one turn of the minute hand producing a one-twelfth turn of the hour hand. Perhaps we have a model of a clock made from a paper plate, and a child has moved the minute hand into position to represent some time by rotating it from the hour hand. If we think dynamically for a moment, considering the hour hand as the initial side of an angle and the minute hand as the terminal side, and one complete revolution as a unit, how much rotation has occurred when the hands name the hour 3:00 o'clock? 6:00 o'clock? 12:00 o'clock? Of course several interpretations might be given to these questions, but let us for a moment keep them "easy." Suppose we say that in the third case there has been no revolution, in the second case there has been half a revolution, and in the first case there has been one quarter of a revolution. Perhaps you should think of some other possible interpretations.

Figure 6-1

Now the question is this: What shall we use as our basic unit? The last chapter made the point that the choice of a unit is quite arbitrary, but often influenced by convenience. Suppose we try one quarter of a revolution as a unit. This would suffice for the angles just mentioned, but what about the angle suggested by the hands at 3:20? (See Fig. 6-1.) (Remember that the hour hand is not pointing directly at the "3" at 3:20 o'clock.) Surely we would find a finer subdivision more convenient. The unit that seems most common for elementary work is the *degree*. If we consider one complete revolution as being made up of 360 small angles, then one of these angles is said to have a measurement of one degree; thus, 1 is the measure of $\frac{1}{360}$ of a revolution, or the measurement of one revolution is 360 degrees, written 360°. Then what is the measure of the angle suggested by the hands of the clock at 6:00 o'clock? Surely half a revolution $= \frac{1}{2}(360) = 180$, so we say the measure of the angle is 180, or its measurement is 180°. What is the measure of an angle pictured by the hands at 12:00 o'clock? at 3:00 o'clock? at 3:20 o'clock? (Did you get 20 or 30 for this last one? Which is correct?)

6-3 ANGLE CLASSIFICATION

Although we quickly dispensed with the quarter-revolution unit, this unit is actually used in a rather different fashion to define other angles. Let us agree that an angle that represents no rotation from the initial side is a **zero angle** and its measure is 0. An angle that represents one quarter of a revolution is a **right angle**. Then what is the degree measure of a right angle? Would you say that this kind of angle occurs frequently around us? Name a model on this page.

Let us limit ourselves to nonnegative angles, angles with positive numbers or zero for measures. An angle other than the zero angle whose measure is less than that of a right angle is an **acute angle**. What is the range of values of the measure of an acute angle? An angle that is larger than one right angle and less than two right angles is an **obtuse angle**. What is the range of values of the measure of an obtuse angle? An angle whose measure is equal to that of two right angles is a **straight angle**. How much of a revolution is this? What is the degree measure? An angle that represents a rotation of more than a straight angle and less than two straight angles is a **reflex angle**. Indicate the set of numbers that is used to name the measure of a reflex angle. An angle of one complete revolution is a **perigon**.

Some pairs of angles, too, depend upon measure for their classification. If the sum of the measures of two angles is equal to the measure of a right angle, then the angles are **complementary angles** and each angle is the **complement** of the other. Likewise, if the sum of the measures of two angles is equal to the measure of a straight angle, the angles are **supplementary angles** and each is the **supplement** of the other. Then what is the sum of the measures of two supplementary angles? If an angle has a degree measure of 40, what is the degree measure of its complement? Of its supplement?

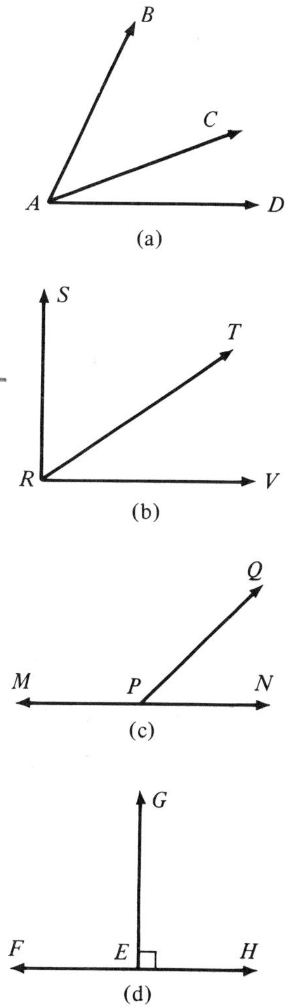

Figure 6-2

Of special interest with respect to complementary and supplementary angles is the situation in which the angles are adjacent. Do you recall what adjacent angles are? Name two pairs of adjacent angles in Fig. 6-2. Which pairs appear to be complementary? What makes you think so? Which pairs seem supplementary? In the case of the adjacent complementary angles, what do you notice about the outermost sides? Do they not assume a position much like the edges at the corner of a box, or the corner edges of a sheet of paper? Such lines are said to be perpendicular to each other. And, in general, two lines, rays, or segments (or pair thereof) are **perpendicular** if they intersect[1] and determine at least one right angle. In the case of the adjacent complementary angles, what is the degree measure of the sum of the angles? How do you know? An angle that has this measure is what kind of angle? Have we not just shown that the exterior sides of two adjacent complementary angles are perpendicular?

Now consider the cases in which the angles are adjacent and supplementary. In Fig. 6-2(c) and (d), what property must we know, in addition to the angles being adjacent and supplementary, in order to conclude that

[1] Our interpretation of perpendicular lines, rays, segments, and so on, requires a nonempty intersection of the sets involved. Note that some authors require only that the carriers of these sets intersect to form right angles.

carriers determined in the figure are perpendicular? What is the degree measure of a straight angle? In Fig. 6-2(d), if $m(\angle FEG) = m(\angle HEG)$, then what is the degree measure of $\angle FEG$? Then we can conclude that $\overleftrightarrow{GE}$ is perpendicular to $\overleftrightarrow{FH}$, and we write "$\overleftrightarrow{GE} \perp \overleftrightarrow{FH}$." Thus we have shown that if two adjacent angles are supplementary and congruent to each other, the carriers of their common ray and either of the other sides are perpendicular. As in Fig. 6-2(d), perpendicularity and right angles are indicated by ∟.

Notice that we started this section with a discussion related to angles, and we now find ourseleves drawing conclusions with respect to lines. One never knows where a particular discussion is going to lead in this geometry!

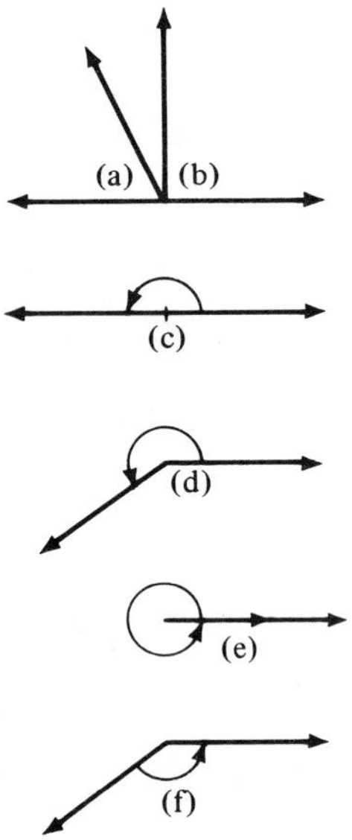

Figure 6-3

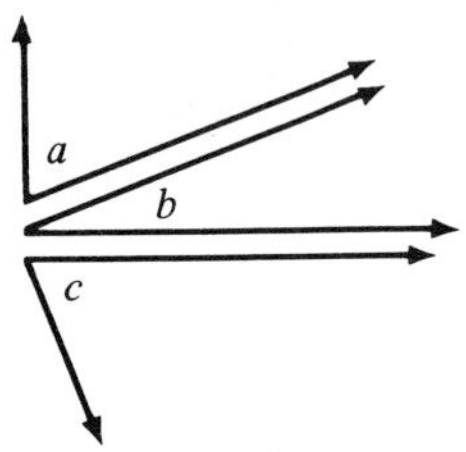

Figure 6-4

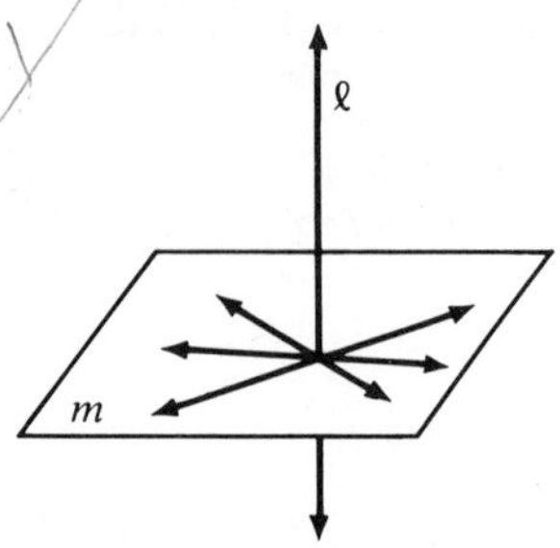

Figure 6-5

EXERCISE 6-3

1. (a) How many revolutions does the minute hand make from 12:00 noon to 3:00 P.M. on the same day? What would be the measure of this same angle in quarter-revolutions? What is the degree measure of this angle?
   (b) What is the degree measure of the angle suggested by the hands of a clock (a child's clock again) at 4:45 o'clock?
2. In Fig. 6-3(a) through (f), list the word name or names that apply to each angle.
3. How does the concept of angle measure as discussed in this chapter differ from that of Chapter 5?
4. List the word names and make sketches of all of the classifications we have had for angles considered as pairs. You are not limited to this chapter.
5. In Fig. 6-4, $\angle a$ is a complement of $\angle b$, and $\angle c$ is a complement of $\angle b$. Then what is the sum of the measures of angles a and b? Of angles c and b? What conclusions can be drawn concerning $\angle a$ and $\angle c$?
6. If two angles are supplements of the same angle, what relationship exists between the two given angles?
7. Using the idea of supplementary angles, discuss how you might justify that what appears in a drawing as two rays not in a line, really is a line. [*Hint:* What relationship exists between the exterior sides of two adjacent and complementary angles?]
8. Consider the definition of two perpendicular lines in a plane and suggest a definition of a **line perpendicular to a plane**. (See Fig. 6-5.)
9. The idea of distance is often developed in terms of "shortest path" (see Sec. 5-2). That is, in a plane the distance between two points is represented by the segment joining them. On a sphere a minor arc of a great circle represents the distance on the sphere between two points on the sphere (not the ends of a diameter—see Exercise 1-7, 6, and Secs. 6-9 and 11-6.) If we continue with this understanding about distance, what would you think would represent each of the following distances?

(a) In a plane, the distance from a point to a line. (Does it matter if the point is on the line?)
(b) In space, the distance from a point to a line.
(c) In space, the distance from a point to a plane.

6-4 TRIANGLES

Let us see whether we might make some use of angles in relation to polygons. Of all of the polygons we have considered, which would you think of as being most basic? Of special interest is the set of triangles. We plan to use the triangle as a building block in further discussion of polygons. In fact, we shall demonstrate several ideas about triangles and then see how we can use these ideas to draw conclusions ("Deductions, my dear Watson") about other polygons.

Carry out the following experiment on a separate sheet of paper. Draw with a straight edge (a ruler will do) a triangle of somewhat the same shape as that shown in Fig. 6-6(a) and label it as indicated. Do not concern yourself with any measurement (after all, measurements are approximate), but make it large enough to take up most of an $8\frac{1}{2}''$ by $11''$ paper. Cut out this triangle. Then tear it into three pieces, somewhat as indicated in Fig. 6-6(b). Be sure the tears leave the sides of the angles intact. On another sheet of paper place these sections in a position along one edge as suggested by Fig. 6-6(c). What does this demonstration lead you to believe about the sum of the measures of the angles of a triangle? Do you suppose the same result will occur if you use a different triangle? Try it. In the next Exercises you will demonstrate this fact without depending upon the placement of a "torn" triangle and without making measurements.

Now suppose that one of the angles of a triangle is an obtuse angle. What can you say of the sum of the measures of the other two? What kind of angle must each of the remaining angles be? Such a triangle is called an **obtuse triangle.** [See Fig. 6-7(a).] Must the acute angles be mentioned in the description, or it is sufficient to say that it contains one obtuse angle?

Likewise consider a triangle in which one of the angles is a right angle. Such a triangle is called a **right triangle**. The side opposite the right angle is called the **hypotenuse** and the other sides, **legs**. [See Fig. 6-7(b).]

Do you suppose we can have a triangle that contains two right angles? Two obtuse angles? Two acute angles? [See Fig. 6-7(c), (d), (e).] Suppose a triangle has angles that are neither right nor obtuse. What kind of angle must each be? Could one of them be the zero angle? Would this figure be a polygon if it has a zero angle? A triangle in which all three angles are acute is an **acute triangle**. Would it be sufficient to say that a triangle that contains an acute angle is an acute triangle? Although we could formulate

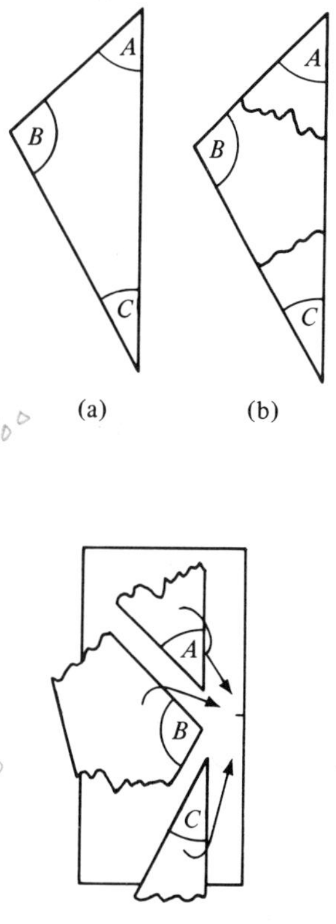

(a) (b) (c)

Figure 6-6

our description in this way, why do you suppose others require all three angles to be acute?

Thus, every triangle must be either acute, right, or obtuse. Suppose each of the angles of a triangle has the same measure. What must the degree measure be? Such a triangle is called **equiangular**. Use your protractor and draw a triangle in which the measure of each angle is 60. Now measure the sides of this triangle. Draw another equiangular triangle in the same way but this time make the sides longer. Measure these sides. Would your observations lead you to believe that an equiangular triangle would have sides that have the same (equal) measure(s)? As you might expect, a triangle all of whose sides have the same measure is called **equilateral.** Do you suppose that in other polygons in which we have angles with equal measures we will have congruent sides? How about vice versa? Do not treat these last two questions hastily. Consider many quadrilaterals, some with angles of the same measure, others with sides of the same measure. We shall investigate this more carefully in the next chapter and reach conclusions without using measuring instruments.

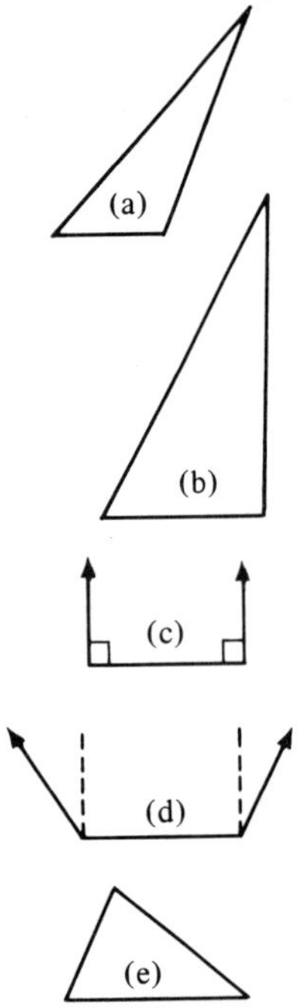

Figure 6-7

Again we started a discussion about angles and have been led to questions concerning lines or segments. Having mentioned a classification of triangles by properties of their sides, let us point out that a triangle with two congruent sides is called **isosceles**; and, a triangle having no two sides with the same measure is called **scalene**.

Since the sum of the degree measures of the angles of a triangle in our geometry is 180, do you suppose there is any limit to the sum of the measures of the sides? Surely not. This number, the sum of the measures of the sides of a triangle, is called the **perimeter**, and the same term is used for the sum of the measures of the sides of any polygon. What would be the physical interpretation of the perimeter of a rectangular field? Some people use the term "perimeter" to refer to the polygon, but its strict mathematical sense is that of a measure. Although there is no limit to the lengths of the sides, and hence to the perimeter, there are some restrictions as to the relative lengths of the sides; these will be investigated in the exercises.

EXERCISE 6-4

1. (a) In a right triangle if an acute angle has a degree measure of 60, what is the degree measure of each of the other two angles?

(b) In any right triangle what is the sum of the degree measures of the acute angles?

(c) Using part (b), then we say each acute angle of a right triangle is the _____ of the other, and they are _____ angles.

(d) If the hypotenuse of an isosceles right triangle has a measure of 14, and the measure of one leg is 10, what is the perimeter?

2. For (a) through (e) of Fig. 6-8 list all of the vocabulary we have developed that relates to each figure. You may use your ruler and protractor.

3. (a) If the perimeter of a triangle is 7 and the measures of two sides are 2 and 3, respectively, what kind of triangle is it?
   (b) Suppose we have a figure that is the union of segments whose measures are 7, 6, and 13. Make drawings to illustrate the possible combinations of a figure. Use statements upon which we have agreed to justify the fact that in this case there is no triangle with these segments as sides.
   (c) Make a general statement about the relationship between the sum of the measures of two sides of a triangle and the measure of the third side based upon your justification of part (b).

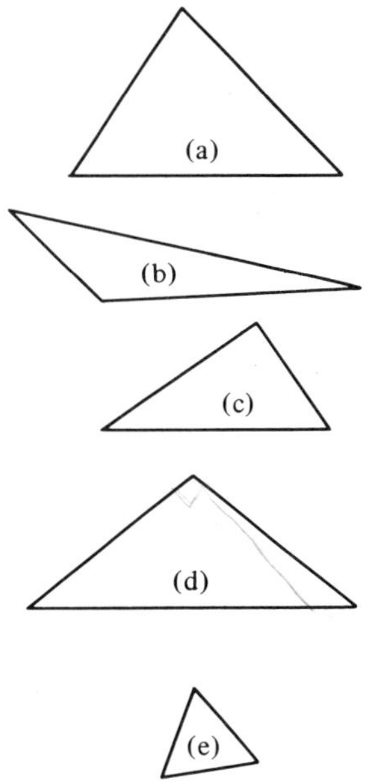

Figure 6-8

4. In Fig. 6-9 the measures of $\angle A$ and $\angle B$ are known to be 47 and 62, respectively.
   (a) What is the degree measure of $\angle ACB$?
   (b) If point P is selected on $\overrightarrow{AC}$ on the opposite side of C from A, then $\angle BCP$ is called an **exterior angle** *of the triangle*. What is the degree measure of $\angle BCP$? Explain how you arrived at this answer.
   (c) Compare $m(\angle BCP)$ with the sum of the measures of $\angle A$ and $\angle B$.
   (d) Suppose we say $m(\angle A) \overset{\circ}{=} a$, $m(\angle B) \overset{\circ}{=} b$. Using the ideas of parts (a), (b) and (c), express the degree measure of $\angle BCP$ in terms of a and b. (Note: "$\overset{\circ}{=}$" is a common symbol for "degree measure.")
   (e) Make a general statement based upon part (d).
   (f) On the basis of part (e) state whether or not you would agree to the next statement and explain: An exterior angle of a triangle has a degree measure equal to the sum of the degree measures of the *nonadjacent interior angles of a triangle*, and its measure is greater than the measure of either of these interior angles. (The nonadjacent interior angles are sometimes called the **remote interior angles** of a triangle with respect to a given exterior angle of a triangle.)

5. In Fig. 6-9:
   (a) What is the degree measure of $\angle PCA$?
   (b) Does $m(\angle A) + m(\angle B) \overset{\circ}{=} m(\angle BCP)$?
   (c) But $m(\angle BCP) + m(\angle ACB) \overset{\circ}{=}$ _____.
   (d) Thus $m(\angle A) + m(\angle B) + m(\angle ACB) \overset{\circ}{=}$ _____.
   (e) Make a general statement justified by parts (a) through (d).

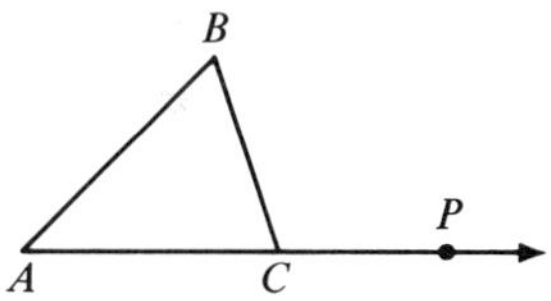

Figure 6-9

6-5 REGULAR POLYGONS

As with the triangle, we speak of polygons in general as being equilateral or equiangular according as their sides or angles, respectively, are equal in measure. A polygon that is both equilateral and equiangular is called **regular**. Actually, this is how most people envision a polygon. When asked to illustrate a polygon they will frequently draw one in which the sides have the same length and the angles the same measure. Why do you suppose this is? You might draw a pentagon that has sides of equal length. Then try to draw one that has angles all of the same measure and sides of different measures. Also try to draw one in which the sides have equal measures but the angles do not.

6-6 QUADRILATERALS

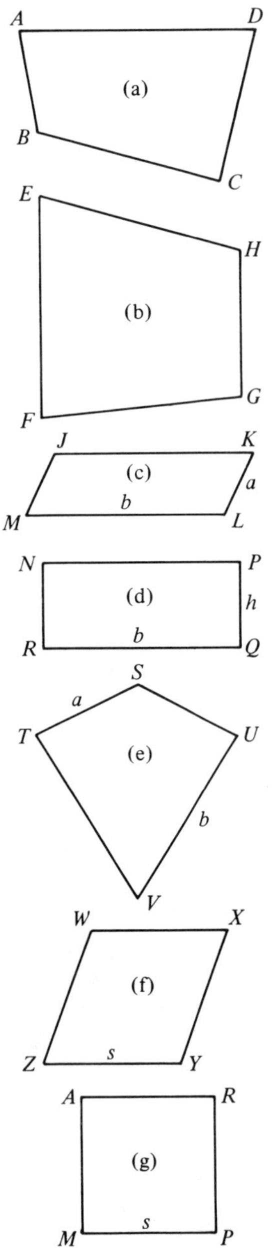

Figure 6-10

We are now able to make further classifications within the set of quadrilaterals. Recall that we distinguished among the trapezium, trapezoid, and parallelogram using only the concept of parallelism. Figure 6-10 reproduces Fig. 3-13. Measure the angles of each of these drawings. Notice the similarities. Figures (d) and (g) represent rectangles. What would you consider suitable conditions (criteria) to establish a figure as a rectangle? To be a **rectangle** a figure need only be a parallelogram with one right angle. We shall investigate other possible sets of conditions that are sufficient to establish a figure as a rectangle, but we shall use the previous statement as the basic one. Measure the sides of each of these quadrilaterals and notice the similarities. Each of figures (f) and (g) is a rhombus (plural: rhombi or rhombuses). To establish a figure as a **rhombus** it is sufficient to know that it is a parallelogram having two adjacent sides of the same measure.

Before completing our classification of quadrilaterals, let us note the following. It is not necessary at the initial stages of development of a concept that a person be able to give a precise definition. (See Sec. 13-3.) If he can recognize that if a parallelogram has right angle*s* then it is a rectangle; if he can see that if the side*s* of a parallelogram are congruent then it is a rhombus; this is sufficient for building these concepts. However, you should realize that a good definition contains no unnecessary information. Therefore we point out in the exercises that if a parallelogram has a right angle, all of the other angles must also be right angles; hence, in the definition of a rectangle it is sufficient to require that only one angle be right.

Perhaps it would be helpful at this time to distinguish between a *working* definition and a *mathematical* or *logical* definition. Most of the descriptions we have given thus far in this text would be classified as working definitions. They enable as to distinguish among the various geometric configurations. Most of the statements used in texts for elementary school children would be of this type. When a formal system is being developed, then a logically precise definition is desired, including a minimal set of conditions. On the other hand, several of the descriptions we have agreed to are also good definitions under this requirement. Other properties of a mathematical definition are discussed in Sec. 13-3.

How, then, might the student distinguish among (c), (d), (f), and (g) of Fig. 6-10? If a quadrilateral is both equiangular and equilateral, it is a

square. What other word states these same facts? What additional fact, then, would you find sufficient to conclude that a given rectangle is also a square? Could we define a square as a rectangle with this property? Is it sufficient to say that a square is a rectangle with two adjacent sides that are congruent? What additional fact would be sufficient to conclude that a given rhombus is really a square? Again we find ourselves in a position to describe a single concept in any of several ways. The usual practice is to define a square as a rectangle with two adjacent sides of equal measure, or congruent.

To complete our classification of the quadrilaterals in Fig. 6-10 we need to name and give a description to figure (e). Although the common term is not widely used in mathematical literature, it is becoming more popular, perhaps because it is most descriptive. Common usage would have us describe a **kite** as a quadrilateral that has two distinct pairs of adjacent sides with the same measure; that is, two sides have the same measure, and another two sides have the same measure, perhaps different from the measure of the previous two. Notice that this means that a kite may be a trapezium or a rhombus. Could a kite be a square? Could a given kite be *both* a trapezium and a rhombus? The introduction of this term indicates how common language can influence technical vocabulary.

Now that the classification of quadrilaterals is complete, you should carefully summarize these statements and consider the relationships among the various classifications.

EXERCISE 6-6

1. Associated with many polygons, quadrilaterals in particular, are formulas that express their perimeters.
   (a) Assume we know that the opposite sides of a parallelogram have the same measure (we show this in Chapter 7). Refer to Fig. 6-10 and write a formula for the perimeter of each of figures (c), (d), (f), (g), and (e). Notice that in this diagram we are using a lower-case letter to represent the measure of a side; thus $m(\overline{ML}) = b$.
   (b) If the measure of one side of a regular polygon is s and the polygon has n sides, what is its perimeter?
   (c) Suppose in Fig. 6-10(a) that $m(\overline{AD}) = 2$, $m(\overline{BC}) = 1\frac{11}{16}$, $m(\overline{AB}) = \frac{5}{8}$, and $m(\overline{DC}) = \frac{7}{8}$. What is the perimeter of trapezium $ABCD$? If a physical model is made with these measures (perhaps in inches), dismantled, the sides placed in a straight line, and the segment that results is measured, what would you expect this measure to be? Why?
2. In polygon $ABCD$, $\angle ACB$ is an angle of the polygon. A S N
3. A rectangle is a square. A S N
4. A rhombus is a rectangle. A S N

5. A trapezoid is a parallelogram. A S N
6. An equilateral triangle is isosceles. A S N
7. An isosceles triangle is obtuse. A S N
8. A regular quadrilateral is a rhombus. A S N
9. In Fig. 6-11 we are given parallelogram $ABCD$, in which $\angle B$ is a right angle.
   (a) What is the degree measure of $\angle B$?
   (b) Draw $\overline{AC}$. What name applies to this kind of segment in any polygon?
   (c) What relationship exists between $\angle 1$ and $\angle 3$? Then:
   $$m(\angle 1) + m(\angle 3) \overset{\circ}{=} \_\_\_\_.$$
   (d) In problem 19 of the Exercises at the end of Chapter 5 we agreed intuitively that if parallel lines are cut by a transversal, then the alternate interior angles are congruent. (We justify this in problem 18 of the Exercises at the end of Chapter 6.) Name two pairs of alternate interior angles in Fig. 6-11. Then what pairs of angles are congruent in parallelogram $ABCD$?
   (e) Since the degree sum of $\angle 1$ and $\angle 3$ is 90, what is the degree sum of $\angle 2$ and $\angle 4$? Then what is the degree measure of $\angle D$? What kind of angle is it?
   (f) Show that $\angle BAD$ and $\angle BCD$ are right angles.
10. Read Ref. 112, pp. 729–730.

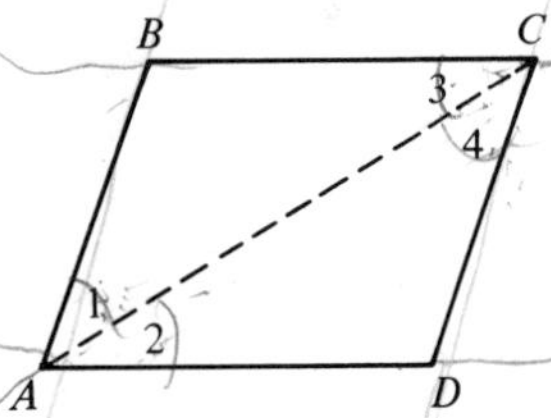

Figure 6-11

6-7 POLYGONS—ANGLE SUMS

Having concluded that in our geometry the sum of the degree measures of the angles of a triangle is 180, we should expect to reach some conclusions regarding the sum of the degree measures of the angles in other polygons. Consider quadrilateral $ABCD$ with a diagonal drawn as in Fig. 6-12. Would you agree that the sum of the measures of all the angles of all the triangles is equal to the sum of the measures of all the angles of the quadrilateral? What, then, is the sum of the angles of this quadrilateral? If segments are drawn connecting some interior point P of Fig. 6-13 to each vertex of the quadrilateral, how would you answer the last two questions? Be sure you take into account the angles with P as a vertex. Are you willing to agree that one of these two methods will surely apply to any quadrilateral? If you are, then we may conclude that the sum in degrees of the measures of the angles of any quadrilateral is 360.

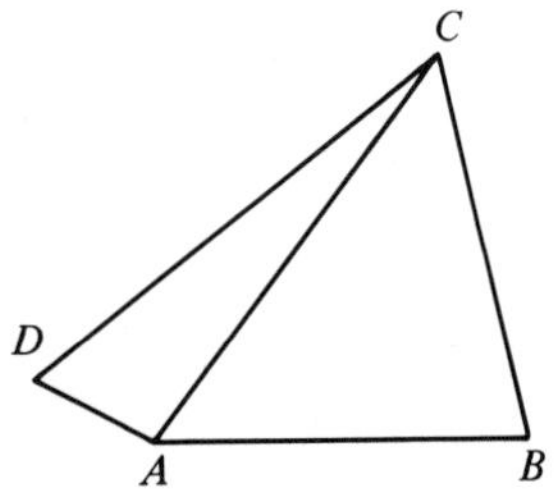

Figure 6-12

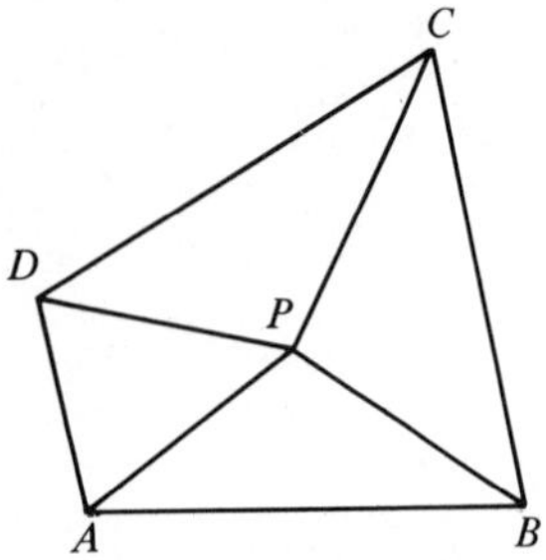

Figure 6-13

Do you suppose this technique can be generalized to a pentagon? A hexagon? An n-gon? You should have answered yes, but with care. Consider Fig. 6-14 and check to be sure one of the techniques applies. On scrap paper make sketches similar to those on this page, and also try one concave polygon with more than five sides. Explain any difficulties. For help in reaching a conclusion concerning the n-gon, consider the following:

(1) Draw diagonals from only one vertex.

(2) Try to predict the number of triangles that will be formed and state this number in terms of n. Take into account (in the second technique) the angles with P as vertex.

Having made a conjecture about the sum of the measures of the angles of any polygon in terms of the number of sides, measure the angles of several polygons with your protractor and calculate the sum.

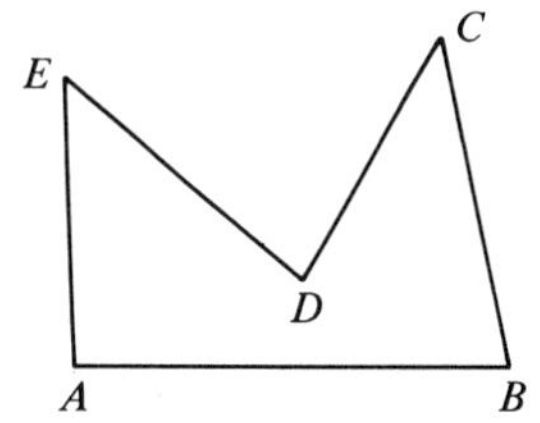

Figure 6-14

If one side of a polygon is extended through a vertex, the angle determined by this extension and the adjacent side of the polygon is an **exterior angle** *of the polygon*. In Fig. 6-15, $\overline{AB}$ is extended as indicated by $\overrightarrow{BP}$. Then $\angle PBC$ is the angle determined by $\overline{BC}$ and $\overrightarrow{BP}$ and is thus an exterior angle of this polygon. If $\overline{BC}$ is extended through C, and Q is a point on this extension, then $\angle DCQ$ is an exterior angle of hexagon $ABCDEF$. If $\overline{CB}$ is extended through B, as indicated by $\overrightarrow{BR}$, then $\angle ABR$ is an exterior angle of this hexagon. What about $\angle RBP$? Describe how the exterior angles would be formed at E.

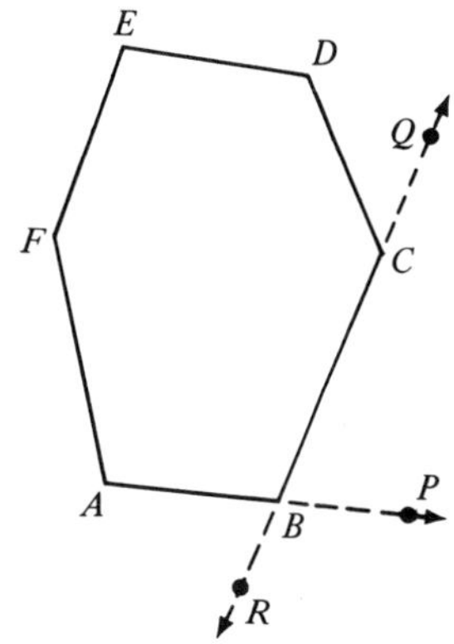

Figure 6-15

Perform the following experiment. Place a pencil along $\overline{AB}$ with the point toward B. Now move the pencil along $\overline{AB}$ until the eraser is at B. Rotate the pencil to correspond to the counterclockwise rotation of $\overrightarrow{BQ}$ from $\overrightarrow{BP}$, thus representing $\angle PBQ$. Now move the pencil along $\overline{BC}$ until the eraser is at C and rotate the pencil counterclockwise through exterior angle QCD. Continue around the polygon until you have rotated the pencil through all of the exterior angles formed in this way, the last being determined by extending $\overline{FA}$ through A. Now how does the position of the pencil compare with its starting position? Would you be willing to say that the sum of the exterior angles of a hexagon, formed by extending the sides in one direction at each vertex (thus excluding $\angle ABR$) is equal in measure to 360 in degrees? Would the measure change if there were more vertices, perhaps for a decagon? Suppose the sides were extended in the opposite direction; that is, form the exterior angle by extending $\overline{BA}$ through A, $\overline{AF}$ through F, and so on. What then is the sum of the measures of *all* the exterior angles of any convex polygon?

EXERCISE 6-7

1. Find the degree measure in each interior angle and each exterior angle of a regular pentagon.
2. For a discussion of the sum of the measures of angles of a concave polygon read Ref. 93, pp. 403–406.

6-8 THE CIRCLE

Is a circle a polygon? Why, or why not? Could it be considered as a polygon? Although a circle is not a polygon, several of the ideas related to a polygon have analogies in a circle. Several segments and lines related to the circle are given names, since they occur frequently. A **radius** (plural: radii) of a circle is a segment that joins the fixed point (center) and any point of the circle. If two points of a circle are joined by a segment, this segment is a **chord** of the circle. If a chord contains the center, then this chord is a **diameter** of the circle. How does the measure of the diameter compare with the measure of the radius of any given circle? Which of these segments might be considered analogous to a diagonal of a polygon? Apply each of these definitions and name the segments of Fig. 6-16.

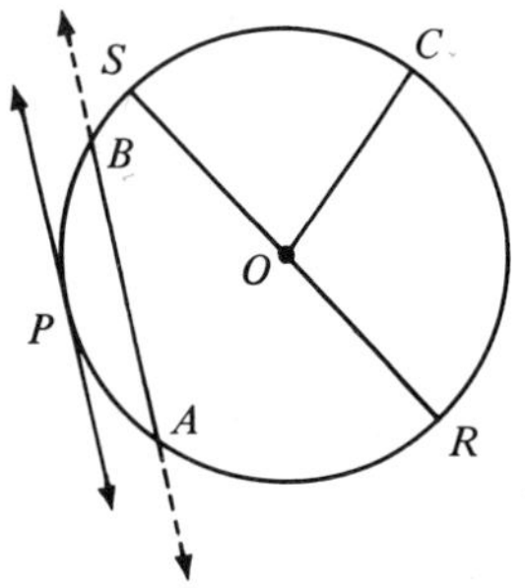

Figure 6-16

Two other closely related sets of points will be considered. The carrier of a chord of a circle is a **secant** of the circle. Thus any line that contains two points of a circle is a secant of that circle. Name six secants determined by the named points of Fig. 6-16. Think dynamically now as we look at A and B of secant AB. Suppose A and B approach each other along the circle until they meet at point P; that is, points A and B coincide at point P of the circle. This line now has but one point in common with the circle. A line in such a position with respect to a given circle is called a **tangent** to the circle, and the single point is called the **point of tangency**. On Fig. 6-16 picture line MCT as a tangent to the circle at C. The point of tangency is also known as the **point of contact**. Although the previous development is a convenient way to think of a tangent to a circle, conventional uses of the words "tangent" and "secant" do not include the tangent as being a special secant—somewhat like the trapezoid-trapezium situation. Thus a tangent to a circle is not a secant of the circle. One other point: we say that if a line is tangent to a circle, the circle is also tangent to the line, or the line and circle are tangent to each other at the point of contact.

6-9 CIRCUMFERENCE

Recall that when we add the measures of the sides of a polygon we get a number that is the perimeter, the measure of the distance around the physical model of a polygon. With respect to a circle we cannot really discuss this from the same point of view. (Where are the segments?) However, it is possible to form an analogy between the perimeter of a polygon and the distance around a circle—or, more precisely, the length of a circle—if we go to a physical model. When we form the sum of the measures of the sides of a model of a polygon, we can actually see the dismantling of the model (polygon) and the laying out of the sides end to end in a line. Then the measure of this segment is also the perimeter of the polygon (since length measure is additive); see Exercise 6-6, 1(c).

Suppose we take a string in the form of a circle (perhaps wrapped around some circular object), cut the string at some point and lay it out along a line. Surely this segment would represent the distance around the circular object.[2] Although some people call this distance the perimeter of the circle, it is not quite the same as for a polygon and is more frequently called **circumference**. Thus the circumference of a circle is the distance around a circle starting and stopping at the same point—or, more simply, the length of the circle. In practice the term "circumference" is used in two rather different ways: (1) as the distance around, hence as a measure, and (2) as the set of points that comprise the circle. Thus some people speak of points "on the circumference." The reader should note that a similar situation exists with regard to the use of "radius" and "diameter." The words are used to indicate either the measure (or measurement) of the segment, or the segment itself. The context usually indicates the intent of the writer (or speaker).

Consider the set of points of the circle in Fig. 6-16 between S and C. (What meaning are we now giving to "between"?) Clearly there are two such sets of points. We must agree upon a way to indicate which set is intended. Any curve that is a portion of a circle is called an **arc** of the circle. Thus an arc of a circle is a special subset of the set of points that comprise a circle. We usually indicate an arc by naming its endpoints and drawing an arc symbol over the letters, much as we name a segment. Thus we might refer to arc SC and write $\overset{\frown}{SC}$. But which arc is intended? There are two common ways to distinguish between the two arcs with endpoints S and C. General agreement would require the symbol $\overset{\frown}{SC}$ to name the shorter arc; if the longer arc is intended, a point on the arc between S and C is named and this name written between S and C, such as $\overset{\frown}{SRC}$. How would $\overset{\frown}{SRC}$ compare with $\overset{\frown}{SBC}$? With $\overset{\frown}{SAC}$? The other convention introduces a modifying term. Thus you refer to the arcs as *major arc SC* for $\overset{\frown}{SRC}$, or *minor arc SC* for $\overset{\frown}{SC}$. An arc whose measure is equal to half the circumference of a circle is called a **semicircle**, or **semicircular arc**. Is a semicircle a major arc or a minor arc? To help you decide consider the following: Is a right angle acute or obtuse?

Very early in the historical development of a formal geometry a certain relationship was discovered to exist between the diameter and circumference of a circle. To "discover" this relationship, complete the following table by measuring the circumferences and diameters as indicated and calculating the other entries. Use appropriate units. C = circumference, d = diameter.

[2] Arc length is really a profound concept that cannot be systematically discussed without the equally profound concept of limit. The present treatment is a gross oversimplification, but it suffices as an intuitive notion for arc length of a circle and other elementary curves. The mathematically mature reader is referred to books on the calculus under such topics as *length of arc* or *rectifiable arc*.

Item (*substitute for unavailables*)	C	d	$C + d$	$C - d$	$(C)(d)$	$C \div d$
Circle with 8″ radius						
Coin (larger the better)						
Large lampshade						
Manhole cover						
Other						

Archimedes (287?–212 B.C.) showed that for any circle the ratio of the circumference to the diameter is more than $3\frac{10}{71}$ and less than $3\frac{10}{70}$. Convert each of these fractions to decimal numerals, showing four places. Observe the interval of tolerance. This ratio, $\frac{C}{d}$, is named by the Greek letter pi, and written π. Its exact value cannot be expressed as a decimal numeral, since there is no unit in common to both the circumference and the diameter (they are said to be incommensurable), hence it cannot be represented by a rational number. You are invited to review some of the literature relating to rational and irrational numbers expressed as decimal representations. (Almost any recent methods text for elementary school teachers should suffice.) There are ways, however, to calculate a decimal representation of π to any desired degree of precision. Thus, rounded to nine decimal places the value of π is 3.141592654. It has been computed now to several thousand places. Values that are often used as approximations are 3.14 and $3\frac{1}{7}$.

The formula definition, $\pi = \frac{C}{d}$, can be used to approximate the circumference of a circle from the radius or diameter. What is the relationship between the radius and diameter of a circle? Then we can express the above formula as $\mathbf{C} = \boldsymbol{\pi}\mathbf{d}$, or $\mathbf{C} = 2\boldsymbol{\pi}\mathbf{r}$. Notice that we can also use these formulas to compute approximations to the radius or diameter if the circumference is known.

EXERCISE 6-9

1. Draw a large circle and on it illustrate each of the following and label: radius, center, diameter, arc, semicircle, major arc, tangent, secant, point of tangency, chord.

2. A proper subset of the points of a circle is an arc of the circle. A S N

3. (a) If the degree measures of five of the angles of a hexagon are 170, 65, 80, 146, and 165, find the measure of the sixth angle.
(b) What is the degree measure of one angle of a regular dodecagon? Of an exterior angle? Show two methods.

4. (a) Using $\pi = 3\frac{1}{7}$, what is the circumference of a circle that has a radius measurement of 21 in.? What is its diameter?

(b) Using $\pi = 3.14$, calculate to two decimal places the diameter of a circle that has a circumference of 12.7.

5. Discuss the following statement: Since $\pi = \frac{22}{7}$, π can be expressed as a rational number, hence is rational.

6-10 CIRCLE—RELATED ANGLES

Among angles frequently discussed in connection with the circle, perhaps the most common is the central angle. A **central angle** of a circle is an angle determined by two radii of a circle. What is the vertex of a central angle of a circle? Thus, as with the polygon we have angles "determined by" sides of the polygon, in a circle we have angles "determined by" radii, and called central angles "of" the circle. However, let us be reminded that the sides of an angle are rays.

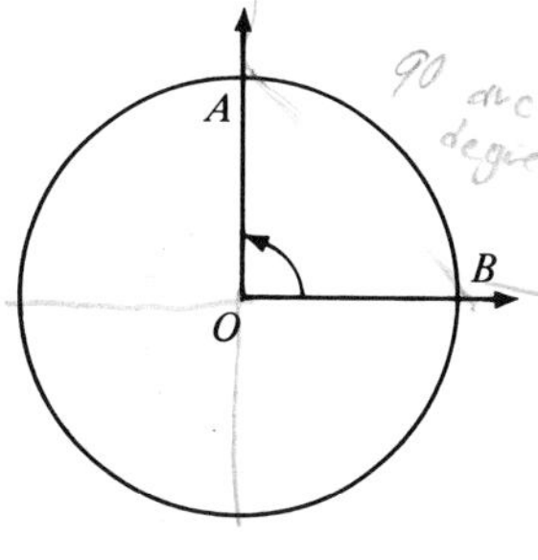

Figure 6-17

In Fig. 6-17 central angle AOB represents a counterclockwise rotation of one quarter of a revolution of $\overrightarrow{OA}$ from $\overrightarrow{OB}$. Then there are how many degrees in the measure of $\angle AOB$? Now think of drawing in the figure, with O as a vertex, all 90 of the one-degree angles that make up $\angle AOB$. Would you not agree that the portion of the circle between A and B, $\widehat{AB}$, is also divided into 90 smaller arcs of equal measure? How many such little arcs are there in the circle? When we think of the number of these smaller arcs in the circle, we can think of the number of unit central angles (whose angular measure is 1) that are determined by the endpoints of these arcs. How many such arcs are there in any circle? Thus we speak of the circumference as containing 360 *arc degrees*. Do not confuse arc degree with angle degree. Surely an angle does not equal an arc! However, in Fig. 6-18 in one complete revolution of $\overrightarrow{OC}$ from $\overrightarrow{OB}$, $\overrightarrow{OC}$ will rotate through 360 angle degrees and point C will traverse 360 arc degrees. How many arc degrees are there in $\widehat{AB}$ in Fig. 6-17?

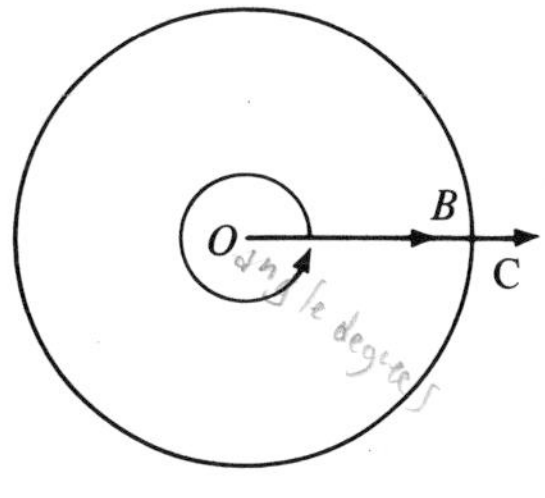

Figure 6-18

Suppose we have a central angle that represents a rotation of one sixth of a revolution. What fractional part of the circumference is **intercepted**, cut off, by the sides of this angle? Then how many arc degrees are there in the measure of this intercepted arc? What is the angular degree measure of this central angle? We are led to the conclusion that a central angle of a circle has the same measure in angle degrees that its intercepted arc has in arc degrees. Explain how the arc-degree concept applies to the angle-degree concept in using a protractor.

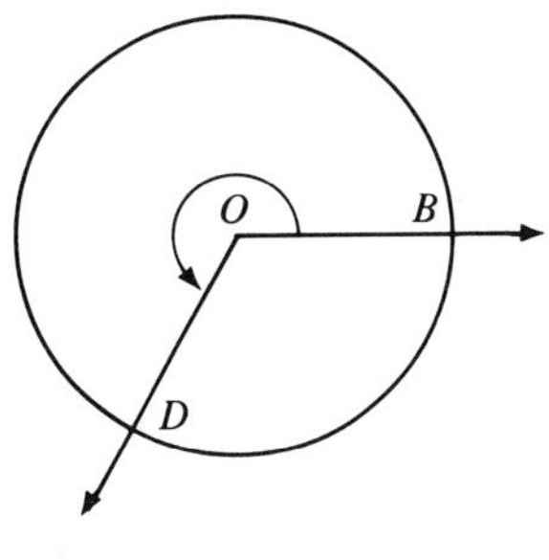

Figure 6-19

Do arcs whose measurement is 1 arc degree always have the same or equal linear measurement? Suppose we have a central angle that represents a rotation of two thirds of a revolution, as in Fig. 6-19. Then the central angle has a measure of $(\frac{2}{3})(360)$ or 240, in degrees. The intercepted arc also has a measure of 240 in arc degrees. Notice that we reached these conclusions about the number of angle degrees and arc degrees without reference to the measure of the circumference, radius, or diameter. But what about the linear measure of the arc? Surely it is two thirds of the circumference. But what happens to the circumference if the radius is changed? Suppose this angle is in a circle with a radius of 7″. How many inches are in the measure of the intercepted arc? How would this compare with the number of inches in the intercepted arc if the same central angle were in a circle of radius $3\frac{1}{2}''$? What about the number of arc degrees? From this we can see that an arc of 1 arc degree has different linear measure, depending on the circumference of the circle of which it is a part.

Several other angles associated with a circle will be used in the Exercises. One of these is the inscribed angle. If an angle is determined by two chords of a circle that intersect at a point of the circle, then the angle is called an **inscribed angle** of the circle. You will see that it has a measure equal to one half the number of arc degrees in its intercepted arc.

SUMMARY OF PRINCIPLES

The exterior sides of two adjacent complementary angles are perpendicular to each other.

Complements (supplements) of the same angle (or congruent angles) are congruent.

The exterior sides of two adjacent supplementary angles lie in a straight line.

The sum of the degree measures of the angles of a triangle is 180.

The acute angles of a right triangle are complementary angles.

The sum of the measures of two sides of a triangle is greater than the measure of the third side.

An exterior angle of a triangle has a degree measure (1) equal to the sum of the measures of the remote or nonadjacent interior angles, and (2) greater than the measure of either remote interior angle.

If an angle of a parallelogram is a right angle, all of the angles of the parallelogram are right angles.

The sum of the measures of the angles of an n-gon in degrees is given by $(n - 2)180$, where n is the number of sides.

The sum of the degree measures of the exterior angles of any polygon, taking one at each vertex, is 360.

A central angle of a circle has the same measure in angular degrees as its intercepted arc has in arc degrees.

1. Two adjacent sides of a polygon have the same carrier. A S N

2. A square is a rectangle. A S N

3. A trapezoid is a trapezium. A S N

4. An isosceles triangle is equilateral. A S N

5. If two lines are coplanar, their intersection contains how many points? (1) 1, (2) 2, (3) none, (4) cannot tell.

6. The sides of an angle of a triangle are segments. A S N

7. A closed curve in a plane is a polygon. A S N

8. A decagonal prism has how many vertices? How many faces?

9. If $m(\angle A) \stackrel{\circ}{=} 70$, then the measure of its complement is _____.

10. The polyhedral angles of a tetrahedron are called _____ angles.

11. Prove that if two lines intersect, the angles in a pair of vertical angles are congruent.

12. To **bisect** means to separate into two congruent parts. By using a central angle with a measurement of 180°, show that a diameter of a circle bisects the circle.

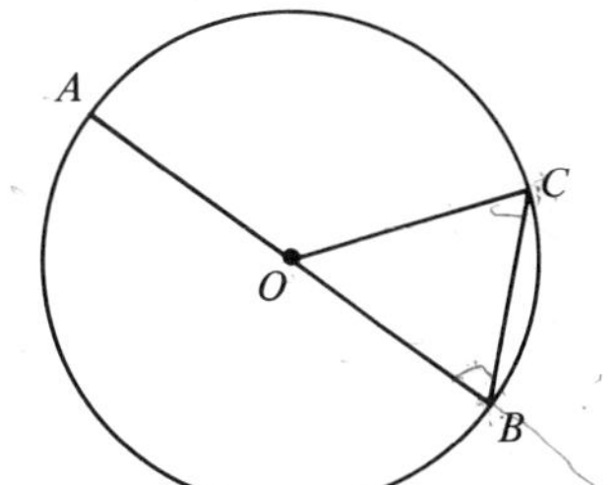

I

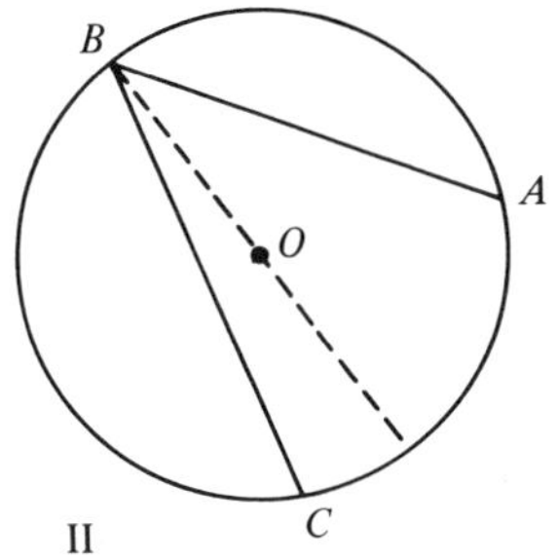

II

13. Accept the statement: "If two sides of a triangle are congruent, the angles opposite these sides are congruent." (This is justified in Exercise 7-6, 6.) Then show that an inscribed angle has one half the degree measure that its intercepted arc has in arc measure. Consider three cases.
   I. Given circle O, inscribed $\angle ABC$, $\overline{AB}$ a diameter.
   To show: $m(\angle ABC) \stackrel{\circ}{=} \frac{1}{2}[m(\widehat{AC})]$.
   [*Hint:* Draw radius OC, use exterior angle of $\triangle OBC$.]
   II. Given circle O, inscribed $\angle ABC$, O lies in the interior of the angle.
   [*Hint:* Apply case I, by drawing diameter AO.]
   III. Given circle O, inscribed $\angle ABC$, O lies in the exterior of the angle.

14. In Fig. 6-20 points A and B are selected on the circle such that $\overline{OA}$ is a radius and $m(\widehat{AB}) = m(\overline{OA})$. Then $\angle AOB$ is said to have a measurement of one *radian*. Point C is selected on the circle such that $m(\overline{AC}) = m(\overline{OA})$. Then what kind of angles are formed in triangle COA?
   (a) What is the degree measure of $\angle COA$?
   (b) What is the formula for the circumference of a circle in terms of the radius?
   (c) How many times can the radius be "marked off" on the circumference? Then what is the number of radians in one complete revolution? Thus a measure of 360 in degrees is equal to what measure in radians?
   (d) An angle with a degree measure of 60 has what radian measure?
   (e) An angle with a radian measure of $\frac{\pi}{8}$ has what degree measure?

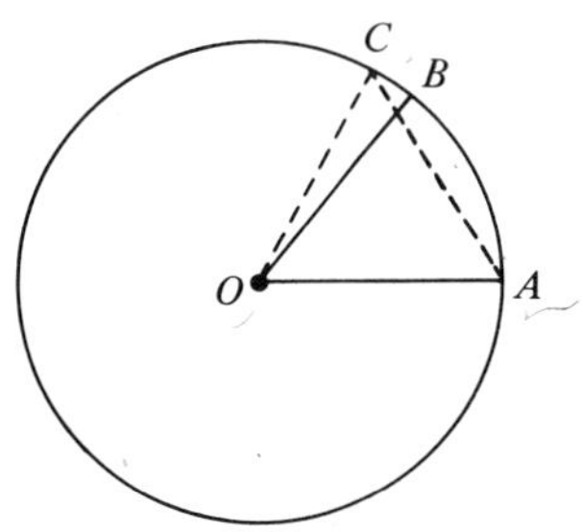

Figure 6-20

15. In Fig. 6-21 points A, B, C, and D are on circle O such that $m(\widehat{AB}) \stackrel{\circ}{=} 110$, $m(\widehat{CD}) \stackrel{\circ}{=} 80$, and $m(\angle C) \stackrel{\circ}{=} 50$, $\overline{AOD}$ is a diameter. Find the degree

measures of (a) $\angle CAD$, (b) $\widehat{AC}$, (c) $\angle DOB$, (d) $\angle CAB$, and (e) $\widehat{BC}$. (f) Prove $\overline{CE} \perp \overline{AD}$. That is, justify it.

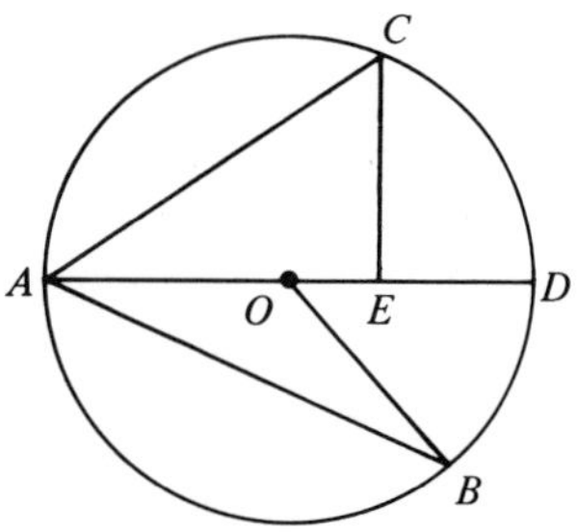

Figure 6-21

In Exercise 6-4 problem 4, we outlined steps that showed (1) an exterior angle of a triangle has a degree measure equal to the sum of the degree measures of the two remote interior angles. In doing this we used the fact that (2) the sum of the degree measures of the angles of a triangle is 180. Then, in problem 5, we "showed," using statement (1), that statement (2) is true. Thus we are "guilty" of circular reasoning. However, in outlining these steps in the Exercises, every statement we make is true. Notice that we say: "on the basis of" or "justified by." Thus our conclusions are true based on statements made in those exercises. In developing a logical system this error of circular reasoning would be serious—we have really proved very little, other than that the statements are not contradictory. (However, we have shown that the two statements are logically equivalent. See Sec. 13-4.) In this introductory treatment our procedural error is not serious, and it is easily correctable, using material in Chapter 7 that is independent of any of this material. Specifically, using the concept of congruence we can prove that statement (1) above is true; then statement (2) follows, as we outlined in problem 5. We may then use these statements in the proofs that follow here. You may find it helpful to reread the introductory comments in problem 7 in the Exercises at the end of Chapter 3 before proceeding.

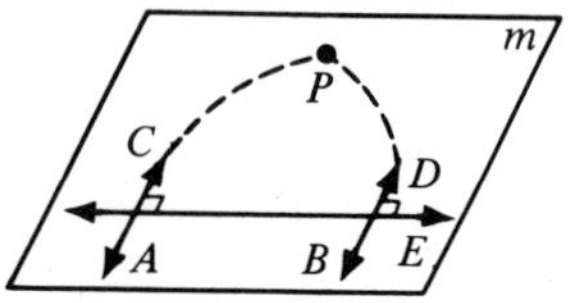

16. The steps outlined below justify that if in a plane two lines are perpendicular to the same line, these lines are parallel.

GIVEN: $\overleftrightarrow{AC}$, $\overleftrightarrow{ABE}$, $\overleftrightarrow{BD}$ all in plane m, $\overleftrightarrow{AC} \perp \overleftrightarrow{ABE}$, $\overleftrightarrow{BD} \perp \overleftrightarrow{ABE}$.

TO PROVE: $\overleftrightarrow{AC} \parallel \overleftrightarrow{BD}$.

STEPS	EXPLANATION
(1) Assume $\overleftrightarrow{AC} \cap \overleftrightarrow{BD} = \{P\}$.	(1) One of two contradictory statements.
(2) Then ABP is a triangle.	(2) ?
(3) $\angle CAE$ and $\angle DBE$ are right angles.	(3) ?
(4) $\angle DBE$ is exterior angle of $\triangle ABP$.	(4) ?
(5) $m(\angle DBE) > m(\angle CAE)$.	(5) An exterior angle of a triangle is greater in degree measure than . . .
(6) But $m(\angle CAE) \stackrel{\circ}{=} m(\angle DBE)$.	(6) ?
(7) Since (5) and (6) cannot be, $\overleftrightarrow{AC} \cap \overleftrightarrow{BD} = \varnothing$.	(7) ?
(8) Hence $\overleftrightarrow{AC} \parallel \overleftrightarrow{BD}$.	(8) ?

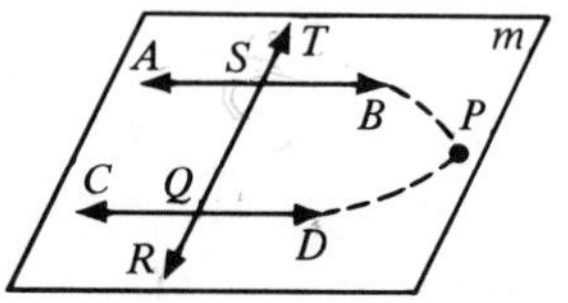

17. This illustration shows that if two lines are cut by a transversal, making a pair of alternate interior angles with the same measure, then these lines are parallel.

GIVEN: $\overleftrightarrow{AB}$, $\overleftrightarrow{CD}$, $\overleftrightarrow{RT}$ in plane m, $\overleftrightarrow{TR} \cap \overleftrightarrow{AB} = \{S\}$, $\overleftrightarrow{TR} \cap \overleftrightarrow{CD} = \{Q\}$, $m(\angle ASR) \stackrel{\circ}{=} m(\angle DQT)$.

TO PROVE: $\overleftrightarrow{AB} \parallel \overleftrightarrow{CD}$.

STEPS	EXPLANATION
(1) Assume $\overleftrightarrow{AB} \cap \overleftrightarrow{CD} = \{P\}$.	(1) ?
(2) Then SQP is a triangle.	(2) ?
(3) $\angle ASR$ is exterior angle of $\triangle SQP$.	(3) ?
(4) $m(\angle ASR) > m(\angle TQD)$.	(4) ?
(5) But $m(\angle ASR) \overset{\circ}{=} m(\angle TQD)$.	(5) ?
(6) Since (4) and (5) are impossible, $\overleftrightarrow{AB} \cap \overleftrightarrow{CD} = \varnothing$.	(6) ?
(7) Hence $\overleftrightarrow{AB} \parallel \overleftrightarrow{CD}$.	(7) ?

18. If two parallel lines are cut by a transversal, then a pair of alternate interior angles have the same measure.

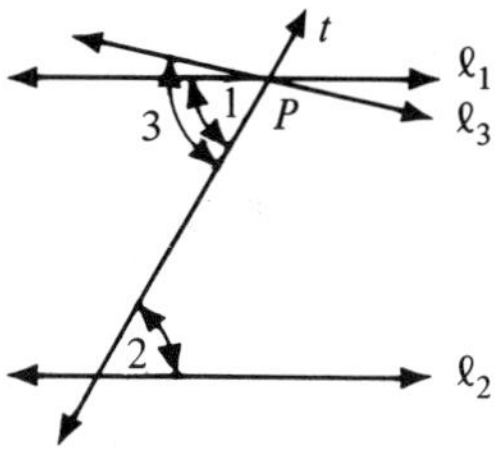

GIVEN: $\ell_1 \parallel \ell_2$, transversal t.

TO PROVE: $m(\angle 1) \overset{\circ}{=} m(\angle 2)$. (Query: Why no plane m?)

STEPS	EXPLANATION
(1) Assume $m(\angle 1) \neq m(\angle 2)$.	(1) ?
(2) Then at P, on t, there is an angle such that $m(\angle 3) \overset{\circ}{=} m(\angle 2)$.	(2) Agreement, Chapter 5.
(3) Let ℓ_3 be the terminal side of $\angle 3$, then $\ell_3 \parallel \ell_2$.	(3) See problem 17 above.
(4) It is impossible that: (a) $\ell_1 \parallel \ell_2$ *and* (b) $\ell_3 \parallel \ell_2$, therefore ℓ_1 and ℓ_3 are the same line.	(4) ?
(5) Thus $\angle 1$ and $\angle 3$ are the same angle, $m(\angle 1) \overset{\circ}{=} m(\angle 2)$.	(5) ?

19. In a plane a line perpendicular to one of two parallel lines is perpendicular to the other parallel line.

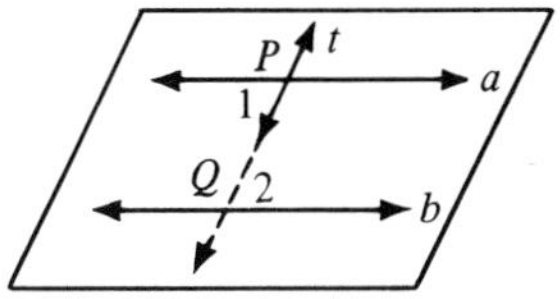

GIVEN: Lines a, b and t in plane m, $t \perp a$ at P, $a \parallel b$.

TO PROVE: $t \perp b$. Note: This is a direct proof as compared with problems 16–18, which are indirect.

STEPS	EXPLANATION
(1) $t \cap b$ in some point, Q.	(1) See problem 7(a) in the Exercises at the end of Chapter 3.
(2) $\angle 1$ and $\angle 2$ are alternate interior angles.	(2) ?
(3) $m(\angle 1) \overset{\circ}{=} m(\angle 2)$.	(3) ?
(4) $\angle 1$ is a right angle, hence $m(\angle 1) \overset{\circ}{=} 90$.	(4) ?
(5) $m(\angle 2) \overset{\circ}{=} 90$, hence $\angle 2$ is a right angle.	(5) ?
(6) Hence $t \perp b$.	(6) ?

20. Read (a) Ref. 25, pp. 229–244, (b) Ref. 43, pp. 126–134, and (c) Ref. 32.

CHAPTER 7

Transformations: Similarity and Congruence

7-1 INTRODUCTION

Although we shall not investigate in great detail the basic concepts involved in transformations in general, we wish to use this idea to develop a basic understanding of similarity. We shall then be able to discuss congruence as a special case of similarity. These two concepts are basic to the material of Chapters 8 through 11.

7-2 ONE-TO-ONE CORRESPONDENCE

In introducing the concept of counting, we use the idea of one-to-one (1 : 1) correspondence to establish the cardinality of sets. (See Appendix A, Sec. A-6.) Thus in Fig. 7-1 we can see several of the 1 : 1 correspondences between the two sets illustrated, and we say the sets have the same cardinality. We then assign the same cardinal number to each set—in this case 3.

Likewise, we can make a 1: 1 correspondence between the sets N and T: $N = \{n \mid n \text{ is a natural number}\}$, $T = \{t \mid t = 3n \text{ and } n \in N\}$. Thus,

$$\begin{array}{l} N = \{1, 2, 3, \ldots, 21, 22, 23, \ldots, \;\; n, \ldots\}, \\ \quad\;\; \updownarrow \; \updownarrow \; \updownarrow \; \updownarrow\updownarrow\updownarrow \;\; \updownarrow \;\; \updownarrow \;\; \updownarrow \; \updownarrow\updownarrow\updownarrow \; \updownarrow \; \updownarrow\updownarrow\updownarrow \\ T = \{3, 6, 9, \ldots, 63, 66, 69, \ldots, 3n, \ldots\}. \end{array}$$

To decide whether a 1: 1 correspondence exists, we consider the following questions: (1) For every element in N, is there a single corresponding element in T? (2) For every element in T, is there a single corresponding element in N?

Let us introduce a useful notation.[1] Let "$T(1)$," which we read "T of 1" or "the image of 1," mean the corresponding element from T that is matched with "1" of N. Thus $T(1) = 3$, $T(2) = 6$, $T(n) = 3n$, and let us call these **images**. Hence the image in T of 5 that is in N is $T(5) = 15$. What is $T(22)$? Likewise, the image in N of 15 in T is $N(15) = 5$. Of course, the context must make clear which sets are involved. If there seems to be some ambiguity, perhaps a subscript could be used. Thus "$T_N(5) = 15$," read "T sub N of 5 is 15," means the image in T of 5 that is in N is 15. But we are making a real effort here not to become overinvolved with notation. The important point at this time is that for every element of either set there is a unique image element in the other set, the basic consideration of a 1: 1 correspondence.[2]

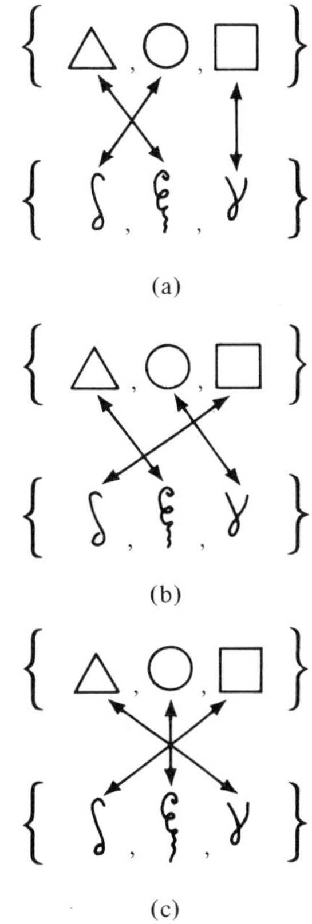

Figure 7-1

7-3 TRANSFORMATION

Let us consider questions (1) and (2) of the previous section as they relate to Fig. 7-2. Select $\overline{AB}$ and $\overline{RS}$ such that they lie in the same plane, they do not intersect, and $\overleftrightarrow{AR}$ and $\overleftrightarrow{BS}$ are not parallel. $\overline{AB}$ and $\overline{RS}$ need not be parallel (but they may be). Now P is selected such that $\{P\} = \overleftrightarrow{RA} \cap \overleftrightarrow{SB}$. To help our analogy with Sec. 7-2, let us say that $\overline{AB} = N$ and $\overline{RS} = T$. Then we can say: $T(A) = R$, $T(B) = S$, $N(R) = A$. What is $N(S)$? We could also say $P_N(A) = P$, $P_N(B) = P$, $P_T(R) = P$, and so on, but at this point we consider this a trivial case. (Notice that it is unique in only one direction; this type of matching is acceptable in transformations but does not constitute a 1: 1 correspondence.[2]) Notice that we have actually set up a mapping (only one of many possible), among sets N, T, and $\{P\}$.

[1] Actually, while we use "$T(1) = 3$," in advanced texts a transformation is usually symbolized "$(1)T = 3$" or "T: $1 \to 3$." Our usage agrees with the more commonly known functional notation, "$f(1) = 3$."

[2] 1: 1 correspondence requires unique pairing—both ways—all elements.

In this geometric type of mapping, any point in the original figure together with its image point in the image figure (set) are called **corresponding points.**

Now how could we describe this mapping geometrically? Suppose we say that to find in T, the image or image point of P_n that is in N, consider $\overleftrightarrow{PP_n}$. Then define $T(P_n)$ as the element in the intersection of $\overleftrightarrow{PP_n}$ and $\overline{RS}$; thus $\{T(P_n)\} = \overleftrightarrow{PP_n} \cap \overline{RS}$, and $T(P_n) = P_t$. Would this properly describe the correspondence we have already set up with points A, B, R and S? In Fig. 7-2 select several points of $\overline{AB}$ and find the image points in $\overline{RS}$. How would we describe the prócedure for finding in N the image points of elements in T? Write out this procedure, select several points of $\overline{RS}$, and find the image points in $\overline{AB}$. Use the notation developed.

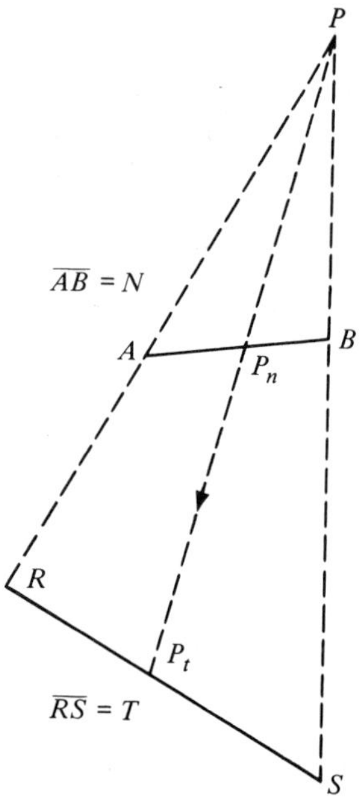

Figure 7-2

Now refer to questions (1) and (2) of Sec. 7-2. Have we a 1: 1 correspondence between N and T? We say we have a **geometric transformation** in a plane that preserves relative positions of points. Figure 7-3 shows a more complicated plane transformation.

If you think of a slide projector, colored slide, and the projection of the slide on the screen, you have an illustration of a space transformation.

EXERCISE 7-3

1. As in Fig. 7-1, show two different possible 1: 1 correspondences between $A = \{x \mid x \text{ is a prime} < 6\}$, and $B = \{\text{red, blue, orange}\}$.

2. Using the braces and ellipsis notation as in Sec. 7-2, show a possible 1: 1 correspondence between $S = \{n \mid n \text{ is a natural number}\}$ and $Q = \{q \mid q = n^2, n \in S\}$. If not possible, explain.

3. In Fig. 7-3 is a drawing that might be considered as either a plane transformation or a space transformation with $\triangle ABC$ and $\triangle RSX$ in different planes.

(a) Considered as a space transformation and defining $\overline{AB} \cup \overline{AC} \cup \overline{BC} = T$ and $\overline{RS} \cup \overline{RX} \cup \overline{XS} = N$, complete the following:
(1) $N(E) =$ _____. (2) $N(C) =$ _____.
(3) $T(Z) =$ _____. (4) $T(Q) =$ _____.
(5) $N[T(Q)] =$ _____.

(b) Considered as a plane transformation in which $\overline{AB}$ maps to $\overline{RS}$, $\overline{AC}$ maps to $\overline{RX}$, and $\overline{BC}$ maps to $\overline{SX}$, defining N and T as in (b):
(1) $N(K) =$ _____. (2) $N(J) =$ _____.
(3) $T(U) =$ _____.

(c) Explain a possible plane transformation that maps $\overline{AB}$ to $\overline{RX} \cup \overline{XS}$, and $\overline{AC} \cup \overline{CB}$ to $\overline{RS}$. Define N and T as before and find some corresponding point pairs.

4. See Ref. 34, pp. 302–311.

In the previous section we actually set up a much more general idea of transformation than we will be using. Figure 7-4 and the illustration of the slide projector exemplify the types in which we will be interested. As with corresponding points, a distance between points in the original figure and the distance between the corresponding image points are called **corresponding distances**. In addition to preserving relative positions of points, if we require that a constant ratio be maintained between every distance in the original figure and the corresponding distance in the image figure, then we have a **similarity transformation** and the figures are said to be **similar**. The relativity ratio between the corresponding distances is known as the **ratio of similitude**, abbreviated r.s.

If in Fig. 7-4 we map $\overline{QR}$ to $\overline{Q'R'}$ so that $\overline{QR} \parallel \overline{Q'R'}$, $\overline{QS}$ to $\overline{Q'S'}$ so that $\overline{QS} \parallel \overline{Q'S'}$, and $\overline{RS}$ to $\overline{R'S'}$, keeping $\overline{RS} \parallel \overline{R'S'}$, we have a parallel mapping. It turns out that any parallel transformation that preserves relative position will retain relative measures for corresponding distances, as in Fig. 7-4. More simply stated, we map $\triangle QRS$ through P onto parallel lines forming $\triangle Q'R'S'$. Then $\triangle Q'R'S'$ is the image of $\triangle QRS$, and we say that $\triangle QRS$ is similar to $\triangle Q'R'S'$, or vice versa, and we write: $\triangle QRS \sim \triangle Q'R'S'$. The order of the letters should be noted. More recent writings require that corresponding vertices be named in the same order, but there is not complete agreement on this. The convenience in this convention, however, is certainly most desirable and we shall adopt this pattern. Either triangle may be considered as the image of the other through the parallel mapping through P.

Since $m(\overline{QR}) = 1\frac{1}{4}$ and $m(\overline{Q'R'}) = 2\frac{1}{2}$, both in inches, the ratio of similitude, r.s., in this instance is 1: 2 from $\triangle QRS$ to $\triangle Q'R'S'$. Common agreement insists that the order in this ratio is the same as the order in the statement of similarity; thus, $\triangle QRS \sim \triangle Q'R'S'$ requires us to write "r.s. = 1: 2" rather than "r.s. = 2: 1." So you see, even if no diagram is given, if you know the ratio of similitude between $\triangle QRS$ and $\triangle Q'R'S'$ is 1: 2, then you know that $\triangle Q'R'S'$ is the larger figure.

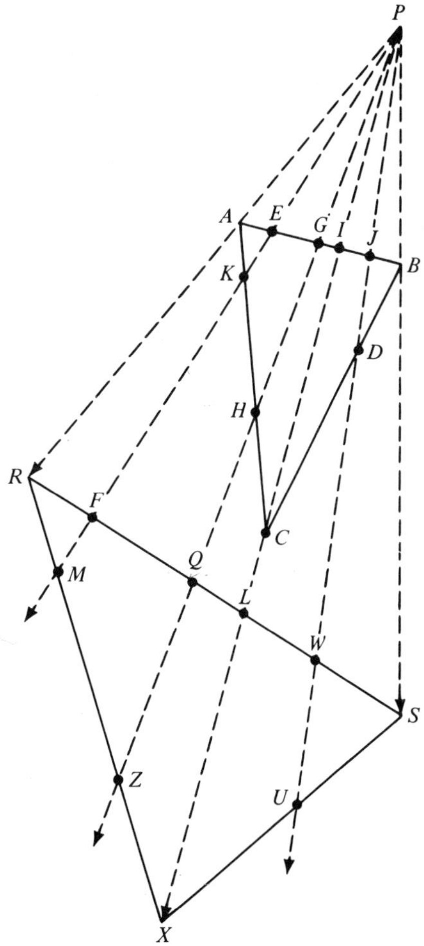

Figure 7-3

Notice also from our understanding of a linear similarity transformation that r.s. is 1: 2 between $\overline{QR}$ and $\overline{Q'R'}$. If you did not know the measure of $\overline{Q'S'}$, but knew all of the above information regarding the parallelism and that $m(\overline{QS}) = 7$, then you could conclude that $m(\overline{Q'S'}) = 14$. In fact, in general if you know r.s. between two similar figures, knowing any measure in one figure enables you to compute the corresponding measure in the other figure by using proportions. This is even true of other than linear measures, but it requires a more careful analysis, which we discuss in Chapters 8, 9, and 10.

Corresponding angular measures will be developed in the exercises, but think for just a moment: since parallelism is involved, what might you expect to be true of corresponding angular measures?

Two warnings:

(1) If you compare two distances in the original figure, the ratio between these is *the same as* the ratio between their corresponding distances in the image figure; r.s. is applied only when comparison is made between a distance in the original figure and its image distance. See Exercise 7-4, problem 2.

(2) We are not limited to parallelism in order to obtain similarity transformations, and some special ones will be considered in Sec. 7-5. Also do not conclude that we must deal only with triangles. Any geometric configuration may be transformed.

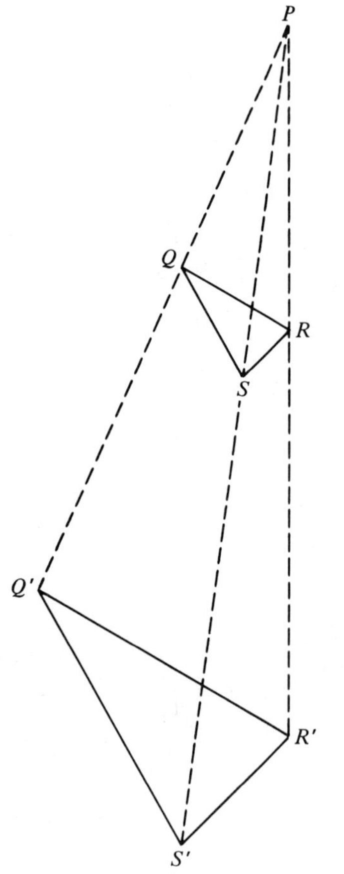

Figure 7-4

We have been illustrating a special type of transformation called **exterior point transformation.** Let us briefly investigate **interior point transformation** and limit ourselves to a similarity transformation. Figure 7-5 compares interior and exterior mappings as follows: using P, $\overline{AB}$, and $\overline{RS}$, we have an exterior point mapping in which $\overline{AB} \sim \overline{RS}$ and r.s. = 2: 3; using P, $\overline{R'S'}$, and $\overline{AB}$ we have an interior point transformation in which $\overline{R'S'} \sim \overline{AB}$ and r.s. = 3: 2; using P, $\overline{RS}$, and $\overline{R'S'}$ we have an interior point transformation in which $\overline{RS} \sim \overline{R'S'}$ and r.s. = 1: 1. In this last case be careful not to confuse the three uses of "1: 1": (a) for the one-to-one correspondence of points, (b) for the ratio between distances and between corresponding distances, and (c) for the ratio of similitude between the original figure and the image figure.

A convenient and familiar model of an interior point similarity transformation is the well-known pinhole or box camera. Notice in Fig. 7-6 that the orientation appears to be changed, and in terms of right-left considerations it is; but otherwise, the image points are relatively placed. Actually we do not consider this to be a case of change in orientation, but more on this later.

EXERCISE 7-4

1. In two sets of points that comprise a similarity transformation, measures a, b, and c in the first set correspond to e, f, and g in the second set, respectively. If $a = 9$, $b = 10$, $e = 12$, and $g = 16$, find c and f.

2. Given that r.s. $= m:n$ and two corresponding distances are h, k, and another pair are j and t, where h and j are in the original figure. Show that $h:j = k:t$. [See "warning" (1) in Sec. 7-4.]

3. In Sec. 7-3 in the first paragraph we placed the restriction with respect to Fig. 7-2 that $\overleftrightarrow{AR}$ and $\overleftrightarrow{BS}$ not be parallel. If we remove this restriction, that is, let $\overleftrightarrow{AR} \parallel \overleftrightarrow{BS}$, how can we set up our mapping?

4. Make a sketch of Fig. 7-4 on paper such that you have room to show $\overline{QR}$ and $\overline{Q'R'}$ extended and A lies on $\overrightarrow{RQ}$ but not on the R side of Q, B lies on $\overrightarrow{QR}$ but not on the Q side of R, C lies on $\overrightarrow{R'Q'}$ but not on the R' side of Q', D lies on $\overrightarrow{Q'R'}$ but not on the Q' side of R', and E lies on $\overrightarrow{QQ'}$ but not on the Q side of Q'.
   (a) Is $\angle RQQ'$ congruent to $\angle QQ'C$? Explain.
   (b) Is $\angle CQ'Q$ congruent to $\angle R'Q'E$? Explain. What of $\angle RQE$ and $\angle R'Q'E$?
   (c) In like manner show that $\angle SQQ'$ is congruent to $\angle S'Q'E$.
   (d) Then is $\angle RQS$ congruent to $\angle R'Q'S'$? Explain.
   (e) In a similar manner show that $\angle QRS$ is congruent to $\angle Q'R'S'$.
   (f) Look carefully at steps (a) and (b) and you will see that we have shown another important statement: If two parallel lines are cut by a transversal the _____ are congruent. (Recall we proved the alternate interior angles congruent in Chapter 6.)

5. (a) Using your protractor, draw two triangles such that the angles of one are congruent to the angles of the other, respectively. Do not draw the special case in which the angles of both are all the same measure—that is, angles with degree measure of 60. Also do not make the sides of one congruent to the sides of the other. In other words, draw two scalene triangles with corresponding angles congruent. Then measure the sides of each triangle and notice the ratios of the measures of corresponding sides. (Use appropriate precision.) From this experiment—performed, perhaps, several times—would you be willing to agree that knowing that the three angles of one triangle are congruent to three angles of another triangle, respectively, is sufficient for you to conclude that the triangles are similar? (As a matter of fact this statement is true, but your experiments do not prove it. They merely fail to disprove it!)
   (b) As in part (a) you are to draw two triangles. This time draw one scalene triangle two sides of which have "nice" measures, perhaps 6 and 8. Then draw another triangle in which the sides corresponding to these are in proportion, perhaps 3 and 4, and the angles determined by these sides and

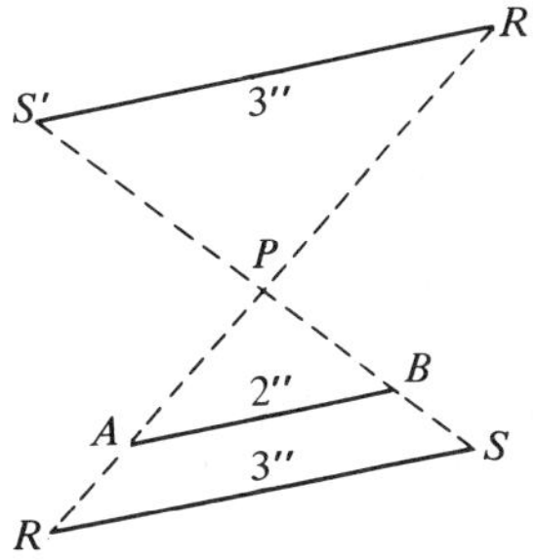

Figure 7-5

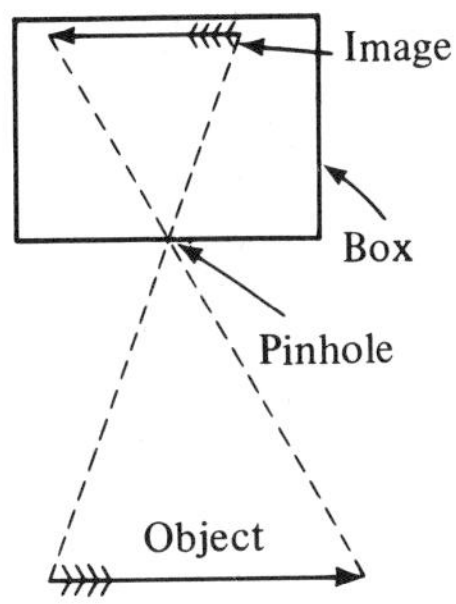

Figure 7-6

the original sides are congruent. Then measure the third sides in each triangle and observe the ratio. Also measure each of the other two pairs of angles. What do you observe? Make a general statement similar to that in part (a). (This statement is also true, although the experiment only reinforces your belief in its validity.)

6. For comments on interesting properties of central projection, including the possibility that the image plane is parallel to one of the projecting rays, see Ref. 6, pp. 41–42, 47–51.

7-5 CONGRUENCE

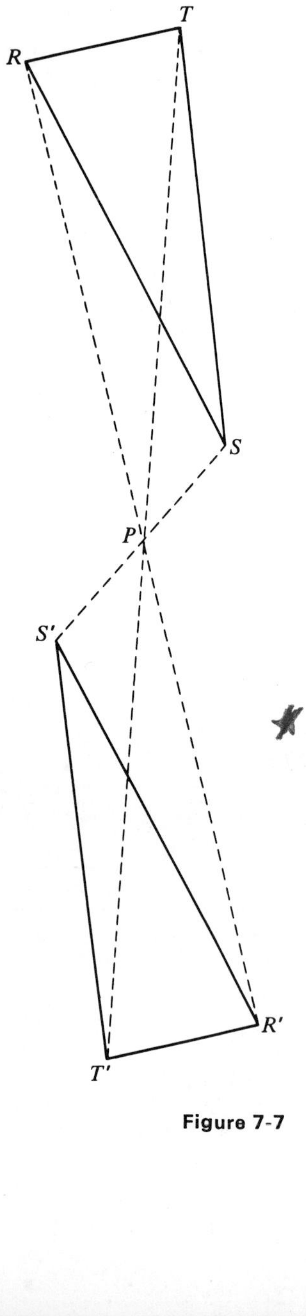

Figure 7-7

In the special case indicated for Fig. 7-5, in which r.s. $= 1\!:1$ or 1, we say the figures are **congruent**. Notice that in this case a segment in the original figure has the same measure as the corresponding image segment. How does this concept of congruence compare with our understanding of congruent segments and congruent angles in Chapter 5? (See Ref. 30, pp. 202–207.)

In general any transformation that preserves relative position and preserves distance betwen pairs of points and pairs of image points is a **congruence transformation**, also known as **Euclidean transformation**, or **isometry**. In Fig. 7-7, $\triangle RTS$ is congruent to $\triangle R'T'S'$ through the interior point congruence transformation, and we write "$\triangle RTS \cong \triangle R'T'S'$," "$\overline{RT} \cong \overline{R'T'}$," and so on. You might note that the symbol "$\cong$" is a combination "$=$" and "$\sim$" sign, for equal and similar. What about angular measure? Since congruence is a special case of similarity, what can we conclude about corresponding angular measures? Suppose the triangles were cut out; do you suppose you could superimpose one upon the other? Actually, some authors define congruence in this way, but we shall say more on this later.

Mechanically there are at least three ways in which Euclidean transformations can be carried out with physical models: (1) rotation, (2) reflection, and (3) translation. To do this we think dynamically. We move points; indeed we move complete figures (at times a most difficult accomplishment.)

In Fig. 7-7 we might think of $\triangle R'T'S'$ merely as being $\triangle RTS$ after it has been rotated (turned) in the plane through an angle of 180° about point P. Now it would appear that the orientation is reversed; indeed, that which was to the left is now to the right. But we prefer to think of orientation in the following way. If we think of a point traversing each triangle, it could proceed in either a clockwise or counterclockwise manner (or sense). Notice that if the point travels clockwise about $\triangle RTS$ starting at R, it passes through R, T, and then S. Also, if it goes clockwise around $\triangle R'T'S'$ starting at R', it passes through R', T', and then S'. Thus we say the **rotation** about this point (called **reflection in a point**) is sense-preserving. Reminder: We have already agreed that these transformations are distance-preserving. Thus we have described **point reflection**, or reflection in a point, as a 180° rotation in the plane. The word "reflection" here is an unfortunate convention, owing to the possible confusion with our second method—reflection. Figure 7-8(a) illustrates a rotation of other than 180°. Try to locate a point that could be used to carry out this rotation.

Suppose we think of flipping $\triangle TRS$ 180° out of the plane about a line in the same plane. For the present consider only those cases in which the line does not intersect the triangle, as in Fig. 7-9(a). As will be seen, this limitation is temporary and merely avoids complication. Then, in the figure, we can consider $\triangle R''T''S''$ as $\triangle RTS$ after flipping it about m until it is again in the plane suggested by the paper. If a small mirror is placed along m, we observe in it a reflection of $\triangle RTS$ Hence, flipping as we have described it is called **reflection in a line**. Is it necessary that m be parallel to one side of the triangle? Try placing the hand mirror in different positions. Let us consider orientation with respect to this reflection. If a point travels clockwise about $\triangle RTS$ starting at R, it would pass through R, T, and then S. In $\triangle R''T''S''$ it would pass through R'', S'', and then T''. Therefore, we say the orientation or sense has been reversed under this reflection.

Now consider $\triangle ABC$ in Fig. 7-8(b). What happens to this triangle as this page is turned to expose another page? Where is the line of reflection? What can you say of the distance[3] from A to this line throughout this experiment? (Refer to the description of a circle given in Sec. 5-7.)

By our description of reflection, in Fig. 7-9(b) $\overline{TT''}$ is perpendicular to m. Let $\overline{TT''} \cap m = \{P\}$. In this situation $\overline{TP} \cong \overline{T''P}$. Hence, m is perpendicular to and bisects $\overline{TT''}$. (Thus m is called the **perpendicular bisector** of $\overline{TT''}$.) In such a case T and T'' are called **symmetric points.** Is it not true that m would be perpendicular to and bisect each segment joining any two corresponding points? Would this be true of m and $\overline{KK''}$? Then are K and K'' symmetric points? What point is symmetric to (or symmetric with) Q?

Notice that the preceding statements will be true in general when a reflection has been carried out, and we say (1) the figures are **symmetric with respect to a line**, (2) they are **symmetric** *to each other about the line*, or (3) the figure possesses **line symmetry.** The line of reflection is called the **axis of symmetry**. What is the axis of symmetry for the experiment with Fig. 7-8(b)?

Now refer to Fig. 7-7. As $\triangle RTS$ is rotated about P, does not each point of the figure maintain its distance from P thoughout the rotation? Specifically, is not $\overline{PS} \cong \overline{PS'}$? What can you conclude of $\overline{PR}$ and $\overline{PR'}$? Select

[3] Distance from a point to a line is discussed in Exercise 6-3, problem 9.

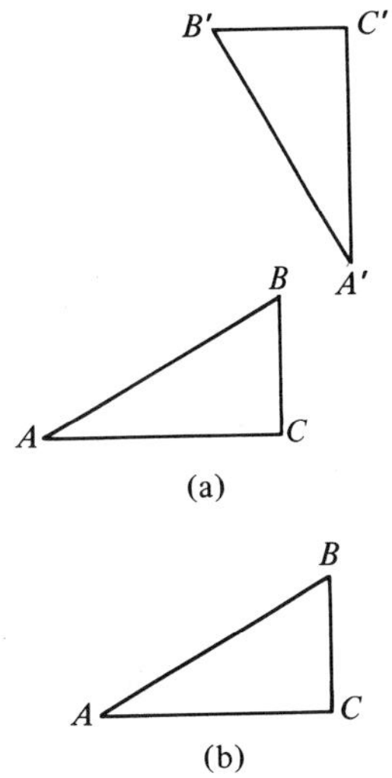

Figure 7-8

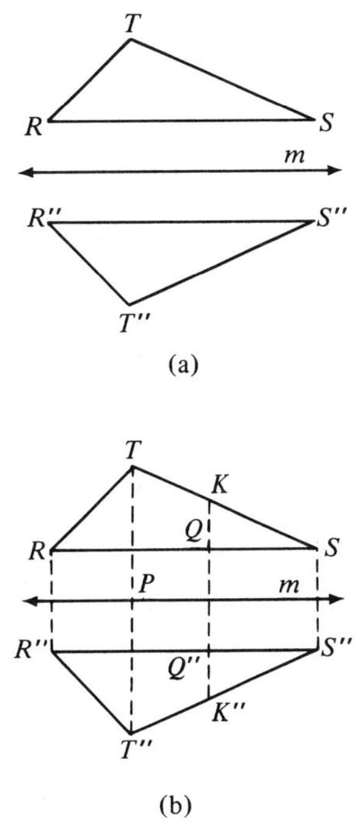

Figure 7-9

any point of the triangle RTS; translate through P in a line until you reach the corresponding point of triangle $R'T'S'$. Measure with a compass the distance from the original point to P and compare this with the distance from the image point to P. As in the case for which we have corresponding points equidistant from the same line we say we have line symmetry, in this last case we have corresponding points equidistant from a point, and we say (1) the figures are **symmetric with respect to a point**, (2) they are **symmetric** *to each other about a point*, or (3) the figure possesses **point symmetry**. Think of some of the figures we have discussed and decide whether they possess point and/or line symmetry.

The third mechanical model of an isometry is the **translation**, or slide, illustrated by Fig. 7-10. It should be noted that a translation is sense-preserving.

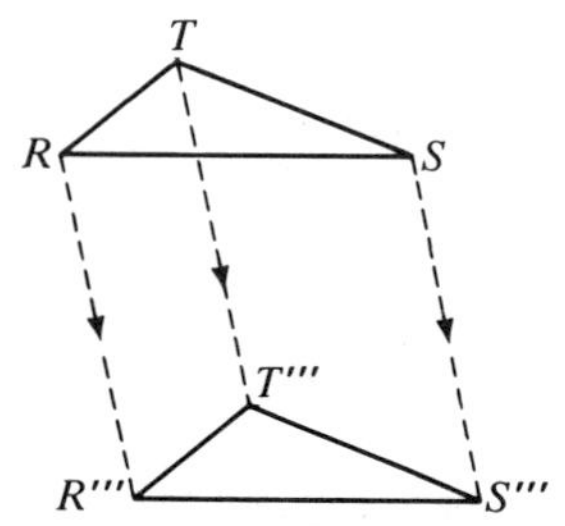

Figure 7-10

Surely by this time you are wondering, "But why all the fuss!" The "fuss" is because congruence is basically a static concept. So also is the idea of a mathematical transformation. Thus, those who object to a concept of superposition (used by Euclid) or coincidence (sometimes stated: a geometric figure can be moved without changing its size or shape) would like a transformation approach. And these people should not be dismissed lightly. After all, if we move a physical model of a circle, for instance a piece of string, we might be hard-pressed not to change some of its properties. On the other hand, other models, perhaps wire for a circle, would not present much difficulty. The point is that such movements do or can exist. At any rate, we shall discuss conclusions that can be deduced under the assumption that a figure can be moved without altering its measurement properties. The concept of transformation (greatly oversimplified in this discussion) accomplishes this.

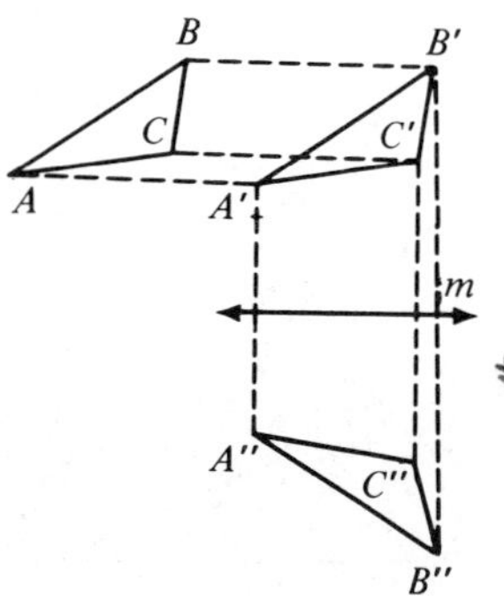

Figure 7-11

We say then, in summary: an isometry, or distance-preserving transformation, also known as a congruence or Euclidean transformation, does exist and we can think physically of:

(1) Rotation—turn—preserves sense.
(2) Reflection—flip—reverses sense.
(3) Translation—slide—preserves sense.

It should be clear that every isometry of a figure in a plane can be accomplished by one of the methods above, or perhaps by a combination of two of them, as in Fig. 7-11. (And the figures need not be merely segments and triangles, which we have used in our illustrations.)

EXERCISE 7-5

1. Which figures in Fig. 7-12 possess (a) line symmetry, (b) more than one axis of symmetry, (c) point symmetry, (d) both point and line symmetry?

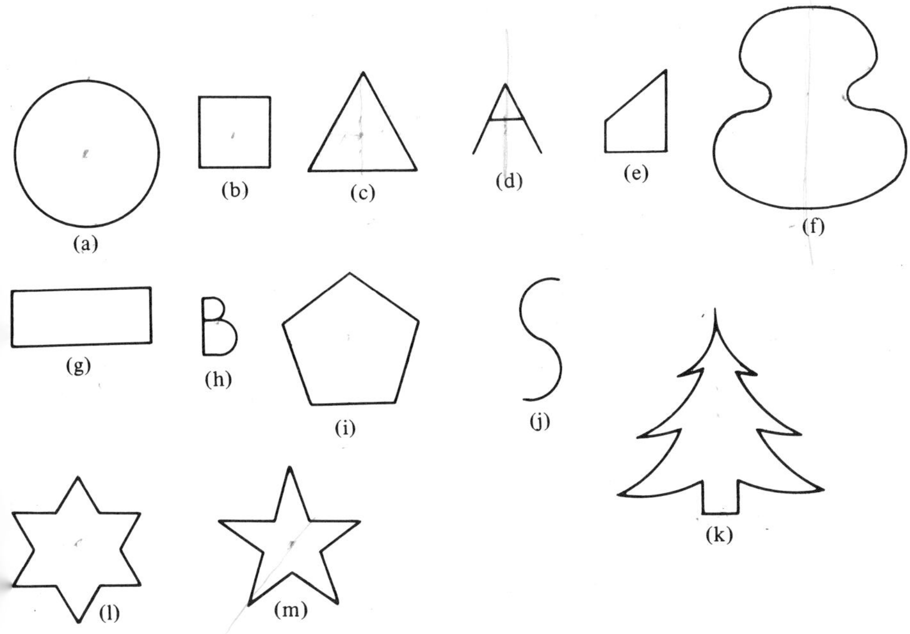

Figure 7-12

2. For which figures in Fig. 7-13 can part 2 be obtained from part 1 by a single (a) translation, (b) rotation, (c) reflection about a line?

3. Collect pictures that display the properties mentioned in problems 1 and 2 above.

4. Collect three-dimensional objects which possess symmetry with respect to (a) a point, (b) a line, (c) a plane.

5. This exercise consists of reference readings related to the previous section. The first two items are, perhaps, the most informative and readable; the last is mathematically technical. The readings in part (a) are especially good for work relating to symmetry.
   (a) (1) Ref. 86, pp. 448–452; (2) Ref. 87, pp. 115–125.
   (b) Ref. 113, pp. 354–369.
   (c) Ref. 90, pp. 230–232.
   (d) Ref. 116, pp. 123–125.
   (e) Ref. 11, pp. 3–5, 29–31, 39–49.
   (f) Ref. 29, Chapters 1–3.

7-6 BASIC CONSTRUCTIONS—SEGMENT, TRIANGLE, ANGLE

Having defined congruence as a similarity transformation in which r.s. = 1, let us see whether we can make use of this in terms of establishing the existence of certain triangles. Hopefully, we shall find that triangles

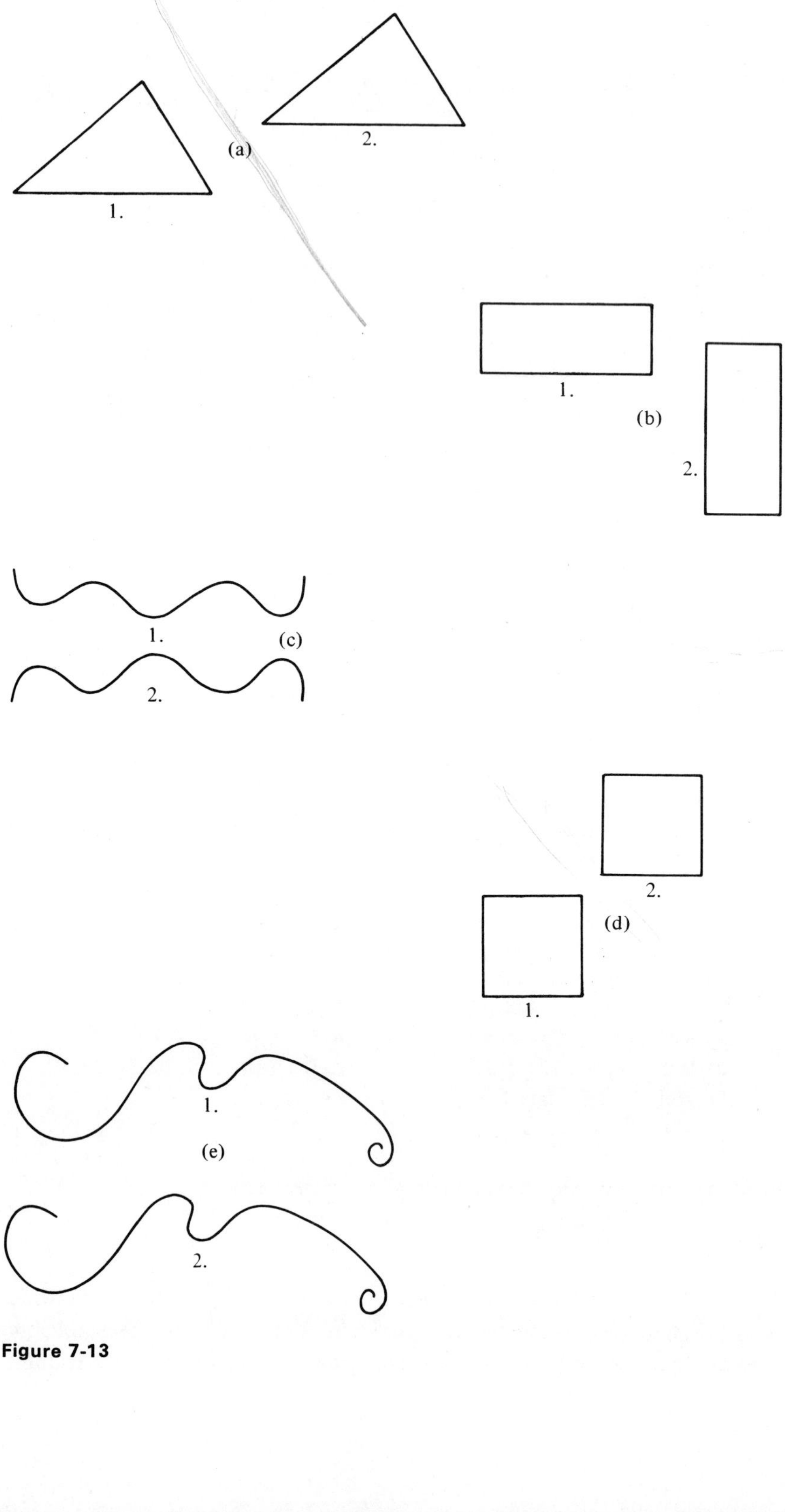

Figure 7-13

exist under certain conditions, and we shall be able to determine any limitations involved in these conditions.

Perhaps you have wondered why the triangle appears so often in situations involving construction. To help your thinking along this line make physical models, one of a triangle and another of a parallelogram. In Appendix C you will find patterns for these. Cut out copies of these patterns, fastening at the corners, perhaps with brass fasteners (not with staples). Set the models on some flat surface and, by pushing at the corners, try to change their shapes. Which model retains its shape—that is, remains rigid? Because of this (and our background in isometries) we are willing to agree that if three sides of a triangle are known (that is, we are assuming the triangle exists), then any triangle that has sides that are congruent to these known sides, respectively, is congruent to the given triangle. In practice this is often abbreviated: If $SSS \cong SSS$, then the $\triangle$'s are $\cong$. Or: If SSS, then the $\triangle$'s are $\cong$. Occasionally this is shortened to: SSS.

Let us use a compass and a straight edge to construct a triangle, knowing only the three sides or their measures. Notice: (1) A construction in which you are permitted to use only an unmarked straight edge and a compass is called a **Euclidean construction**. When we say "construct," we mean you are limited to this type of construction. (See Ref. 9, pp. 152–155; Ref. 30, pp.2 14–219, 227–228, 238–244; Refs. 18 and 40.) (2) We use the wording "know the sides" rather than "measure" because, on occasion, we might be given segments that represent the sides, and you will see that we can construct the triangle without knowing the measures.

We must first agree that it is possible to "transform" a given segment onto a given line. That is, in Fig. 7-14 you must agree that there exists on line w a point X such that $\overline{PX} \cong \overline{AB}$. Surely the discussions in Secs. 5-8 and 7-5 convince you that one of the physical models (rotation, reflection, translation), or a combination of them, will result in such a segment. Often a translation will suffice. In practice we use a compass to perform such transformations without regard to the particular combination of methods required to justify them. It will be worth your while, on occasion, to consider how this practice could be justified.

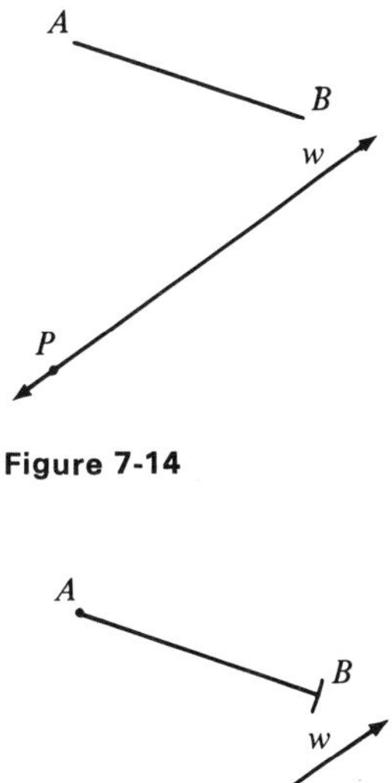

Figure 7-14

Agreeing that such a point (X) exists, we can use the following procedure to find its approximate location. (We are once again using physical embodiments to help us to "see" what exists in the abstract world.) Place the point of your compass at A. Open it so that the pencil point is at B. Without changing the adjustment, place the point at P and swing a short arc of a circle until it intersects w. The intersection of the arc with w is X. (See Fig. 7-15.) The fact that there might be other such segments (in this case determined by X' on the opposite side of P from X) should not arouse

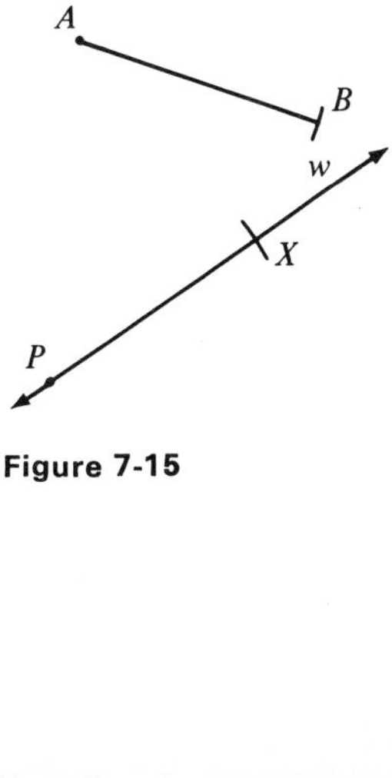

Figure 7-15

any anxiety, for we are merely seeking agreement that at least one such segment exists.

As usual, we are oversimplifying the situation, depending to a large extent upon your intuition; and let us not forget the inherent dangers. However, let us see just what we are asking you to "swallow":

(1) In a plane there exists a circle with a given point as center and a given length as radius.
(2) If two segments have the same measure, they are congruent.
(3) A point separates a line. (This is needed only if we wish to establish uniqueness of point X.)
(4) If two angles have the same measure, they are congruent.
(5) If two triangles have the three sides of one congruent to three sides of the other, respectively, the two triangles are congruent.

Surely you recognize statement (3) from Chapter 2. And our discussion concerning transformations clearly leads us to agree with (1), (2), and (4). We feel this is also true of (5), but let us discuss it further. What must be true for two figures to be congruent? Our concept of transformation requires that any two corresponding (image) points would be relatively placed as in the original, and r.s. $= 1$; that is, any distance in the original figure is the same for the corresponding image distance. Would this not mean, then, that making the sides the same length would guarantee us congruent figures? Do not be too hasty. Do you see that we are asking, "Does knowledge that three distances are invariant suffice to allow us to conclude that all distances are invariant?" We suggest that you are willing to say yes—although it is not as "obvious" as some people make it.

After all, consider the model you made of a parallelogram. What is given basically is four segments. But you were able to change the shape of this parallelogram; hence, other parallelograms exist having these same segments for sides, and they are not congruent to each other. What is it about a triangle that permits us to agree readily to the congruence? We suggest that the answer lies in our experiments, such as those with the triangle and parallelogram, and our discussion of transformations. We just "know" that one of the three methods listed will illustrate congruence for the triangle, and such is not the case with many other figures.

We could have avoided all of this by merely defining congruent triangles as those in which corresponding sides are congruent. Some authors do; but we feel our approach is more meaningful (although perhaps not as simple). We know, as we have developed it, that if we have congruence, any corresponding measures are the same, and if *all* corresponding measures are the same, we have congruence (with proper relative position).

Construction: But let us return now to our original problem—that of constructing a triangle given the three sides as in Fig. 7-16. [Note: For all the constructions of this chapter you should start with a full sheet of paper and make a large reproduction, step by step as stated, answering any questions raised in the discussion.] Let us proceed: Draw a working line w. Select a point on w and label it B. Using your compass, as discussed above, find A on w such that $\overline{BA} \cong c$. Using B as center, draw a circle with a as radius.[4] Using A as center, draw a circle with b as radius.

Discussion: Every point on the circle with B as center is distance a from B. Every point on the circle with A as center is distance b from A. We seek a point, C, or points, satisfying both of these conditions. Where is it? Connect these points, C and C', with A and B (you should have two such points as in the figure). We now claim that (1) either of these triangles satisfies the given conditions, (2) these two triangles are congruent to each other, and (3) any triangles constructed in this manner are congruent each to each.

To gain confidence in these assumptions, you are invited to try several such constructions using these same three segments. Perhaps you should change the order in which you use them; for instance, in the wording of the steps above, replace a with b, replace b with c, and replace c with a. Carefully cut these triangles out, place them on a flat surface, and consider what physical method will make them coincide. You might also compare your triangles with those constructed by someone else.

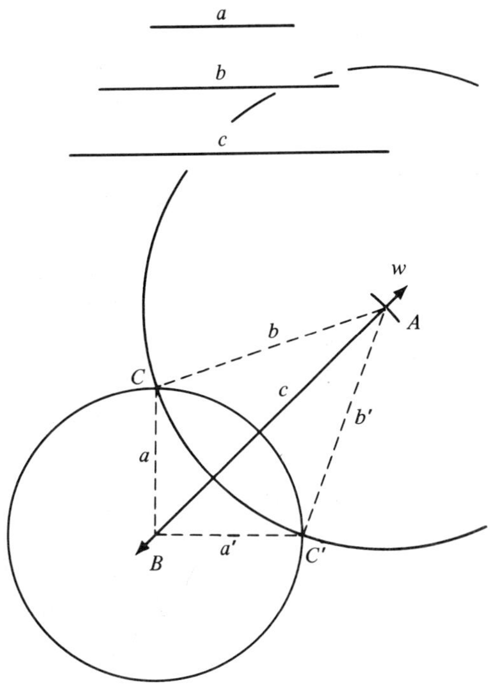

Figure 7-16

Construction: Let us now discuss a very useful construction that can be justified on the basis of this last construction. Let us show that on w in Fig. 7-17 we can contruct $\overrightarrow{PN}$ such that $\angle NPM \cong \angle SQR$. On a working line w select P as suggested in Fig. 7-17. Then, as in Fig. 7-18, using Q as center and any convenient length as radius, draw an arc cutting $\overrightarrow{QR}$ and $\overrightarrow{QS}$ in E and F, respectively. Without changing this radius, and using P as center, draw an arc cutting w; label this point M. The arc should extend in the general direction in which we expect $\overrightarrow{PN}$ to lie. Set the compass so that its points are at E and F, respectively. Using this setting as a radius and with M as center, swing an arc that cuts the previous arc. Label this

[4] As mentioned earlier, a is used both as a measure and as the name of a side. Convention requires that vertex A be opposite the side named a—that is, vertex A lies opposite side a.

intersection N and draw $\overrightarrow{PN}$. Our claim is that the angle determined by $\overrightarrow{PM}$ and $\overrightarrow{PN}$ is congruent to $\angle RQS$.

Discussion: If you reread closely the steps of the construction, you will see that we really constructed $\triangle NPM$ congruent to $\triangle FQE$. But if two figures are congruent, what can we conclude about them? Thus, $\angle SQR = \angle FQE \cong \angle NPM$. And, you notice that this conclusion requires most of the previous agreements; specifically, (1), (2), (4), and (5) are necessary. Furthermore, if we wish to show that this angle is uniquely placed on the N side of w, we need only refer to a statement concerning line-separation of a plane.

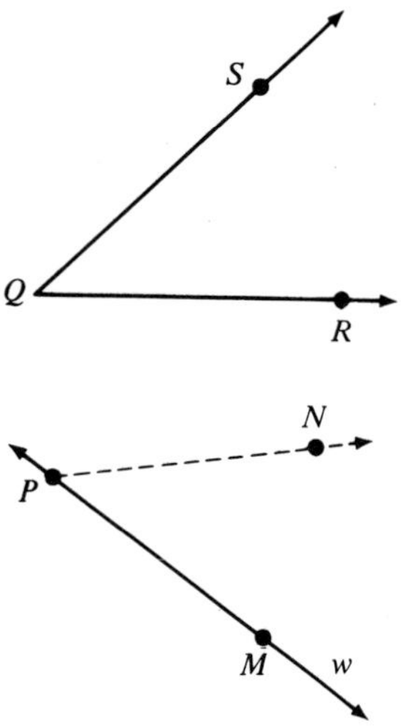

Figure 7-17

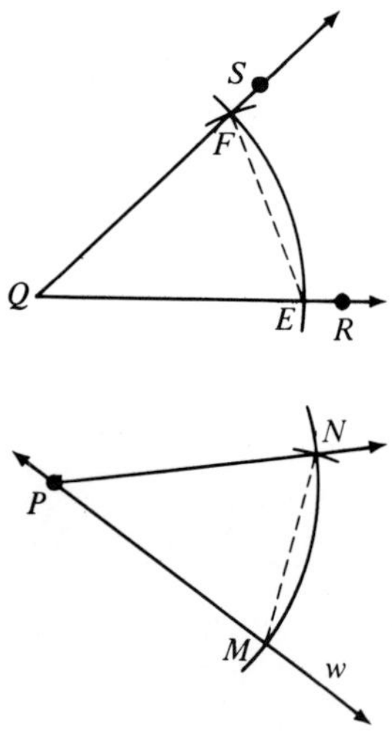

Figure 7-18

EXERCISE 7-6

In Exercise 6-4, 3, you were led to a conclusion regarding the relative lengths of the sides of a triangle. The following constructions should help to verify this conclusion as well as lead us to other minimal conditions that establish congruence of triangles.

1. Using the method above, construct triangles with sides of:

(a) 5″, 3″, 6″. (b) 4″, 2″, 6″.

(c) 4″, 4″, 3″. (d) 6″, 6″, 6″.

After completing the construction in each case, measure the angles with a protractor and make any appropriate general comments about their sizes, compared perhaps with the length of the sides.

2. Using a compass to compare, determine which of the following are pairs of congruent segments.

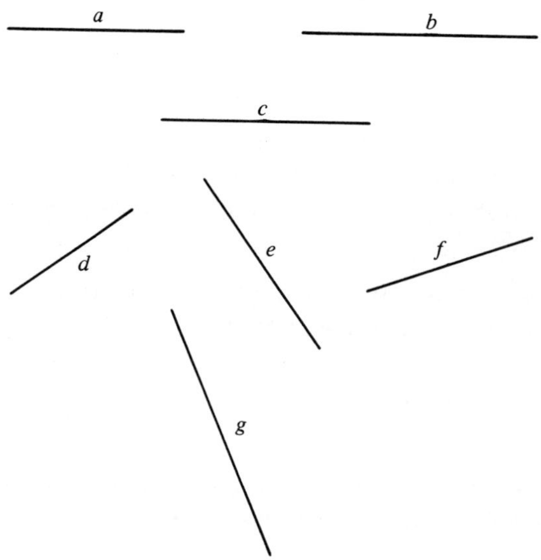

3. For each pair of figures below, determine whether the two are congruent by tracing one and trying to superimpose the copy on the other.

(a)

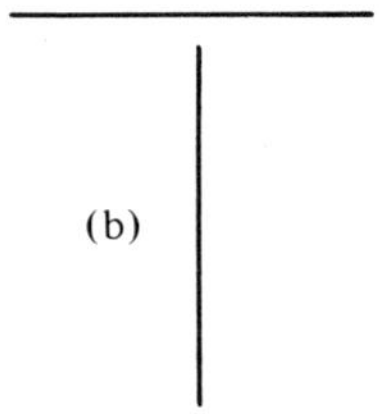

(b)

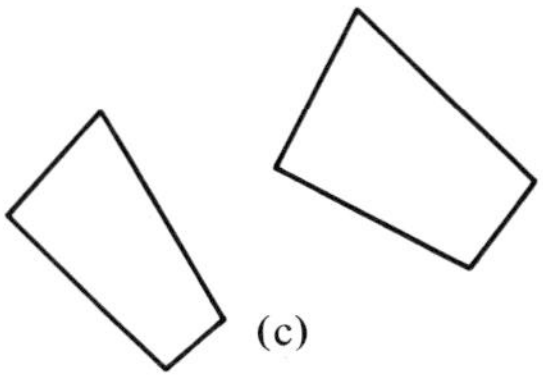

(c)

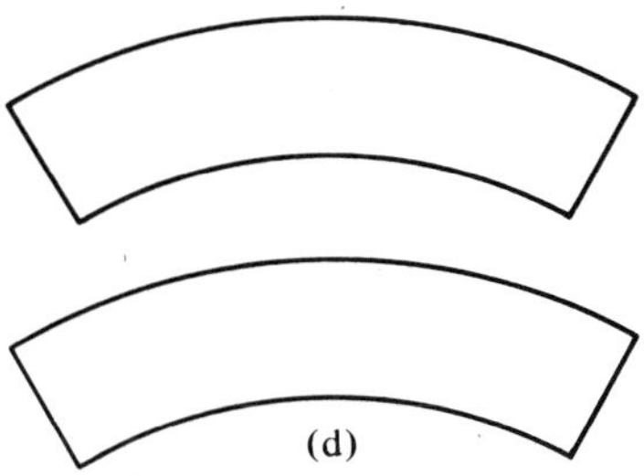

(d)

4. There are two other fundamental methods of showing congruence of triangles. Let us investigate.

(a) Suppose we know that two sides are 4″ and 5″ and the angle determined by them has a degree measure of 35. Question: Do these conditions determine congruent triangles? Test this somewhat as we did our three-side method in Sec. 7-6. Suggestion: Copy the 35° angle onto the working line and then construct copies of the 4″ and 5″ sides on the sides of this angle. The question then is: Do the outer endpoints of these segments determine a unique length? The abbreviation for this method is: If $SAS \cong SAS$, then the $\triangle$'s are $\cong$ (or simply *SAS*). *SAS* means two sides and the included angle, the angle determined by those two sides. What are the limitations on the given parts?

(b) Suppose we know the measures of two angles are 30 and 60 in degrees, and the "included" side has a measurement of 5″. Do these conditions determine congruent triangles? This time start with the 5″ segment and then copy the angles at the endpoints. What is the abbreviation for this method?

(c) A fourth method is based on the third method. Given that the measures of two angles are 35 and 45 in degrees and the side opposite the 45° angle has a measure of 3 in inches. Construct such a triangle. The abbreviation would be: If $AAS \cong AAS$, then . . . , which would be read: If in a triangle two angles and the side opposite one of them . . .

5. To show one application of some of this material consider Fig. 7-19, in which it is known that $\angle R \cong \angle S$ in $\triangle RST$. The problem is to show that in any triangle in which this is true, the sides opposite these angles are congruent. Explain each of following steps: (a) $\angle R \cong \angle S$, (b) $\overline{RS} \cong \overline{SR}$, (c) $\triangle RTS \cong \triangle STR$, (d) hence $\overline{TS} \cong \overline{TR}$.

6. Show that if two sides of a triangle are congruent to each other, the angles opposite these sides are congruent. [*Hint:* See steps of problem 3.]

7. The reader who is familiar with the function concept and the distance formula of a rectangular coordinate system (see Sec. 11-4 and Exercise 11-4, 13) could profit from the discussion in Ref. 24, pp. 1–5, in which the *SSS* agreement is proved.

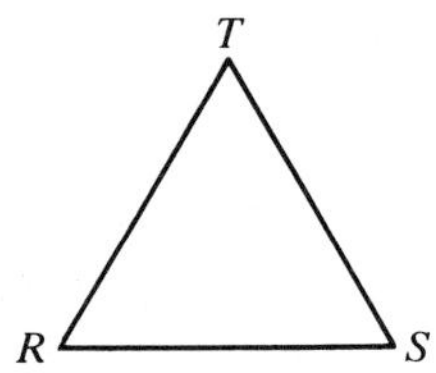

Figure 7-19

7-7 USING BASIC CONSTRUCTIONS

Using the basic constructions and their geometric properties, we are able to develop and justify many other useful constructions. In the development of these constructions we shall use a different format. A construction will be proposed, hints given, or steps listed with a diagram showing the final picture. Occasionally we shall include a discussion as to why these steps yield the desired result. Where necessary or desirable, new vocabulary will be introduced with little or no further comment beyond the defining statement (which also serves as a hint). Be prepared to offer a justification of your construction.

Figure 7-20: To construct the midpoint of a given segment.

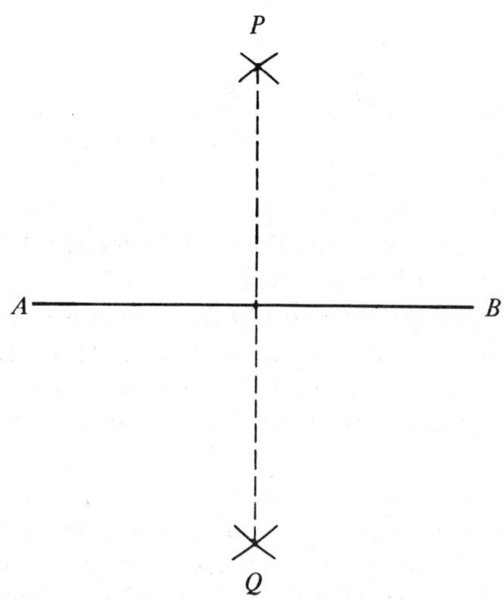

Figure 7-20

Given: $\overline{AB}$.
Construction: Place the compass point at A; swing circle A with a radius greater than $\frac{1}{2}(AB)$. Swing circle B using the same length as radius. Label the intersections P and Q. Draw $\overline{PQ}$. Then $\overleftrightarrow{PQ}$ bisects $\overline{AB}$ at M, the intersection of $\overleftrightarrow{AB}$ and $\overleftrightarrow{PQ}$.

DEFINITION: To **bisect** is to separate into two congruent parts. That which bisects is a **bisector**. A point that bisects is called a **midpoint.**
Discussion: Why take the radius of circle A greater than $\frac{1}{2}(AB)$? Is it necessary to draw complete circles each time? How many points must be constructed to determine a line? How do you construct a point? Draw $\overline{AP}$, $\overline{PB}$, $\overline{AQ}$, and $\overline{BQ}$. What two segments must be proved congruent to prove that $\overline{PQ}$ is a bisector of $\overline{AB}$? Prove $\triangle APQ \cong \triangle BPQ$, $\triangle APM \cong \triangle BPM$. How do you know that $\overline{AM} \cong \overline{MB}$? What can you conclude of $\angle AMP$ and $\angle BMP$?

Figure 7-21: To construct the midray of a given angle.

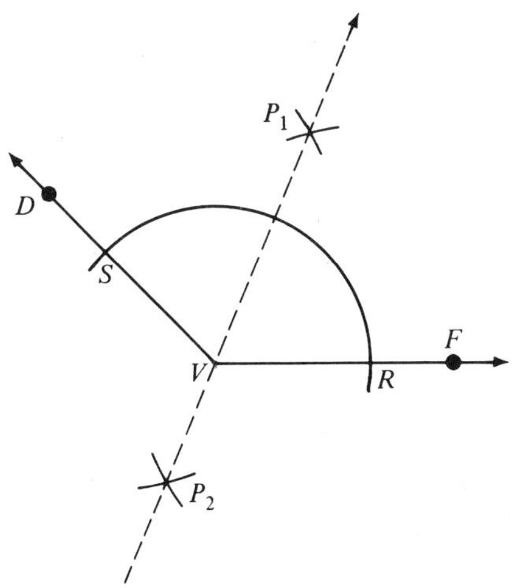

Figure 7-21

Given: $\angle DVF$.
Construction: Place the compass point at V, swing an arc with a convenient radius that intersects $\overrightarrow{VF}$ and $\overrightarrow{VD}$ at R and S, respectively. Select a length greater than $\frac{1}{2}(RS)$ as radius and draw circles at R and S. Label the intersections of these circles P_1 and P_2. Draw $\overrightarrow{VP}$. (Which P?) Then $\angle PVS \cong \angle PVR$.

DEFINITION: A ray that bisects an angle is called a **midray.**

Discussion: Prove $\triangle VSP \cong \triangle VRP$? Do we need both pairs of arcs? How do you decide? [*Hint:* Is there but one $\angle DVF$?]

Figure 7-22: To construct perpendicular lines. (a) To construct a line

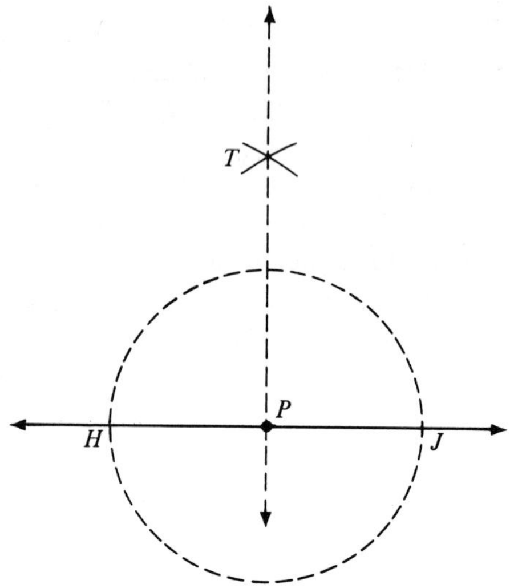

(a)

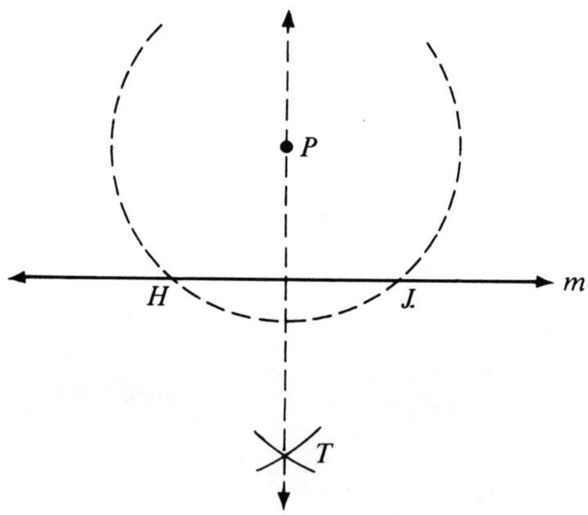

(b)

(c)

Figure 7-22

perpendicular to m, containing P. P in m. (b) To construct a line perpendicular to m, containing P. P not in m. (c) To construct a line perpendicular to a given segment at one of its endpoints.

Construction:

(a) Draw circle P using some convenient radius. Let the intersection of this circle with m be H and J. Using a radius greater than $\frac{1}{2}(HJ)$, draw arcs using H and J as centers. Label their intersection T. Then $\overrightarrow{TP} \perp m$.

(b) Follow the steps listed for part (a). In the first step, the radius must be taken to be greater than the distance from P to m. Although we do not know the distance from P to m, we surely can select a segment whose measure is greater than this. How?

(c) Construct $\overrightarrow{PB}$ perpendicular to $\overline{AB}$.

Figure 7-23: To construct the medians of a given triangle: m_a, m_b, m_c.

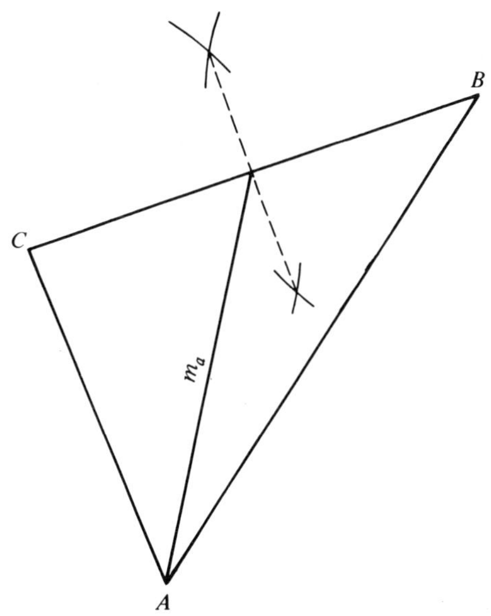

Figure 7-23

Given: $\triangle ABC$.

DEFINITION: A **median** of a triangle is a line segment joining a vertex and the midpoint of the opposite side. In $\triangle ABC$ we have illustrated median m_a, from A to the midpoint of $\overline{BC}$. The point of concurrence of the three medians is the **centroid** of the triangular region determined by the triangle. If a physical model of constant thickness (and density) is made of a triangular region, the figure will balance at the centroid.

Figure 7-24: To construct the altitudes of a given triangle: h_s, h_t, h_r.

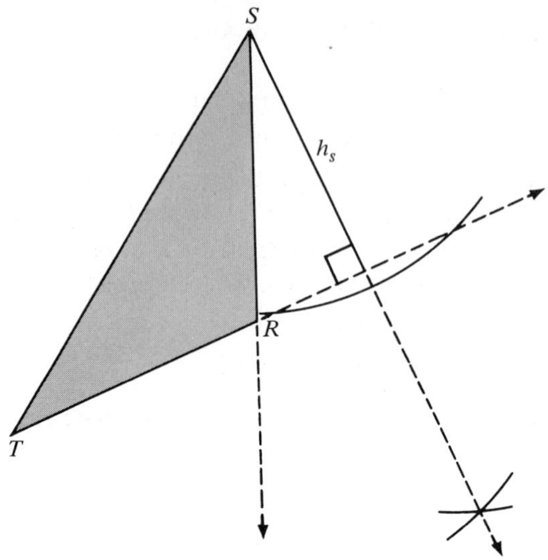

Figure 7-24

Given: $\triangle RST$.

DEFINITION: An **altitude** of a triangle is a segment from a vertex, drawn perpendicular to the carrier of the opposite side. In $\triangle RST$ the altitude, h_s, from S and perpendicular to $\overrightarrow{TR}$ is illustrated. The carriers of the altitudes of a triangle are concurrent in a point called the **orthocenter** of the triangle.

Figure 7-25: To construct the three angle bisectors in a triangle.

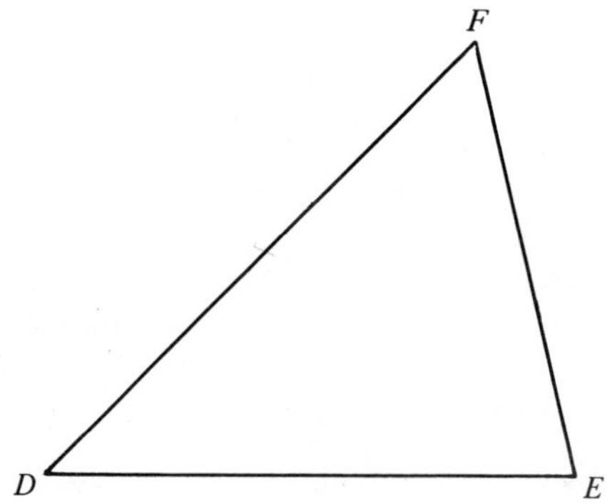

Figure 7-25

Given: $\triangle DEF$.

DEFINITION: The angle bisector of an angle of a triangle is the midray of the angle. (See Fig. 7-21.) The point of concurrence of the three angle bisectors is the **incenter** of the triangle. (See Fig. 7-35 for further insight regarding this name.)

Figure 7-26: To construct the three perpendicular bisectors of the sides of a given triangle.

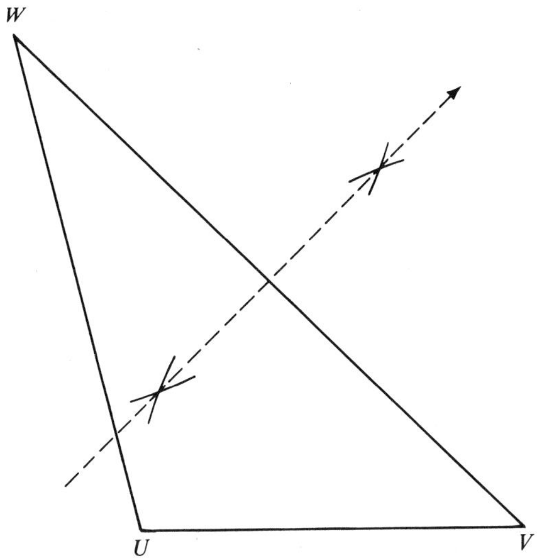

Figure 7-26

Given: $\triangle UVW$.

DEFINITION: A **perpendicular bisector** of a side of a triangle is a *line* that is perpendicular to and bisects a side of the triangle. The point of concurrence of these lines is the **circumcenter** of the triangle. (See Fig. 7-37 for insight with respect to this name.)

Discussion: Does the construction of Fig. 7-20 also yield perpendicular lines?

Figure 7-27: To construct a right triangle knowing the hypotenuse and one leg.

Figure 7-27

Given: c (the hypotenuse)
b (one leg)

[*Hint:* There are at least two solutions. Consider the following: (1) What kind of angle is inscribed in a semicircle? (2) Consider constructing a segment perpendicular to w and congruent to b.]

Figure 7-28: To construct a right triangle knowing the hypotenuse and one acute angle.

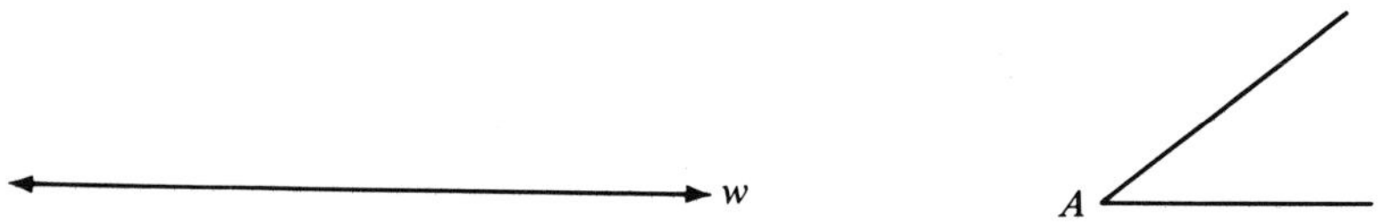

Figure 7-28

Given: c (the hypotenuse)

[*Hint:* There are at least two solutions. Consider the following: (1) If $\angle A$ is constructed on w, where is c? Where is $\angle C$? (2) If you wish to use ASA in your solution, what else do you need?]

Figure 7-29: To construct an isoceles right triangle knowing one leg.

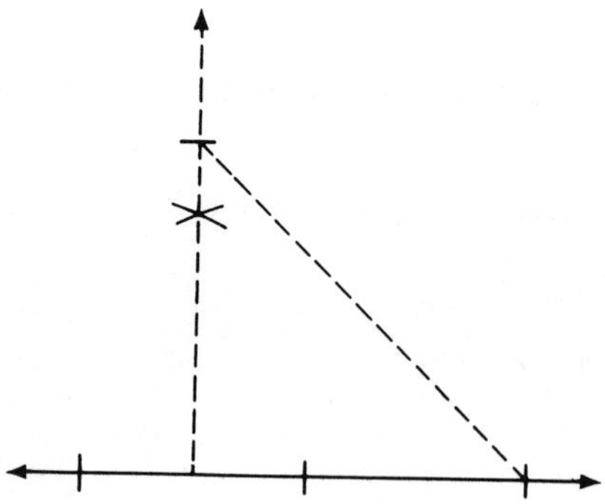

Figure 7-29

Given: b (one leg)

DEFINITION: A **leg** *of an isosceles triangle* is one of the two congruent sides. The **base** *of an isosceles triangle* is the third side. The **vertex angle** *of an*

isosceles triangle is the angle determined by the legs. The other two angles are the **base angles.**
Discussion: Compare the measures of the base angles.

Figure 7-30(a): To construct an equilateral triangle given one side.

Figure 7-30(a)

Given: s (one side)
Discussion: To prove that an equilateral triangle is equiangular, consider the following question: Is $\triangle ABC \cong \triangle BCA$? Do you think every equiangular triangle is equilateral? When are equiangular triangles congruent?

Figure 7-30(b): To construct an equiangular triangle knowing one side.

Figure 7-30 (b)

Given: s (one side)
Discussion: If the measures of all of the angles are equal, what must the measure be? (Answer this without using your protractor.) How then would you construct an angle with a measure of 30 in degrees?

Figure 7-31: To construct a line that is parallel to a given line and containing a point not in the given line.
Given: Line m, point P, not in m.
[*Hint:* Recall that if a pair of lines are cut by a transversal in such a way that certain angles are congruent, then the lines are parallel. What kind of angles does our solution use? What other kinds of angles can be used?]

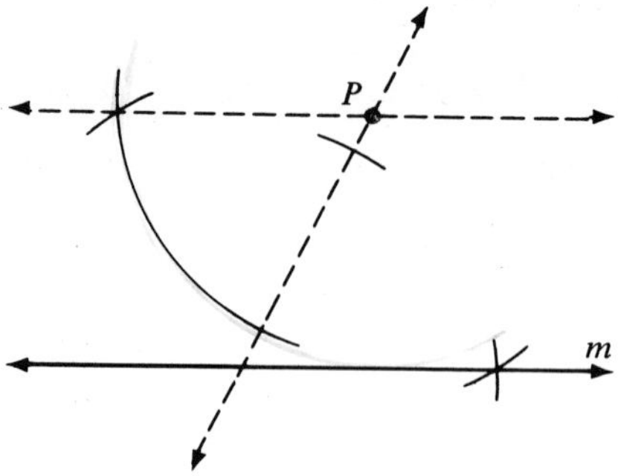

Figure 7-31

Figure 7-32: To construct a rectangle given two adjacent sides.

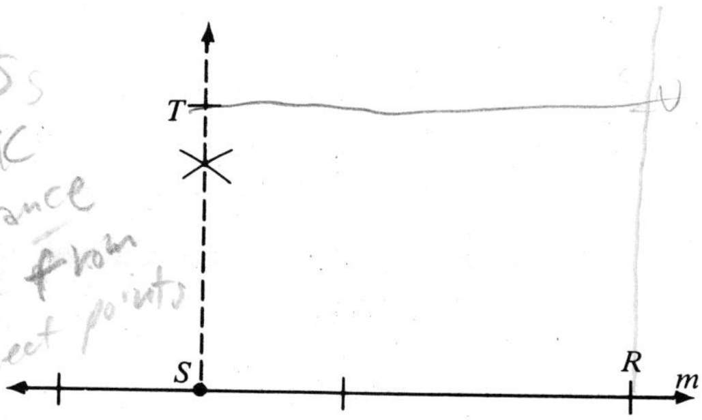

Figure 7-32

Given: S ———————— T
R ———————————— S

[*Hint:* Construct a parallelogram.]

Figure 7-33: To construct a square given one side.

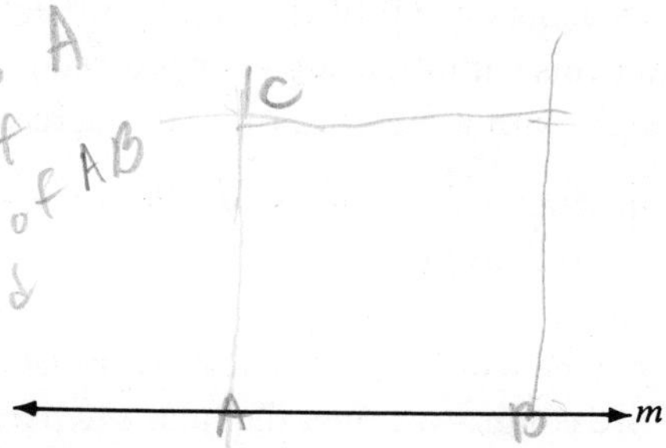

Figure 7-33

Given: s ________________

[*Hint:* Construct a rectangle.]

Figure 7-34: To construct a tangent to a circle at a point of the circle.

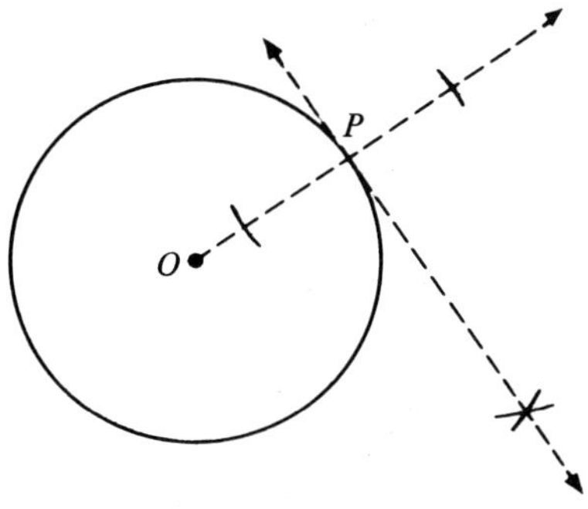

Figure 7-34

Given: Circle O, containing point P.

DEFINITION: (Alternate) In a plane, if a line is perpendicular to a radius of a circle at a point of the circle, the line is **tangent** to the circle. (Note: Frequently a tangent is defined as in Sec. 6-8. Then the statement above is proved. Since either statement can be proved on the basis of the other, the choice of definition is arbitrary.)

Figure 7-35: To construct the inscribed circle in a triangle.

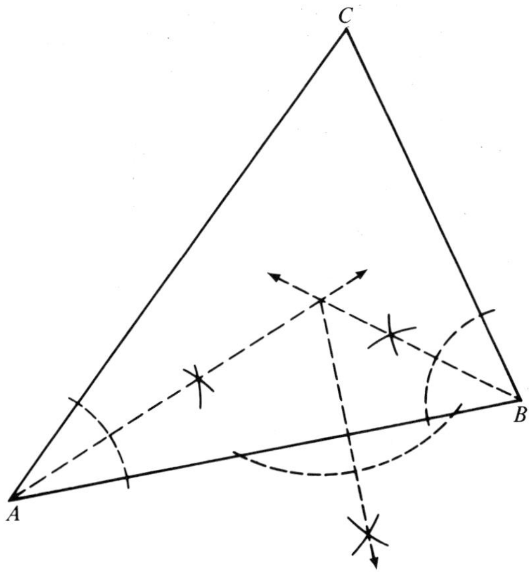

Figure 7-35

Given: $\triangle ABC$.

DEFINITION: An **inscribed circle** or a circle **inscribed** *in* a polygon is a circle that is tangent to each of the sides of the polygon. The polygon is then **circumscribed** *about* the circle. The center of this circle is the **incenter** of the polygon. The circle is also known as the **incircle**. (See Fig. 7-25.)

Discussion: How can you find the distance from the incenter to one of the sides of the triangle?

Figure 7-36: To construct a circumscribed polygon about a given circle at points on a given circle.

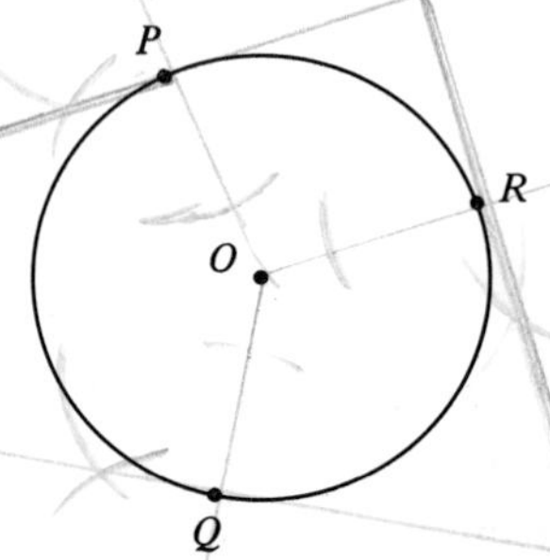

Figure 7-36

Given: Circle O, containing points P, Q, and R.

Figure 7-37: To construct the circumscribed circle about a given triangle.

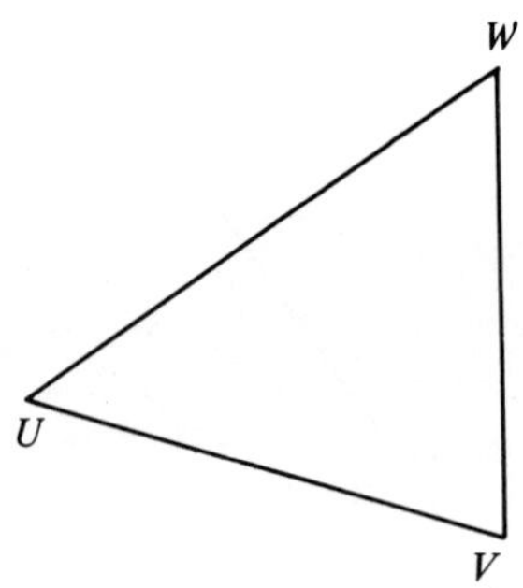

Figure 7-37

Given: $\triangle UVW$.

DEFINITION: A circle is **circumscribed** *about* a polygon if each vertex of the polygon is a point of the circle. The polygon is then **inscribed** *in* the circle.

The center of this circle is the **circumcenter** of the polygon. The circle is known as the **circumcircle**. (See Fig. 7-26.)

Figure 7-38: To construct an inscribed regular hexagon. (a) To construct an inscribed regular hexagon in a circle if the center is given or the radius is known. (b) To construct an inscribed regular hexagon in a circle if the center is not indicated and the radius is not known.

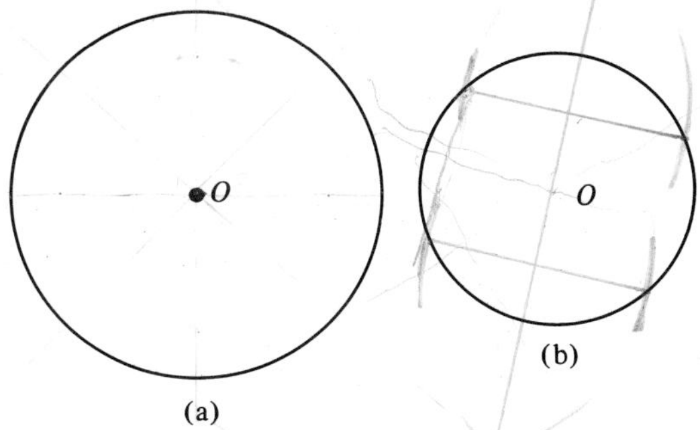

Figure 7-38

Given: Circle *O*.
Construction:

(a) Make a freehand sketch of the circle and the inscribed hexagon. Sketch the central angles determined by the center and vertices and draw conclusions about these central angles.

(b) Consider how you might use the information supplied by problem 14(a) in the Exercises following this chapter. [For this example assume that the statement in problem 14(a) has been proved. You will not be guilty of circular reasoning, since its proof does not depend upon this construction.]

7-8 REGULAR POLYHEDRONS

A **regular polyhedron** is a polyhedron that has regular polygons as faces, and all of whose polyhedral angles are congruent. If you think back to the discussion of congruence in a plane and consider the rather complicated diagrams using transformations, you will appreciate why we limit our comments here to the construction of the only five regular polyhedrons that exist. Although it is rather difficult to prove that five exist, you will not find it hard to construct five that look rather "nice." And it is relatively easy to prove that no more than these five can exist. The proof can be found in almost any text devoted to an extensive treatment of solid or space geometry, such as Ref. 14, p. 408. Other proofs use material in Chapter 12; see Ref. 4, pp. 43–48.

Suffice it for us to conclude that there are five:

Name	*Face*
tetrahedron	equilateral triangle
hexahedron	square
octahedron	equilateral triangle
dodecahedron	regular pentagon
icosahedron	equilateral triangle

EXERCISE 7-8

1. (a) Make models of the five regular polyhedrons using patterns in Appendix C.

(b) For these models, verify the formula in problem 13 of the Exercises at the end of Chapter 4.

(c) How do you construct these patterns using only a straight edge and a compass?

2. (a) See Ref. 53. (b) See Ref. 33.

(c) See Ref. 16.

3. In Exercise 6-3, 8, the question concerning a line perpendicular to a plane is raised. This is quite important in discussing perpendicular planes. A *line perpendicular to a plane* is one that is perpendicular to every line in the plane passing through its foot. With this definition it is possible to prove that if a line is perpendicular to each of two intersecting lines at their point of intersection, then it is perpendicular to the plane determined by these lines.

We have discussed types of dihedral angles and pointed out that the vocabulary and concepts involved are quite analogous to those of planar angles. However, we must devise a way to discuss the measure of a dihedral angle.

(a) Draw $\angle_{\text{dih}}\ R\text{-}AB\text{-}S$, labeling the faces m and n with R in m. With this labeling, what line is the edge? Now in m draw $\overline{RH} \perp \overleftrightarrow{AB}$ at H. Is H in n? In n draw $\overleftrightarrow{HQ} \perp \overleftrightarrow{AB}$. Then the angle determined by $\overline{HR}$ and $\overrightarrow{HQ}$ is called a plane angle of the dihedral angle. Thus, a **plane angle** *of a dihedral angle* is an angle whose sides lie one in each face and perpendicular to the edge of the dihedral angle at the same point. Then the dihedral angle is measured by its plane angle; that is, a dihedral angle has the same measure as any one of its plane angles. What do you suppose we call a dihedral angle if its plane angle is a right angle? What do you suppose is the relationship between its faces? Thus, two *planes* are *perpendicular* if the plane angle of their dihedral angle is a right angle. (Notice we say "the" plane angle, or "a" plane angle. Realize that there are many. Since they all have the same measure, it is common practice to speak of "the" or "a" plane angle.) What is the degree measure of a plane angle of a dihedral angle that represents a rotation of two thirds of a revolution of one face from another?

(b) Now consider the plane determined by $\angle HQR$ above. What is the relationship of this plane to each of the faces of the dihedral angle? What is the relationship of $\overleftrightarrow{AB}$ to plane HQR? This illustrates the fact: If a line ($\overleftrightarrow{AB}$) is perpendicular to a plane (HQR), any plane (either m or n) containing this line ($\overleftrightarrow{AB}$) is perpendicular to that plane (HQR).

(c) If a right prism is defined to be a prism in which a lateral edge is perpendicular to the base, what is the relationship of the lateral faces to the bases? Does it follow that if two planes are perpendicular to the same plane (at different lines of intersection) they are parallel?

4. For some interesting historical notes regarding regular polyhedrons see Ref. 100, pp. 42–44.

SUMMARY OF PRINCIPLES

Corresponding parts of congruent figures are congruent.

In a plane there exists a circle with a given point as center and a given distance as a radius.

A triangle is a rigid figure.

If two triangles have the three sides of one congruent to three sides of the other, respectively, the triangles are congruent. (SSS for $\cong$.)

If two triangles have two sides and the included angle of one congruent to two sides and the included angle of the other, respectively, the triangles are congruent. (SAS for $\cong$.)

If two triangles have two angles and the included side of one congruent to two angles and the included side of the other, respectively, the triangles are congruent. (ASA for $\cong$.)

If two triangles have three angles of one congruent to three angles of another, respectively, the triangles are similar. (AAA for $\sim$.)

If two triangles have two sides of one proportional to two sides of the other, respectively, and the included angles are congruent, then the triangles are similar. (SAS for $\sim$.)

If two angles of a triangle are congruent, then the sides opposite these angles are congruent.

If two sides of a triangle are congruent, then the angles opposite these sides are congruent.

An equilateral triangle is equiangular.

An equiangular triangle is equilateral.

In a plane a line perpendicular to a radius of a circle at a point of the circle is tangent to the circle.

A tangent to a circle is perpendicular to a radius of the circle drawn to the point of contact.

EXERCISES

1. (a) For each illustration in Fig. 7-39 indicate whether it possesses: (1) only line symmetry, (2) only point symmetry, (3) neither point nor line symmetry, or (4) both point and line symmetry. For those that possess symmetry make a statement regarding the axis of symmetry and/or the point of symmetry.

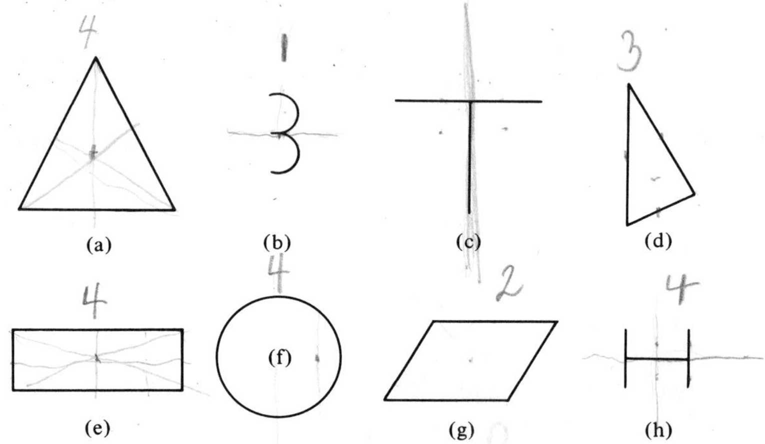

Figure 7-39

(b) Another type of symmetry is of special interest in the fine arts. Locate some references in regard to this and write a brief paragraph of explanation.

2. Using only a straight edge and a compass (no protractor), construct angles of (a) 45°, (b) 30°, (c) 150°, (d) 315°.

3. Construct an equilateral triangle with sides of $5\frac{1}{2}$ in.
   (a) Label the vertices A, B, and C and in this triangle construct (1) h_a, (2) m_b, (3) the midray of $\angle C$, and (4) the perpendicular bisector of $\overline{AB}$.
   (b) Look at constructions 7-23 through 7-26 and comment on any "unexpected" results of part (a) above.
   (c) Is there another type of triangle in which you now might expect something of the same nature to occur? Name the kind of triangle and comment on what you might expect. Construct a triangle of this kind to help verify your conjecture.

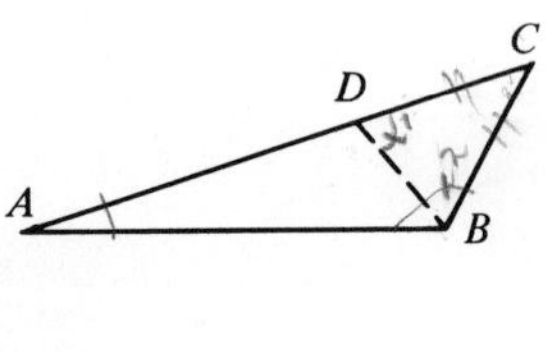

4. We have thus far been interested in congruence or equal measures. However, at times we wish to show that things are not equal. The following is an example.

   GIVEN: $\triangle ABC$ in which $AC > BC$.

   TO PROVE: $m(\angle B) > m(\angle A)$. (Note: This is a direct proof.)

STEPS	EXPLANATION
(1) On $\overline{AC}$, take point D such that $CD = CB$. Draw $\overline{BD}$.	(1) Given $AC > BC$, use agreement from Chapter 5.
(2) Then $\angle CDB \cong \angle CBD$.	(2) ?
(3) But $m(\angle CDB) > m(\angle A)$.	(3) ?
(4) And $m(\angle CBA) > m(\angle CBD)$.	(4) Note that point D of $\overrightarrow{BD}$ lies in the interior of $\angle CBA$.
(5) So we have: $m(\angle CBA) > m(\angle CBD)$; but $m(\angle CBD) = m(\angle CDB)$; but $m(\angle CDB) > m(\angle A)$; hence, $m(\angle CBA) > m(\angle CAB)$.	(5) If three quantities are so related that the first is greater than the second and the second is . . .

Make a general statement summarizing what the above demonstration justifies.

5. Use the indirect method of proof and show that if two angles of a triangle have unequal measures, then the sides that lie opposite to these angles are unequal in measure and in the same order; that is, the side with the larger measure lies opposite the angle with the larger measure. (Warning: In the first step, in which you assume the contradiction of that which you are trying to prove, be sure you consider all possibilities.)

6. Prove that in a parallelogram, (a) a diagonal forms two congruent triangles, (b) the opposite sides are congruent, and (c) the opposite angles are congruent.

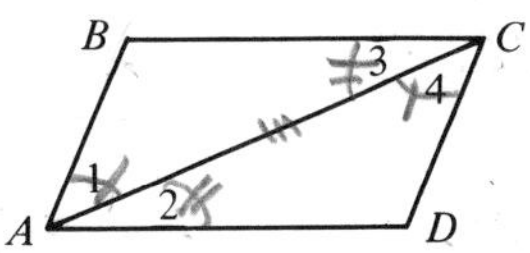

GIVEN: Parallelogram $ABCD$, diagonal AC.

TO PROVE: (a) $\triangle ABC \cong \triangle CDA$.
(b) $\overline{AD} \cong \overline{CB}$, $\overline{AB} \cong \overline{CD}$.
(c) $\angle B \cong \angle D$, $\angle BAD \cong \angle BCD$.

[*Hint:* See definition of bisect with Fig. 7-20. Consider statements we have had that show congruence.]

STEPS	EXPLANATION
(1) $\overline{BC} \parallel \overline{AD}$, $\overline{AB} \parallel \overline{DC}$.	(1) ?
(2) $\angle 1 \cong \angle 4$, $\angle 3 \cong \angle 2$.	(2) ?
(3) $\overline{AC} \cong \overline{AC}$.	(3) Identity. (Reflexive property of "$\cong$".)
(4) $\triangle ABC \cong \triangle CDA$. Hence (a) is done.	(4) ?
(5) $\angle B \cong \angle D$, $\overline{AD} \cong \overline{CB}$, $\overline{AB} \cong \overline{CD}$. Hence (b) is done. Is (c) done? If not, complete the proof.	(5) ?

7. Prove that the diagonals of a parallelogram bisect each other. (You may use results of problem 6 if you wish.)

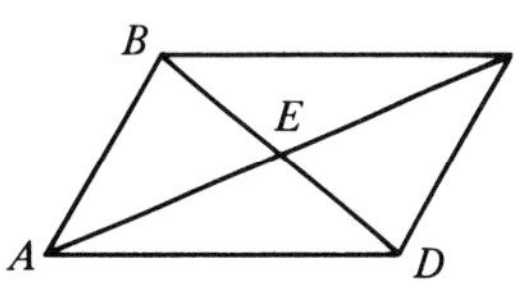

GIVEN: Parallelogram $ABCD$, diagonals AC, BD.

TO PROVE: $\overline{BD}$, $\overline{AC}$ bisect each other.

[*Hint:* What segments must be proved congruent?]

8. (a) What is the definition of a parallelogram?

(b) Prove that if two sides of a quadrilateral are parallel and congruent, then the quadrilateral is a parallelogram.

GIVEN: Quadrilateral $ABCD$, $\overline{AB} \cong \overline{CD}$, $\overline{AB} \parallel \overline{CD}$.

TO PROVE: $ABCD$ is a parallelogram.
[*Hint:* Draw $\overline{AC}$. Prove $\overline{BC} \parallel \overline{AD}$. How can we prove parallels?]

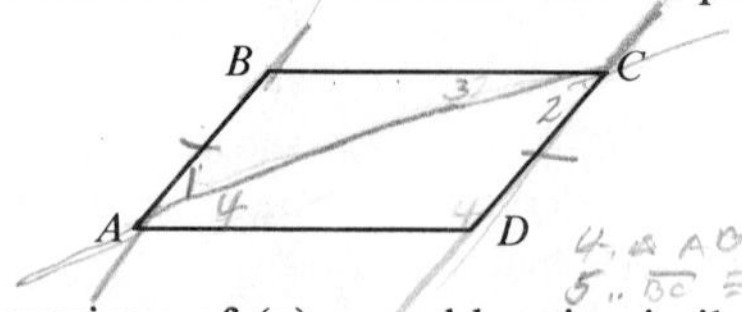

9. See Ref. 9 for discussions of (a) a problem in similarity, pp. 4–6, (b) construction, pp. 149–154, (c) symmetry, pp. 166–170, (d) geometry in daily life, pp. 155–160, (e) geometry in decoration, pp. 164–186.

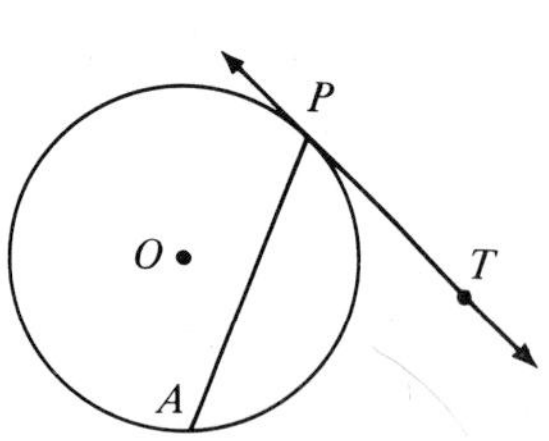

10. Prove that an angle determined by a tangent to a circle and a chord drawn to the point of contact has an angle degree measure equal to one half the arc degree measure of its intercepted arc.

GIVEN: Circle O, tangent PT, chord PA.

TO PROVE: $m(\angle APT) \overset{\circ}{=} \frac{1}{2}m(\widehat{AP})$, or $\angle APT \overset{m}{=} \frac{1}{2}(\widehat{AP})$.
[*Hint:* Draw diameter determined by P and O. See definition with Fig. 7-34.]

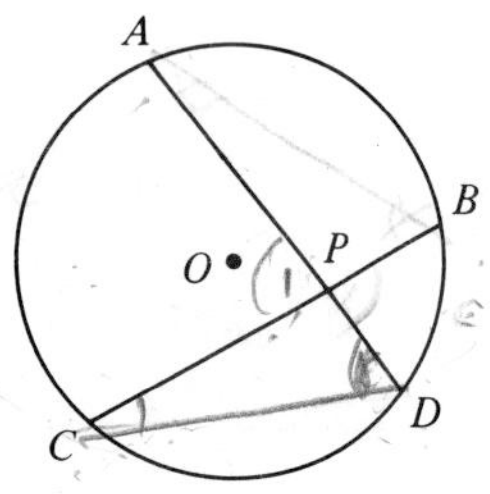

11. Prove that an angle formed by two chords that intersect at an interior point of a circle has an angular degree measure equal to one half the sum of the arc degree measures of the arcs intercepted by it and its vertical angle.

GIVEN: Circle O, chords AD, CB, such that $\overline{AD} \cap \overline{CB} = \{P\}$, P an interior point.

TO PROVE: $m(\angle APC) \overset{\circ}{=} \frac{1}{2}[m(\widehat{AC}) + m(\widehat{BD})]$.
[*Hint:* Draw $\overline{CD}$. Use exterior angle of a triangle.]

12. (a) Prove that the construction of Fig. 7-22(a) is valid.

(b) Prove that the triangles constructed for Fig. 7-27 are right triangles.

(c) Repeat (b) for Fig. 7-28.

(d) The constructions referred to in parts (b) and (c) are special cases of two other rules for proving congruence (one of which is correct only if used in special situations). The first would be abbreviated SSA for two sides and the angle oppsite one of the sides, the second would be abbreviated AAS for two angles and the side opposite one of them. By performing constructions with the parts listed below, decide which one is always valid.

(1) $m(\angle A) \overset{\circ}{=} 30$
$c = 4$
$a =$ (i) $1\frac{1}{2}$
(ii) 2
(iii) 3
(iv) 4
(v) 5

(2) $m(\angle A) \overset{\circ}{=} 35$
$m(\angle B) \overset{\circ}{=} 45$
$a = 3$

13. Define a square to be a rectangle in which two adjacent sides are congruent. Then prove that a square is a regular quadrilateral. [*Hint:* What is a regular polygon?]

14. The following three statements are often useful in more advanced work. Prove each, being careful not to confuse the known (assumed) facts with the conclusions (to be proved).

(a) Any point on the perpendicular bisector of a line segment is equidistant from the endpoints of that segment.

GIVEN: $\overleftrightarrow{CD}$ bisects $\overline{AB}$ at C, $\overleftrightarrow{CD} \perp \overline{AB}$ at C, P on $\overleftrightarrow{CD}$.

TO PROVE: $AP = PB$.

(b) At a point of a line there is only one line that is perpendicular to the given line (in a plane).

GIVEN: $\overleftrightarrow{EF}$, $P \in \overleftrightarrow{EF}$, $\overleftrightarrow{HJ} \perp \overleftrightarrow{EF}$ at P.

TO PROVE: $\overleftrightarrow{HJ}$ is the only line in this plane perpendicular to $\overleftrightarrow{EF}$ at P.

[*Hint:* (1) Assume $\overleftrightarrow{PS}$ also $\perp$ $\overleftrightarrow{EF}$ at P. (2) At Q($Q \in \overleftrightarrow{EF}$, $Q \neq P$) consider $\overleftrightarrow{QT} \perp \overleftrightarrow{EF}$. (3) Use a parallel postulate.]

(c) Any point that is equidistant from the endpoints of a line segment lies on the perpendicular bisector of that segment.

GIVEN: $\overline{RS}$, $RQ = QS$, $\overleftrightarrow{VT} \perp \overline{RS}$ at T, $\overleftrightarrow{VT}$ bisects $\overline{RS}$ at T.

TO PROVE: Q lies on $\overleftrightarrow{VT}$.

[*Hint:* Draw $\overline{QT}$, work with $\triangle RQT$ and $\triangle SQT$.]

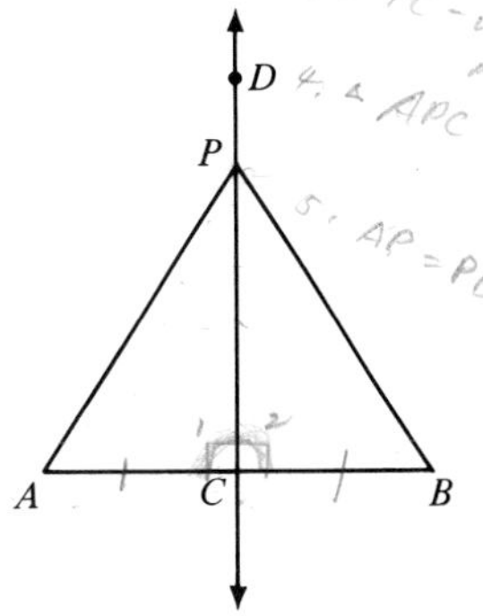

(a)

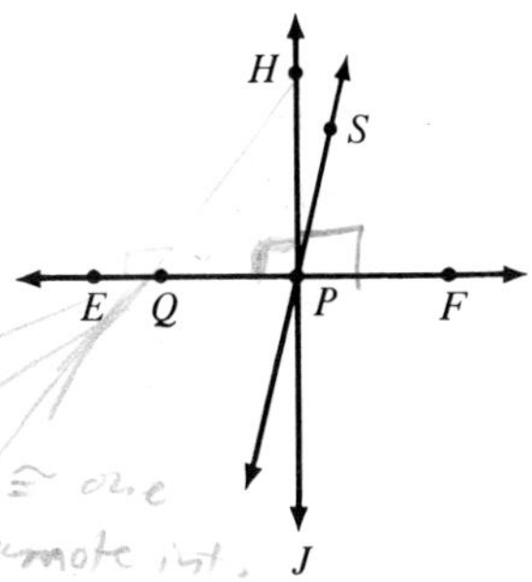

(b)

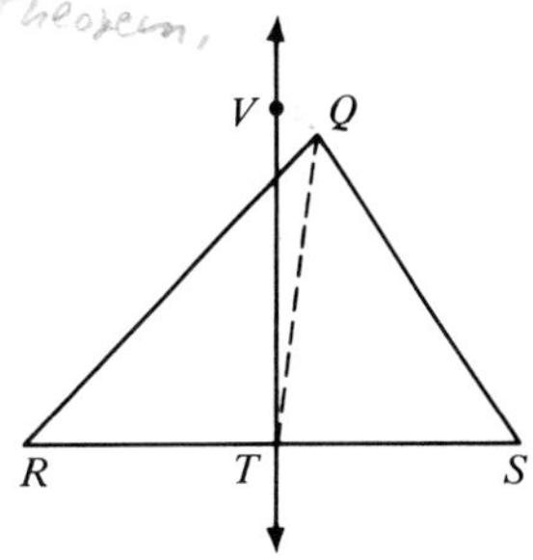

(c)

CHAPTER 8

Area

8-1 INTRODUCTION

In Chapter 5 we analyzed the nature of linear and angular measurement by looking at mathematical abstractions for each situation. In Chapter 6 we investigated some metric properties of polygons. In Chapter 7 we looked at two relations between geometric figures that are intimately related to measurement—similarity and congruence. In the present chapter we extend the analysis of measurement to size of planar regions. Again we shall see that the measuring process is a way of assigning numbers to certain characteristics of geometric figures in order to answer the question: How large?

Determining **area**—the measure associated with the size of a region in a plane—presents some problems not encountered in linear and angular measurement. Most simply, the problem is that such regions have different "shapes" as well as different sizes. This makes the comparison task somewhat more difficult. Notice in Fig. 8-1 that the closed regions have different "shapes." The segments, even though they may have different lengths, all have the same "shape." This makes direct comparison relatively easy. The situation is similar for angles: all angles may be said to have the same shape, but may have different sizes.

However, the problem of different shapes is not unsolvable. Using some of the fundamental properties of congruence, and making a basic assumption about the relationship of area to congruence, we can determine the area for figures of many different shapes, if we can also assume that the area of at least one figure is determinable.

8-2 COMPARISON OF AREAS

One way to begin the analysis is to note how we usually make a direct comparison of two areas. In Fig. 8-2 we would probably say that the rug on the floor has less area than the floor. There is simply more floor space than rug space. But what do we mean in this case by the phrase "more space"? The floor may be thought of as corresponding to a closed planar region, as may the rug. Would you say that the region corresponding to the rug is a proper subset of the region corresponding to the floor? Why? This observation suggests a possible criterion for determining when one planar region has greater area than another planar region: if one closed region contains another closed region as a proper subset, then the first has a greater area than the second. When would two closed regions have equal area? (You may wish to turn to Sec. 5-2 to note the similarity of this discussion for area measurement to that for linear measurement.)

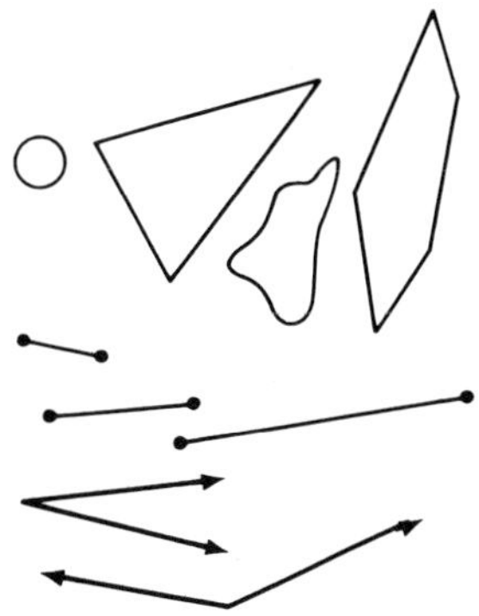

Figure 8-1

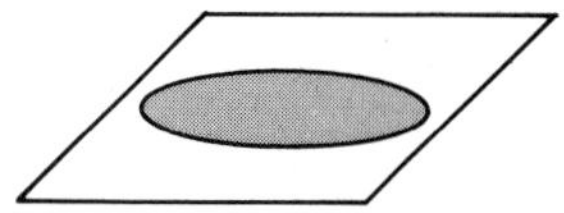

Figure 8-2

Now suppose the rug is in someone else's house and you still want to know which has greater area, your floor or the rug. One solution in the physical situation is to bring the rug to your house to compare the areas directly. But suppose we try to analyze the situation using geometric figures. Let closed planar region $ABCD$ correspond to the rug and closed region $EFGH$ correspond to the floor. (See Fig. 8-3.) What we need to assume is that there are closed regions congruent to $ABCD$ and $EFGH$ in such positions that we can compare the areas directly. Note that we also must assume that the measures of congruent figures are equal. In this case we have the areas of congruent closed regions being equal. If we can find a closed region congruent to $ABCD$ that is also a proper subset of closed region $EFGH$, then $EFGH$ has greater area than $ABCD$. When would $ABCD$ have greater area than $EFGH$? When would it have the same area?

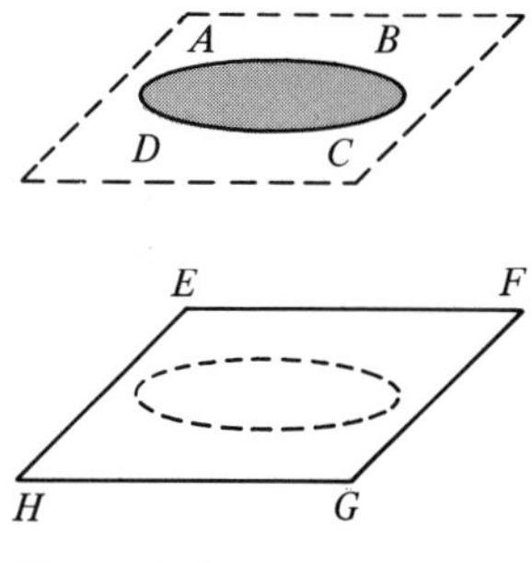

Figure 8-3

We now need to consider the problem of "shape" in direct comparisons of area. Suppose that the closed regions corresponding to the rug and floor are as in Fig. 8-4. Which has greater area? It happens that neither region has a region congruent to it that is also a subset of the other region. The simple criteria given above for determining relative size of area do not work in this case. What might we do to decide which has the greater area? The best we can do, it seems, is to cut the rug (or a paper copy) into smaller pieces and try to fit all the pieces onto the floor without having any of the

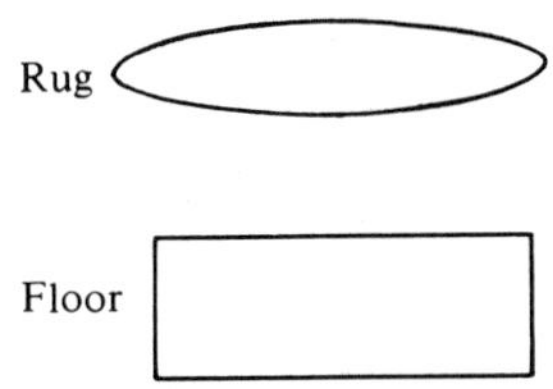

Figure 8-4

pieces overlap. In doing this we would have to assume that the sum of the areas of the smaller pieces is the same as the area of the rug.

That is, we must assume that given a region in a plane and a partition on this region (see Fig. 8-5), the area of this region is equal to the sum of the areas of each of the regions into which it has been separated. (This is the two-dimensional analogue to the additive property of linear measurement.)

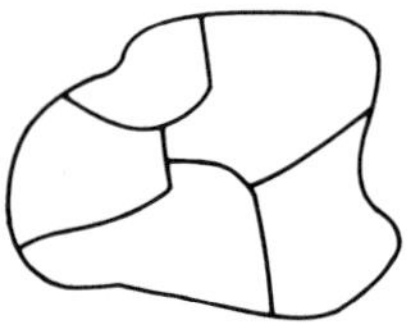

Figure 8-5

To compare the area of closed region x in Fig. 8-6 with that of closed region y, we can make a suitable partition of y and then find regions in x that are congruent to these regions such that the regions do not overlap (may have only boundary points in common).

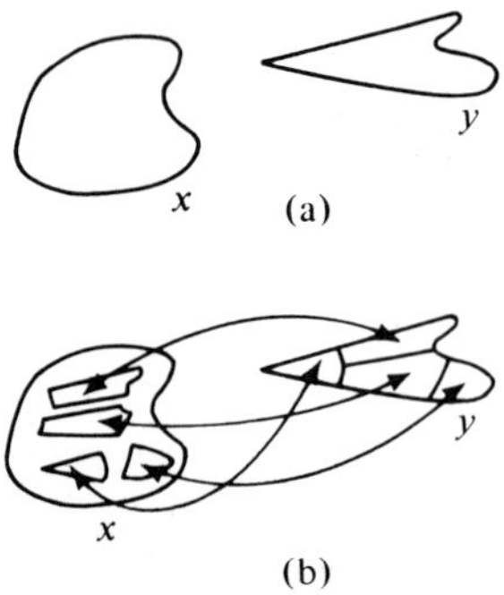

Figure 8-6

EXERCISE 8-2

1. In each part of Fig. 8-7 show that region r has greater area than region s.

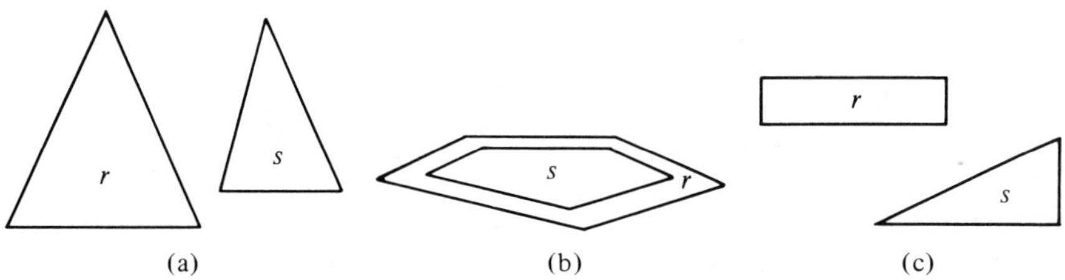

Figure 8-7

8-3 AREA AND COUNTING

As for linear and angular measurement, the comparison procedure above does not tell "how much more" area there is for one region than for another. In order to do this, we require some way of assigning numbers to the regions. One possible approach is to select some unit of area—a closed region in a plane, the measure of whose size we will say is one—and then count the number of units that "fit" into a given region. That is, partition the given region into nonoverlapping regions congruent to the unit region. Since planar regions are continuous sets, as are line segments, the simple counting procedure must be superseded by some assignment system that uses the set of nonnegative real numbers.

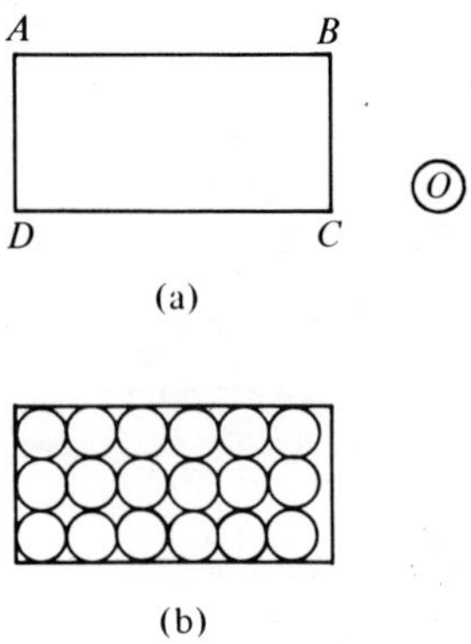

Figure 8-8

In any event, we can get a better idea of the nature of the situation by looking at some examples of the primitive counting situation. We require first a unit of area, then we find some partition of the region to be measured that determines a set of nonoverlapping regions congruent to the unit and in the interior of the region to be measured. The area of the region is given by the number of regions congruent to the unit, provided that the region is completely covered by these nonoverlapping congruent regions. In Fig. 8-8 the unit is the circular region O, and the closed region to be measured is rectangular region $ABCD$. Notice here that any attempt to cover this region with unit regions so that they do not overlap creates some problems. Some parts of the region are not "covered" by any part of this unit region.

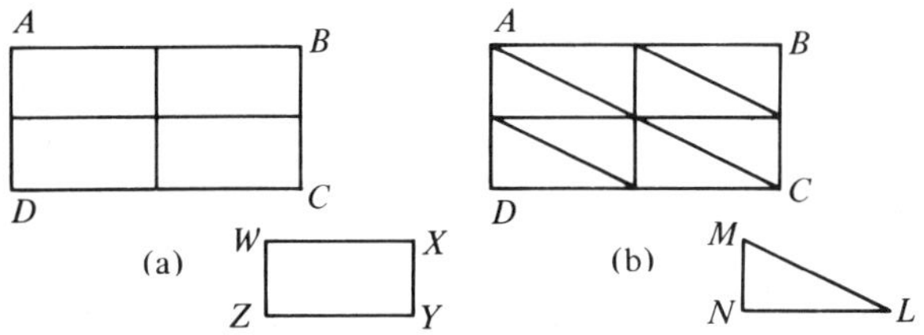

Figure 8-9

Thus, although the actual choice of a unit region is quite arbitrary, there are certain characteristics of "good" units for practical situations. One of these is that the unit regions can be juxtaposed so that no part of the surface to be measured is left uncovered and yet no unit regions overlap. A unit that satisfies these criteria in the case of $ABCD$ is the rectangular region $WXYZ$. [See Fig. 8-9(a).] Note that a right triangular region would also have been suitable. [See Fig. 8-9(b).] What is the area of region $ABCD$ using the rectangular unit? Using the triangular region as unit?

However, if the region to be measured is circular region A of Fig. 8-10, the rectangular unit region presents some difficulties. We do not seem to be able to arrange the unit regions in the circular region so that all points of the region are covered and no part of a unit region is in the exterior of the circular region. So it seems that one unit serves well in one situation but not in another. This is the problem of different "shapes" that makes area measurement somewhat different from linear or angular measurement. However, by constructing our geometry in a particular way we shall be able to show that once we establish a way to find the area of a particular figure based on a particular unit, we shall be able to relate the area of many other figures to the area of this basic figure. Here again we shall see how important are proper assumptions and reasoning to the development of explanations for geometric relations.

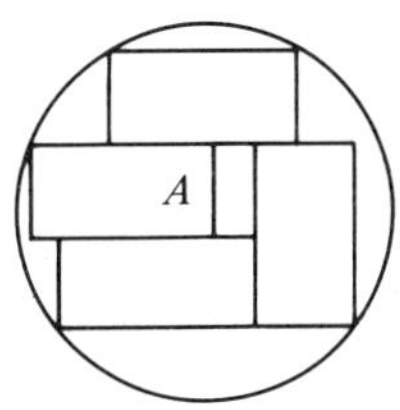

Figure 8-10

EXERCISE 8-3

1. Find the area of each of the closed regions drawn on the grids of intersecting lines in Fig. 8-11. Use the "small" regions determined by the intersecting lines as unit regions.
2. For the regions in Fig. 8-12 simple counting of the units seems "inadequate" to determine the area of the regions. What might we have to "do to" the unit to get a more adequate determination of the area? Find the approximate area of the regions in Fig. 8-12.
3. Use a postage stamp as a model of a unit and find the area of the cover of this book.
4. Make exact copies of the regions shown in Fig. 8-13. Arrange the pieces to form a T-shaped region.
5. See Ref. 82, pp. 668–672.

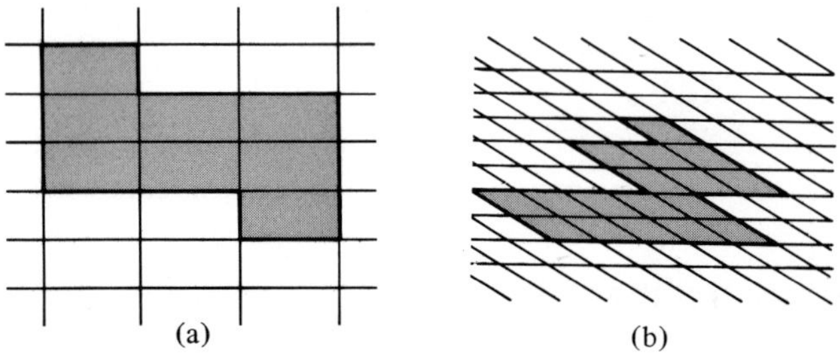

Figure 8-11

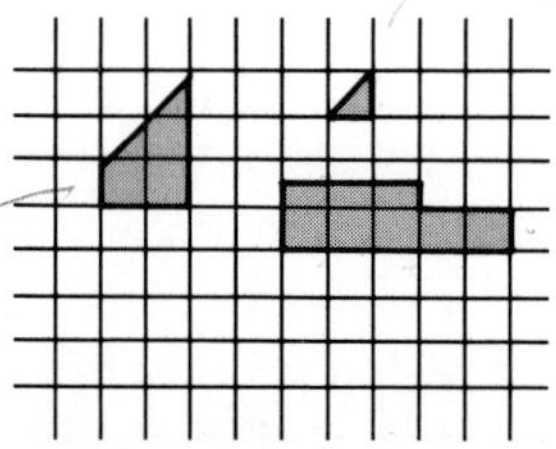

Figure 8-12

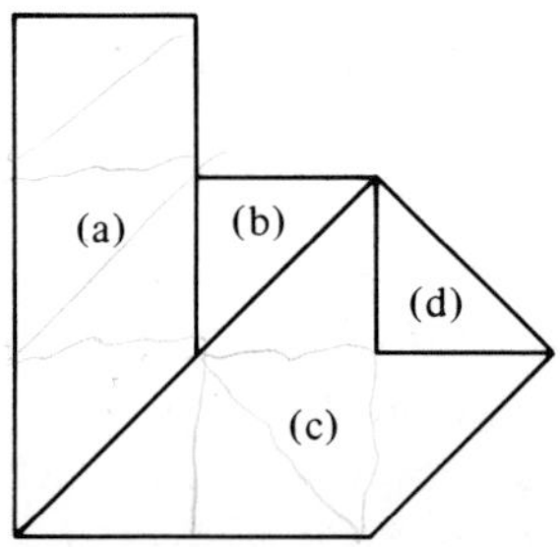

Figure 8-13

8-4 STANDARD UNITS

In actual measuring situations the standard units for area measurement are usually square regions. Figure 8-14 shows that square regions can be arranged so that no part of a region is left uncovered, although we still have the problem of having some parts of unit regions in the exterior of the figure. As is suggested by Exercise 8-3, 2, this problem may be solved by subdividing the unit or by redefining the unit so that it is a smaller region.

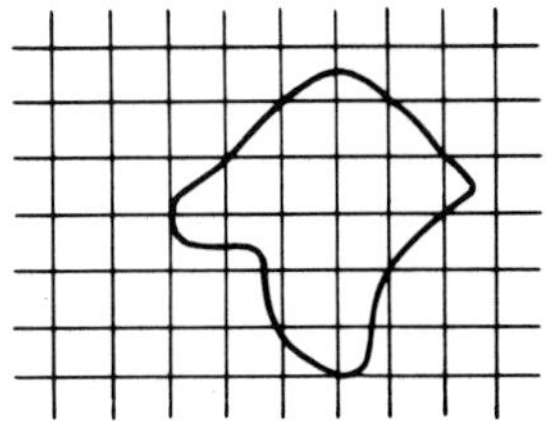

Figure 8-14

Figure 8-15 shows a picture of a square region whose sides have a measurement of one inch. The area measurement of this region is said to be one square inch (usually abbreviated 1 sq in. or 1 in.^{2}). Any region that is congruent to this will have an area measurement of 1 sq in. Any region that has area equal to the area of this region is also said to have an area measurement of 1 sq in.

Figure 8-15

Figure 8-16 shows a picture of a square region whose sides have a measurement of one centimeter. The area measurement of such a region is said to be one square centimeter (usually abbreviated 1 sq cm or 1 cm^2). Approximately, how many square centimeters are equivalent to a square inch?

Figure 8-16

Square regions as units of area seem to be favored because of their usefulness in determining the area of rectangular regions. The sides of rectangle *ABCD* in Fig. 8-17 have measurements as indicated. If we use square regions whose sides have measurement of one unit (linear unit, that is), how many square regions can be placed in the rectangular region along the side whose measure is 5? How many such rows of unit regions fit in this region? Obviously, 3. How many unit square regions are contained in the rectangular region? You could get this number by direct counting. You can get the same number by multiplying 3 by 5. Since these 15 congruent regions completely cover the rectangular region, the measure of the area must be 15. One can say that the area measurement is 15 square units.

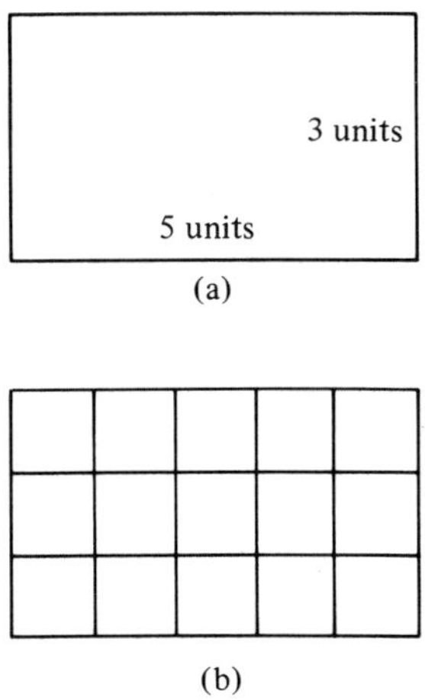

Figure 8-17

What would the area be if the measurements of the sides are 5 units and $3\frac{1}{4}$ units?

Other standard units for area measurement using the English system are square regions whose sides have measurements of one foot (area measurement is 1 sq ft), one yard (1 sq yd), and one mile (1 sq mi). There is also a standard unit of area called the "acre." One acre is equivalent to 43,500 sq ft. The acre, a measure of land used by the English, was originally the amount of land that a yoke of oxen could plow in one day. Get a better yoke of oxen and you will have a larger acre! Hardly standard!

Among the standard units of area used in the metric system are the square regions whose sides have measurements of one meter (area measurement is 1 sq m), one centimeter (1 sq cm), one dekameter (1 sq dkm, also known as 1 are), and one hectometer (1 sq hm, also known as 1 hectare).

Each of the standard units can be subdivided.

EXERCISE 8-4

1. Make paper copies of square regions with the following area measurements.

(a) 1 sq cm. (b) 1 sq in.
(c) 1 sq ft. (d) 1 sq yd.
(e) 1 sq m.

2. How many copies of a square inch will it take to cover a copy of a square foot?
3. How many copies of a square foot will it take to cover a copy of a square yard?
4. How many copies of a square centimeter will it take to cover a copy of a square meter?
5. Which has greater area, the square yard or the square meter?
6. Which square region in problem 1 has measurement area equal to that of one row of nine square regions, each of whose sides measures one foot?
7. Cut the copy of the square inch into four congruent pieces. What is the area of each piece?
8. How many copies of a square inch will it take to cover the faces of a cube, each of whose edges measures one inch?

8-5 STRATEGY

Direct determination of area using an instrument is not common. Although there is an instrument, the planimeter, that gives approximate areas for closed regions, areas are more often calculated by the use of special formulas or certain procedures of the integral calculus. We shall direct our attention to some of the formulas used to determine the area of certain common geometric figures—rectangular regions, parallelogram regions, trapezoidal regions, triangular regions, and circular regions.

Our strategy is to assume that a particular formula works for one of the figures—the rectangular region—and then to show that the formulas for the area of other figures can be derived from this one because of certain properties of the geometric figures, certain properties that we have assumed for area, and certain other mathematical relations.

8-6 BASE AND ALTITUDE

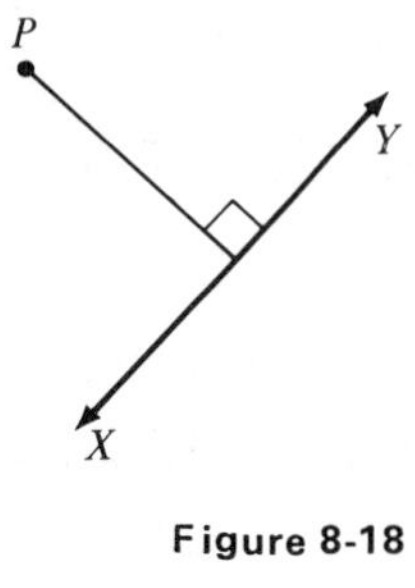

Figure 8-18

Let us first introduce some concepts that will help simplify the subsequent discussion. If we have a point and a line that does not contain the point, as in Fig. 8-18, the distance between the point and the line is the length of the segment that joins the point to the line and that is perpendicular to the line.

When we speak of distance between two parallel lines, we mean the length of a segment that has an endpoint on each line and is perpendicular to one of the lines. (The segment will be perpendicular to the other line. Why?) For example, in Fig. 8-19 the length of $\overline{XY}$ is the distance between $\overleftrightarrow{AB}$ and $\overleftrightarrow{CD}$.

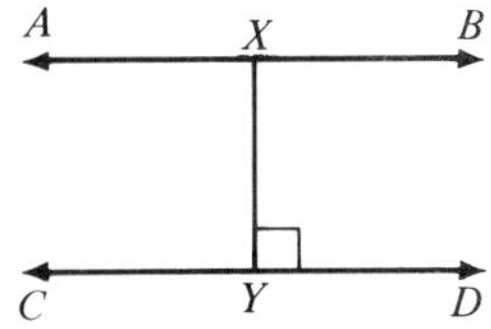

Figure 8-19

Any segment that is perpendicular to both lines and has an endpoint on each line will be congruent to $\overline{XY}$. To see why this is so, consider Fig. 8-20. $\overleftrightarrow{AB} \parallel \overleftrightarrow{CD}$. $\overline{XY} \perp \overleftrightarrow{AB}$. $\overline{MN} \perp \overleftrightarrow{AB}$. Therefore, $\overline{MN} \parallel \overline{XY}$. Why? This means that $XMNY$ is a parallelogram. As we have seen in problem 6 in the Exercises at the end of Chapter 7, the opposite sides of a parallelogram are congruent. Therefore, $\overline{XY} \cong \overline{MN}$, and we have $m(\overline{XY}) = m(\overline{MN})$. (You begin to see the importance of the property that congruent figures have equal measures.) In effect, this tells us that the perpendicular distance between two parallel lines in our geometry is always the same no matter where the perpendicular segment has its endpoints on the lines. This, by the way, is what people mean when they say two parallel lines are "everywhere the same distance apart." (See Sec. 1-2.)

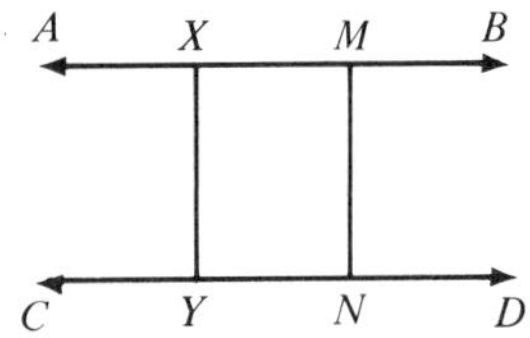

Figure 8-20

Now, if a triangle is drawn so that one side lies on one of the parallel lines and the vertex opposite this side lies on the other parallel line, we call the side on the parallel the **base** *of the triangle*, and the distance between the parallels is the **height** or **altitude** *of the triangle* with respect to the given base. (See Fig. 8-21.) With respect to Fig. 7-24 we noted that an altitude of a triangle is a segment having one endpoint the vertex of the triangle and the other endpoint on the carrier of the side opposite the vertex. The segment is also perpendicular to the carrier. This segment is precisely one of the segments that defines the distance between the two parallels that we have been discussing in this section. Could the altitude of a triangle be described as the distance between a point and a line? Which point and which line?

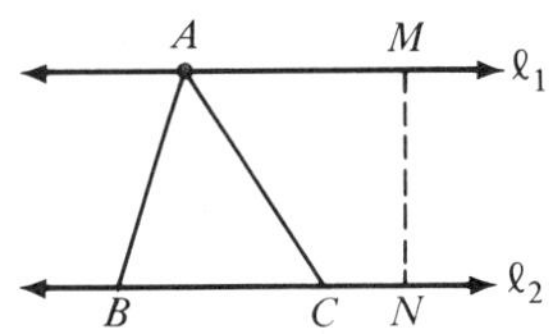

Figure 8-21

Note also that we need not start with the pair of parallel lines and then draw the triangle. We could start with the triangle, one of its sides, and the vertex opposite this side. Will there always be a line through the vertex parallel to the side (in our geometry)?

The words "base" and "altitude," like "radius," sometimes denote a measure, sometimes a measurement, and sometimes a segment. One can usually determine the interpretation from the context.

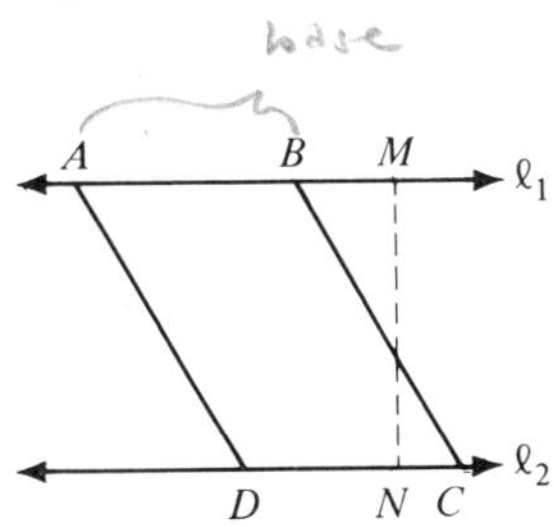

Figure 8-22

What is the relationship between the carriers of a pair of opposite sides of a parallelogram? In a parallelogram, opposite sides are called **bases** *of the parallelogram*, and the distance between the carriers of a pair of bases is called the **height** or **altitude** *of the parallelogram* with respect to these bases. In Fig. 8-22 name the indicated pair of bases and name a segment whose measure is the altitude with respect to these bases.

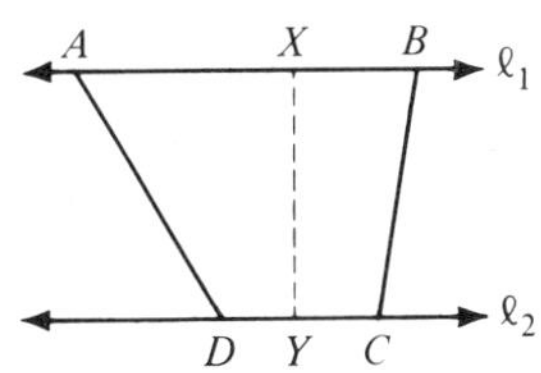

Figure 8-23

The parallel sides of a trapezoid are called the **bases** *of the trapezoid*, and the distance between the carriers of the bases is called the **height** or

altitude *of the trapezoid.* Name the bases and a segment whose measure is the altitude of the trapezoid in Fig. 8-23.

EXERCISE 8-6

1. For each part of Fig. 8-24, name a base or pair of bases and find the altitude of the figure with respect to this base. Use a protractor to draw a segment perpendicular to the base and find the measurement of the segment to the nearest sixteenth inch.

2. For each triangle in Fig. 8-25 construct an altitude to the base $\overline{AC}$. (Use only a compass and straight edge.)

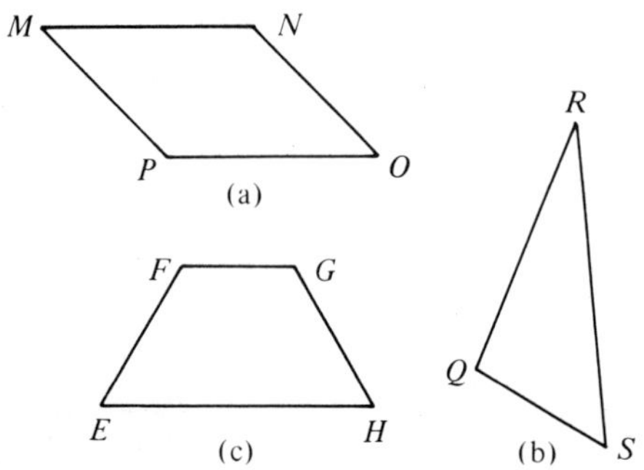

Figure 8-24

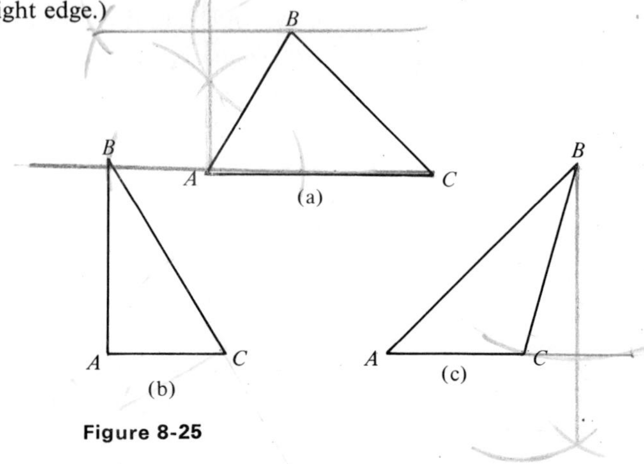

Figure 8-25

3. What is the shortest path between a point and a line? Why is perpendicular distance called "the" distance?

4. Suggest possible meanings for (a) distance between a point and a plane not containing the point, (b) distance between two parallel planes, (c) altitude of a pyramid or cone, (d) altitude of a prism or cylinder.

8-7 AREA OF A RECTANGULAR REGION[1]

In Fig. 8-17(b) we found the number of units of area in the rectangular region by multiplying the number associated with the size of one side by the number associated with the size of an adjacent side. This situation provides the intuitive model for our assumption about the formula for finding the area of a rectangular region.

Considering $WXYZ$ in Fig. 8-26 as representing any rectangle, we know

[1] Some authors use the phrase "area of a rectangle" when they mean area of a rectangular region. And, in general, they call the area of a polygonal region the "area of a polygon." However, we shall maintain the precision in terminology and speak only of the area of a region.

that $\overline{WX}$ and $\overline{ZY}$ are parallel. We also know that $\overline{ZW}$ (and $\overline{YX}$) is perpendicular to $\overline{WX}$ and $\overline{ZY}$. Thus, $\overline{WX}$ and $\overline{ZY}$ may be considered bases of the rectangle, and WZ (or XY) may be considered an altitude of the rectangle. Note that $WX = ZY$, and that $WZ = XY$. Why?

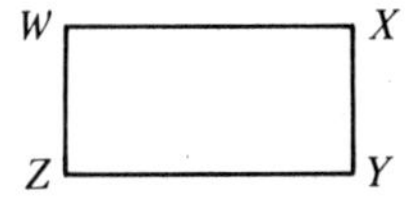

Figure 8-26

We usually represent the measure of the base of any figure by b and the altitude or height of any figure by h. Taking a cue from Fig. 8-17, we shall assume that every rectangular region has a unique area (for a given unit system) and that this area is equal to the product of the measure of the base and the altitude (for the given unit system). Thus, if A represents the area, we have for any rectangular region the formula:

$$\mathbf{A = bh.}$$

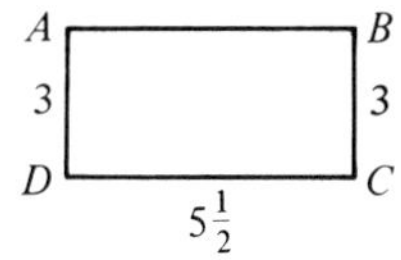

Figure 8-27

Example:[2] In Fig. 8-27, if $DC = 5\frac{1}{2}$ and $AD = 3$, then the area is found by substituting in the formula:

$$A = bh,$$
$$A = (5\tfrac{1}{2})(3),$$
$$A = 16\tfrac{1}{2}.$$

In practical measurement situations we would be able to find the measurement of the base and altitude in terms of a linear unit, and we could then report the area measurement in terms of an area unit. If both dimensions of the rectangle (base and height) are given in terms of the same unit system, then the area measurement is usually reported in terms of the standard unit of area based on that linear unit.

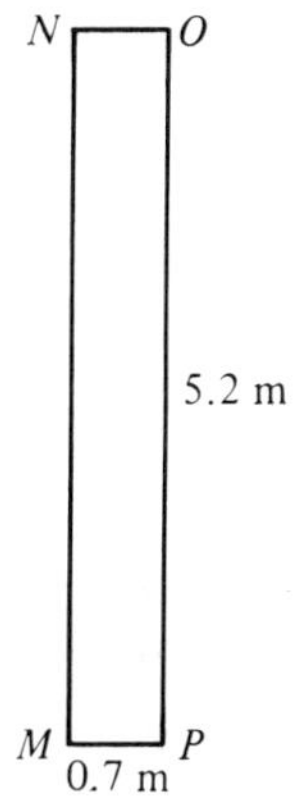

Figure 8-28

Example: For rectangular region $MNOP$ in Fig. 8-28, $m(\overline{PO}) = 5.2$ and $m(\overline{MP}) = 0.7$.

$$A = bh,$$
$$A = (5.2)(0.7),$$
$$A = 3.64.$$

The area measurement is 3.64 sq m.

If the measurements used in the preceding example are obtained from physical measurement, can we consider the measures to be exact? For this reason we should say that the computed value approximates the area. (See Sec. 8-16.)

If the measurements of the base and altitude are not given in terms of the same unit, then it is customary to find the equivalent of one measurement in terms of the unit system of the other before calculating the area.

Example: $PQRS$ is a rectangle. (See Fig. 8-29.) Notice that the measurements are not expressed in terms of the same units. However,

[2] In these sample problems areas will be expressed as exact products. See Sec. 8-16 for a discussion of errors of measurement as they affect the reporting of area measurements obtained from formulas that involve products.

since 1 yd is the equivalent of 3 ft, 3 yd must be the equivalent of 9 ft. Then using feet as the unit of the system, $m(\overline{PQ}) = 2$ and $m(\overline{QR}) = 9$.

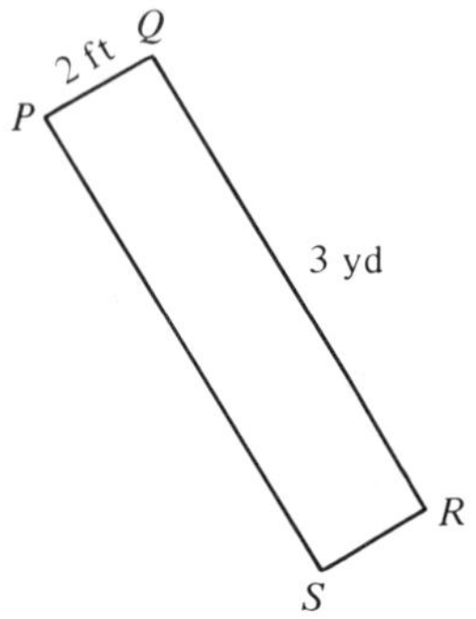

Figure 8-29

$$A = bh,$$
$$A = (2)(9),$$
$$A = 18.$$

The area measurement is 18 sq ft.

What would be the area of the rectangular region in Fig. 8-29 if the unit of area were a rectangular region whose sides had measurements of 1 ft and 1 yd?

We have assumed that every rectangular region has a unique area and that this is given by bh. That this agrees with our experience in covering rectangular regions with unit square regions explains in part our willingness to accept this assumption. From a mathematical standpoint, it is a useful assumption because it allows us to analyze the areas of other closed regions in terms of the area of a rectangular region. This we shall proceed to do.

EXERCISE 8-7

1. Find the area measurement of each of the rectangular regions whose dimensions are given below:
   (a) 6 ft and 10 ft. (b) 11 yd and 15 yd.
   (c) $5\frac{1}{3}$ in. and 4 in. (d) $9\frac{1}{5}$ mi and $3\frac{1}{3}$ mi.
   (e) 2.5 m and 17.3 m. (f) 0.75 in. and 0.33 in.
   (g) 8 yd and 4 ft. (h) 150 cm and 2 m.

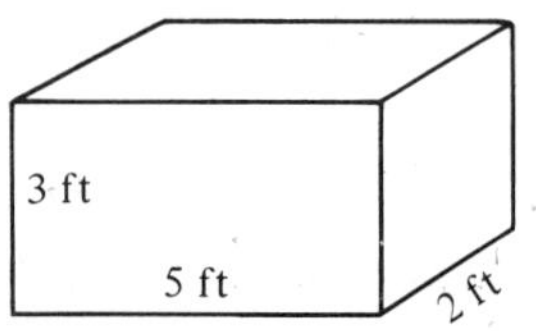

Figure 8-30

2. **A rectangular parallelepiped** (Fig. 8-30) is one in which every face is a rectangular region. If the base rectangle has dimensions of 2 ft and 5 ft, and one of the rectangles in the lateral surface has dimensions of 5 ft and 3 ft, what is the sum of the areas of the faces of the parallelepiped?

3. What is the area of a rectangular region whose base measurement is 15 cm and whose height measurement is 15 cm?

4. Construct a rectangular region whose base and height measurements are $\frac{3}{4}$ in. and $\frac{2}{4}$ in., respectively. Make a unit square region each of whose sides has measurement of 1 in. Cut the unit square region into sixteen congruent regions. How many of these square regions will it take to cover the rectangular region you constructed?

5. (a) Find the perimeter and area of each of the following rectangular regions:

	b	h	p	A
(1)	30	2		
(2)	22	10		
(3)	16	16		
(4)	12	20		
(5)	$\frac{1}{2}$	$31\frac{1}{2}$		

(b) Which rectangular region has the greatest perimeter? Least perimeter?
(c) Which region has the greatest area? Least area?
(d) This exercise seems to indicate that, in general, what kind of rectangular region gives the maximum area for a given perimeter?
(e) If you wish to design a rectangular garden so that its perimeter measurement is 100 ft and its area is a maximum, what shape and dimensions should the rectangle have?

6. (a) Find the perimeter and area of each of the following rectangular regions:

	b	h	p	A
(1)	12	3		
(2)	9	4		
(3)	6	6		
(4)	2	18		
(5)	$\frac{1}{2}$	72		

(b) Which rectangular region has the greatest perimeter? Least perimeter?
(c) Which region has the greatest area? Least area?
(d) This exercise seems to indicate that, in general, what kind of rectangular region gives the minimum perimeter for a fixed area?
(e) If you wish to design a garden so that its area measurement is 100 sq ft and you use a minimum amount of fencing, what shape and dimensions should the rectangle have?

8-8 AREA OF A SQUARE REGION

Every square is a rectangle; therefore, the area of a square region can be found by multiplying the base by the altitude of the square. We noted earlier that the square, by one definition, has the interesting property that all four sides are congruent and thus their measures are equal. If we let s represent the measure of one side of a square, then each side of the square has measure s. Therefore, $b = s$, $h = s$, and the area formula is $A = (s)(s)$ or

$$\mathbf{A = s^2}.$$

EXERCISE 8-8

1. Make several copies of square regions whose sides have measurement of 1 in. (You will need about 75 of them.) By pasting the 1-in. square regions on a piece of heavy paper, make square regions whose sides have measurements of 1 in., 2 in., 3 in., 4 in., 5 in., and 6 in.

8-9 AREA OF A TRIANGULAR REGION

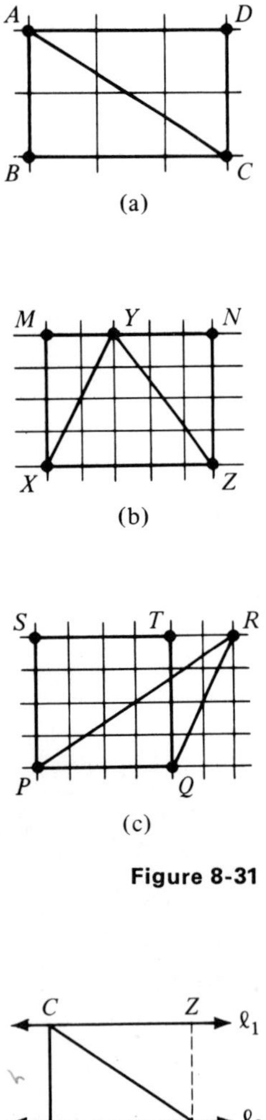

Figure 8-31

We can also relate the problem of finding the area of a triangular region to that of finding the area of a rectangular region. We will see that triangular regions are related to rectangular regions in certain ways. Before we investigate how this relationship is mathematically derived, we can grasp the nature of the relationship by drawing some triangles on squared graph paper. In Fig. 8-31(a) $\triangle ABC$ and rectangle $ABCD$ are drawn. By considering the square regions marked on the graph paper as units, we can find by counting square regions that the area of the rectangular region is 6 and the area of the triangular region is approximately 3. In Fig. 8-31(b) $\triangle XYZ$ and related rectangle $MNZX$ are drawn. By counting square regions, we find that the area of the rectangular region is 20 and the area of the triangular region is 10. In Fig. 8-31(c) we have pictures of $\triangle PQR$ and related rectangle $PSTQ$. The area of the rectangular region is 16, and the area of the triangular region is approximately 8. Notice the relationship of the area of the triangular region to the area of the rectangular region. The area of the triangular region seems to be what fractional part of the area of the rectangular region?

You may wonder why we picked these three triangles. If you consider their altitudes, you will see the reason. The three diagrams summarize the three possible relationships of the altitude segment to its base: (1) the altitude has one endpoint that is a vertex of the base (this makes the altitude one side of the triangle), (2) the altitude has one endpoint between the two vertices of the base (the altitude is in the triangular region but is not a side), and (3) the altitude has neither endpoint on the base (the altitude is not a side and is not in the interior of the triangle).

Figure 8-32

Without getting too involved in mathematical proof, we would like to show that the relationship between the area of a triangular region and the area of a rectangular region with base and altitude equal to those of the triangular region follows from our assumptions and principles about area and relations between geometric figures.

We must consider each of the three types of triangles mentioned above. Remember that this categorization is based on the relative positions of the base and altitude and also on the relationship of the base and altitude to two parallel lines associated with the triangles. (See Sec. 8-6.)

For case 1 (Fig. 8-32) the altitude drawn to the base from vertex C is also a side of the triangle. What kind of triangle is this? We shall agree to call the measure of the base ($\overline{AB}$) b and the measure of the altitude ($\overline{CA}$) h. There is a segment ($\overline{BZ}$) perpendicular to $\overline{AB}$ at B with endpoint on the parallel to $\overline{AB}$ through C. $CZBA$ is a rectangle. Its base $\overline{AB}$ has measure b, and its altitude $\overline{CA}$ has measure h. Therefore, the area of the rectangular region is bh. Now look at $\triangle CAB$ and $\triangle BZC$. The sum of the areas of the triangular regions must be the same as the area of the rectangular region. Why? But $\triangle CAB \cong \triangle BZC$. Why? Since the triangles are congruent, the areas of the trianglar regions must be equal. Then, using "$\blacktriangle$" to denote a triangular region:

$$\begin{aligned} \text{area } (\blacktriangle CAB) + \text{area } (\blacktriangle BZC) &= \text{area (region } CZBA), \\ 2 \times \text{area } (\blacktriangle CAB) &= bh, \\ \text{area } (\blacktriangle CAB) &= \tfrac{1}{2}bh. \end{aligned}$$

We said that $\triangle CAB$ is a right triangle. Thus we see that the area of any right triangular region must be equal to one half the area of a rectangular region that has the same base and altitude as the triangle.

Now that we have developed a formula for the area of a right triangular region, we can use this result to find formulas for the areas of the other two cases. For case 2 (Fig. 8-33) the area of $\blacktriangle DEF$ must be equal to the sum of the areas of $\blacktriangle FGD$ and $\blacktriangle FGE$. Segment FG is an altitude of $\triangle DEF$. What kinds of triangles are $\triangle FGD$ and $\triangle FGE$? In the previous paragraph we concluded that the area of any right triangular region is equal to one half the product of its base and altitude. The base of $\triangle FGD$ is $\overline{DG}$; call its measure d. The base of $\triangle FGE$ is $\overline{GE}$; call its measure e. Recall that b is the measure of $\overline{DE}$ and h is the measure of $\overline{FG}$. The area of $\blacktriangle FGD$ is $\frac{1}{2}dh$, and the area of $\blacktriangle FGE$ is $\frac{1}{2}eh$. Then,

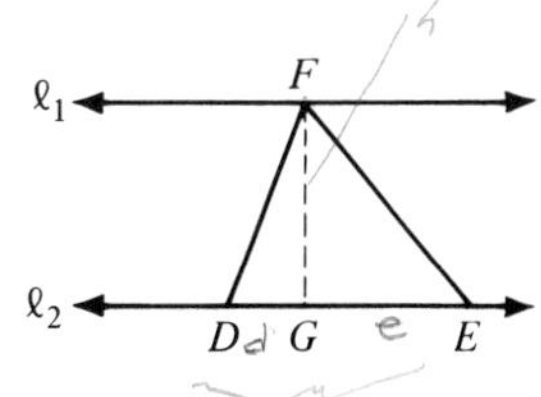

Figure 8-33

$$\begin{aligned} \text{area } (\blacktriangle FDE) &= \text{area } (\blacktriangle FGD) + \text{area } (\blacktriangle FGE), \\ \text{area } (\blacktriangle FDE) &= \tfrac{1}{2}dh + \tfrac{1}{2}eh. \end{aligned}$$

Now, using the distributive property, we see that

$$\text{area } (\blacktriangle FDE) = \tfrac{1}{2}(d + e)h.$$

But $d + e = b$. Why? This gives us

$$\text{area } (\blacktriangle FDE) = \tfrac{1}{2}bh.$$

Therefore, for any triangular region whose altitude drawn to a given base is in the triangular region, the area is one half the product of the base and the altitude.

Finally, for case 3 (Fig. 8-34) we want to find the area of $\blacktriangle HIJ$, where $\overline{JK}$ is the altitude to base $\overline{HI}$. Notice that $\blacktriangle JKH$ and $\blacktriangle JKI$ are right triangular regions whose area we know how to find. Notice also that:

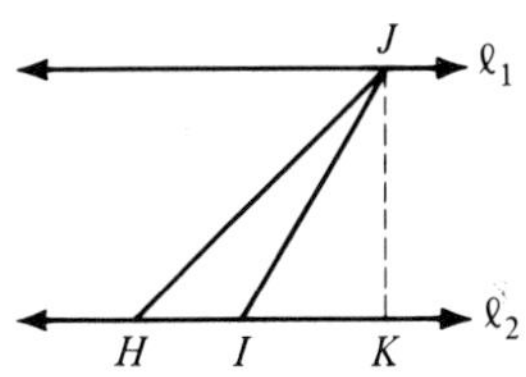

Figure 8-34

$$\begin{aligned} \text{area } (\blacktriangle HIJ) + \text{area } (\blacktriangle JKI) &= \text{area } (\blacktriangle JKH), \\ \text{area } (\blacktriangle HIJ) &= \text{area } (\blacktriangle JKH) - \text{area } (\blacktriangle JKI). \end{aligned}$$

Recalling that $m(\overline{HI}) = b$ and $m(\overline{JK}) = h$, we shall say that $m(\overline{IK}) = c$ and $m(\overline{HK}) = d$. Thus, $b + c = d$, which gives $b = d - c$. Also note that the area of $\triangle JKI$ is $\frac{1}{2}ch$ and that the area of $\triangle JKH$ is $\frac{1}{2}dh$. Thus,

$$\text{area } (\triangle HIJ) = \tfrac{1}{2}dh - \tfrac{1}{2}ch,$$
$$\text{area } (\triangle HIJ) = \tfrac{1}{2}(d - c)h,$$
$$\text{area } (\triangle HIJ) = \tfrac{1}{2}bh.$$

Thus, for all three cases (which exhaust all possibilities) we have found that the area of a triangular region is given by one half the product of the measure of the base and the measure of the altitude. If A represents the area, we have the general formula for finding the area of any triangular region:

$$\mathbf{A} = \tfrac{1}{2}\mathbf{bh}.$$

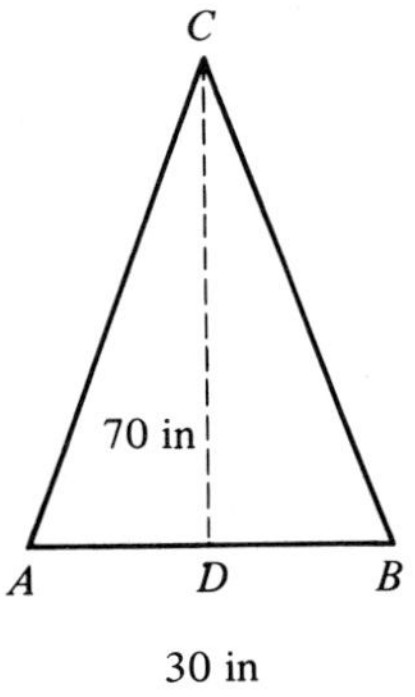

Figure 8-35

Example: In $\triangle ABC$ of Fig. 8-35, $m(\overline{AB}) = 30$ and $m(\overline{CD}) = 70$. What is the area of the triangular region?

$$A = \tfrac{1}{2}bh,$$
$$A = (\tfrac{1}{2})(30)(70),$$
$$A = 1050.$$

The area measurement is 1050 sq in.

EXERCISE 8-9

1. Find the area measurement of each triangular region shown in Fig. 8-36. $\overline{XY}$ is the altitude in each case.

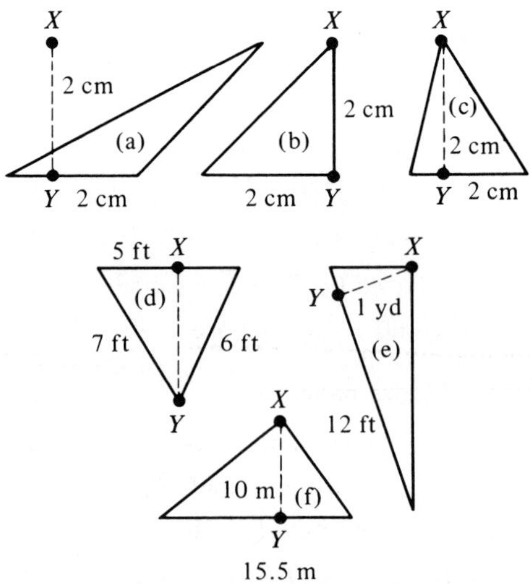

Figure 8-36

2. Draw an acute scalene triangle. Draw an altitude to each base. Measure each base and altitude to the nearest millimeter and compute the area of the region using the three different bases. Do your results agree with each other? Explain why they may not.

3. Make a model of an acute scalene triangular region from stiff paper. Label the three vertices A, B, and D. Find the midpoints of $\overline{AD}$ and $\overline{AB}$. Draw the segment joining the midpoints and call it $\overline{XY}$. Draw the segment from A perpendicular to $\overline{XY}$. Call it $\overline{AZ}$. Cut along $\overline{XY}$ and $\overline{AZ}$ to form the three regions AZX, AZY, and $XYBD$. Rearrange the three pieces to form a rectangular region. How does the base of this rectangle compare with the base of triangle ABD? How does the altitude of this rectangle compare with the altitude of triangle ABD? How does the area of the rectangular region compare with the area of the original triangular region?

4. Make a model of an obtuse scalene triangular region. Make a second copy of the region. How do the areas of the two regions compare? Arrange the two triangular regions to form a parallelogram region. How does the area of the parallelogram region compare with the area of one of the triangular regions? How does the base of the parallelogram compare with the base of the triangle? With respect to the base used above compare the altitude of the parallelogram with the altitude of the triangle.

8-10 AREA OF A PARALLELOGRAM REGION

We shall now see that, knowing how to find the area of a triangular region, we can derive the area of a parallelogram region. (Note: The area of a parallelogram region may be derived directly from knowing how to find the area of a rectangular region.)

Any diagonal of a parallelogram forms with the sides a pair of congruent triangles. Recall problem 6 of the Exercises at the end of Chapter 7. In Fig. 8-37 the parallelogram $MNOP$ has diagonal $\overline{NP}$. $\triangle PMN \cong \triangle NOP$. Thus, the area of parallelogram region $MNOP$ is equal to twice the area of one of the triangular regions. Why? Notice that the measure of the base $\overline{NO}$ of $\triangle NOP$ is the same as the measure of the base of the parallelogram. Notice also that the height of the parallelogram is the height of the triangle. We shall say that $m(\overline{NO}) = b$ and $m(\overline{RQ}) = h$.

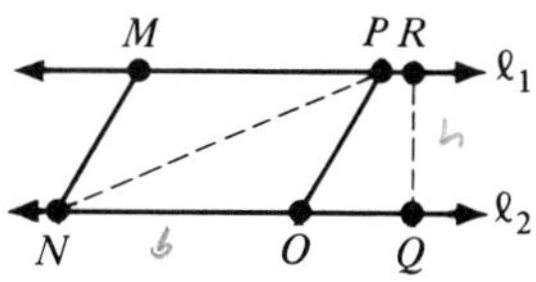

Figure 8-37

Area of parallelogram region $MNOP = 2 \times$ area ($\triangle NOP$). Since the area of $\triangle NOP$ is $\frac{1}{2}bh$, we have

$$\text{area of region } MNOP = 2(\tfrac{1}{2}bh),$$
$$\text{area of region } MNOP = bh.$$

Since $MNOP$ represents any parallelogram, we have the general formula for finding the area of any parallelogram region:

$$\mathbf{A = bh.}$$

Example: For parallelogram $WXYZ$ of Fig. 8-38, $m(\overline{XT}) = 6.8$, $m(\overline{XY}) = 8.0$, and $m(\overline{ZY}) = 2.5$.

$$A = bh,$$
$$A = (2.5)(6.8),$$
$$A = 17.00.$$

The area measurement is 17.00 sq cm.

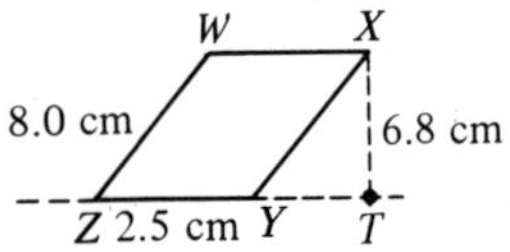

Figure 8-38

Notice that actually we do not need to know the measure of the non-base side of the parallelogram in the previous example to find the area of the parallelogram region. How would we get the perimeter of this polygon?

EXERCISE 8-10

1. Draw pictures of four parallelogram regions, each of which has an area of 20 sq in. and no two of which are congruent.

8-11 AREA OF A TRAPEZOIDAL REGION

There are several different approaches to determining the formula for the area of a trapezoidal region; we have selected but one.

Any diagonal of a trapezoid forms with the four sides a pair of triangles. Unlike the case of the parallelogram, these two triangles are not congruent. Figure 8-39 shows diagonal $\overline{RT}$ of trapezoid $LRCT$, $\triangle LRT$, and $\triangle RTC$. Figure 8-40 shows that the altitudes of the two triangles and the trapezoid are equal. If we call this measure h, b_1 the measure of $\overline{TC}$, and b_2 the measure of $\overline{LR}$, we have the area of $\triangle RTC$ is $\frac{1}{2}b_1h$ and the area of $\triangle LRT$ is $\frac{1}{2}b_2h$. The area of the trapezoidal region, then, is

$$\tfrac{1}{2}b_1h + \tfrac{1}{2}b_2h.$$

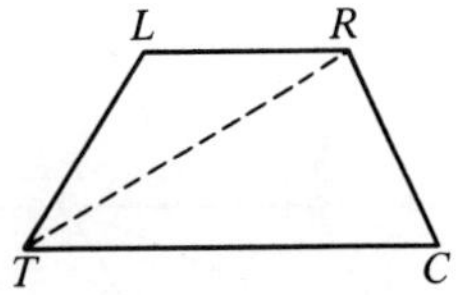

Figure 8-39

Using the distributive property, we have

$$\tfrac{1}{2}h(b_1 + b_2).$$

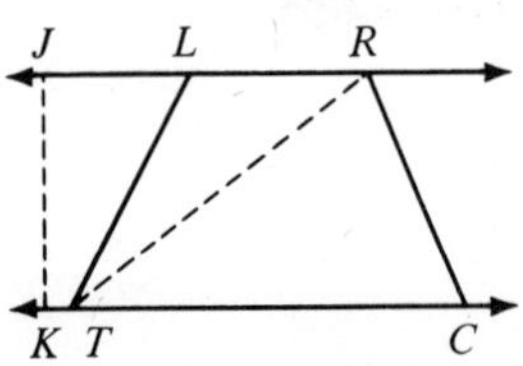

Figure 8-40

In general, the area of a trapezoidal region is equal to one half the product of the measure of the altitude and the sum of the measures of the two bases. If A represents area, b_1 the measure of one base, and b_2 the measure of the other base, then

$$\mathbf{A = \tfrac{1}{2}h(b_1 + b_2).}$$

Example: Trapezoid $ABCD$ has altitude with measure 6 and bases with measures 10 and 15, respectively.

$$A = \tfrac{1}{2}h(b_1 + b_2),$$
$$A = (\tfrac{1}{2})(6)(10 + 15),$$
$$A = 75.$$

The area measurement is 75 square units.

EXERCISE 8-11

1. Find the area of a trapezoidal region whose altitude measurement is 2.5 ft and whose bases have measurements of 6 ft and 8.7 ft.
2. Make a model of a trapezoidal region. Make a copy of this model. What is the relationship between the areas of the two trapezoidal regions? Arrange the two trapezoidal regions so that their union is a parallelogram. How does the altitude of the parallelogram compare with the altitude of one of the trapezoids? How does the base of the parallelogram compare with the bases of one of the trapezoids? How does the area of one of the trapezoidal regions compare with the area of the parallelogram region?
3. The formulas for the area of a region determined by a parallelogram, triangle, square, and rectangle can be considered special cases of the formula for the area of a trapezoidal region. Read Ref. 101, pp. 106–108, for an example of this approach.

8-12 USING BASIC FORMULAS

Before investigating the relationship for the area of a circular region, we shall consider how the formulas for simple figures may help us find areas of other figures. We have noted on several occasions that given any region, its area is equal to the sum of the areas of the regions determined by a partition. This is but a restatement of the fundamental assumption about the additive nature of area.

Given a polygonal region such as shown in Fig. 8-41, we may be able to determine its area if we can find a partition that determines regions each of whose area we know or can find. One possible partition is shown in Fig. 8-41(b). Of what polygonal regions do we need to find the area?

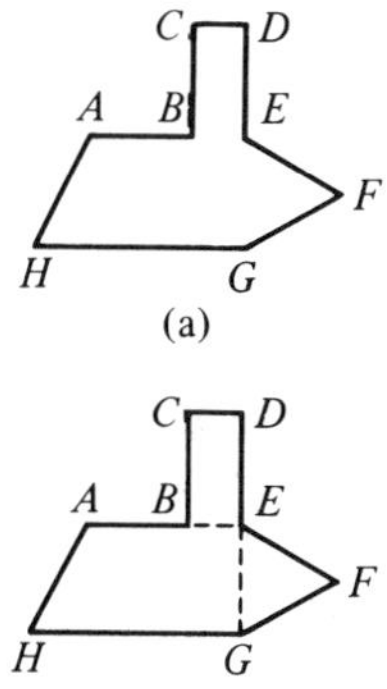

Figure 8-41

Of special interest is a formula that we can use to find the area of a regular polygonal region. Recall that a regular polygon is one that is both equiangular and equilateral. For any regular polygon there is a point in the interior that is equidistant from each vertex of the polygon. (This point, called the **center** of the regular polygon, is the common center of the incircle and the circumcircle.) The sides of the polygon and the segments joining this point to the vertices partition the polygonal region into a set of nonoverlapping congruent triangular regions. How many sides are there in the polygon of Fig. 8-42? How many triangles are formed? In general, if the polygon has n sides, we can form n congruent triangles by

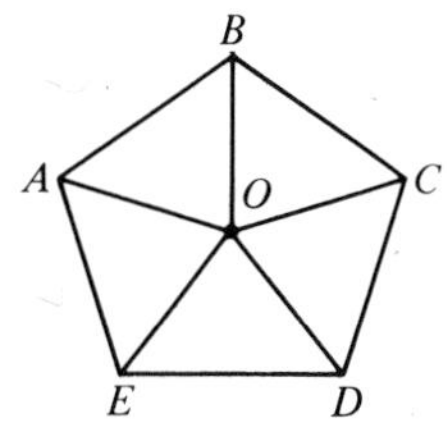

Figure 8-42

selecting the appropriate point in the interior of the polygon. Then, if we know the area of one of the triangular regions, we can find the area of the polygonal region by multiplying the area of this triangular region by the number of sides of the polygon. In Fig. 8-43 what must we know in order to determine the area of $\triangle FOE$? We need the measure of the base, $\overline{FE}$, and the measure of the altitude, $\overline{OS}$. Notice that all the triangles have the same altitude measure. Why?

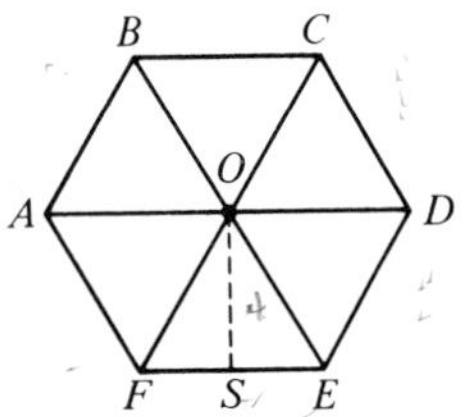

Figure 8-43

The distance that is the altitude of each triangle is called the **apothem** of the regular polygon. (This is also the radius of the inscribed circle.) If we call the measure of one side of the polygon s and the measure of the apothem a, then the area of one triangular region is given by $\frac{1}{2}sa$. The area of the polygonal region is, then, $n(\frac{1}{2}sa)$, where n is the number of sides of the polygon. Note that $n(\frac{1}{2}sa) = \frac{1}{2}a(ns)$. Why? But what is ns? Is this not the perimeter of the polygon? If we replace ns by p, the formula for the area, A, of a regular polygonal region is given by

$$\mathbf{A = \tfrac{1}{2}ap,}$$

where p is the perimeter and a is the apothem of the polygon.

EXERCISE 8-12

1. Find the area of the polygonal region in Fig. 8-44.
2. Draw a circle with radius of 2 in. Draw the inscribed regular hexagon. Each side of the hexagon has measurement of 2 in. Why? The apothem measurement of this polygon is approximately 1.73 in. Find the perimeter of the polygon. Find the area of the polygonal region.
3. The circumscribed circle of the hexagon in problem 2 has a radius of 2 in. Using 3.14 as the approximation of π, find the circumference of this circle. Compare with the perimeter of the polygon.
4. For a regular polygon of 24 sides, if one side has a measurement of 2 in., the apothem measurement is approximately 7.66 in. Find the perimeter of the polygon. Find the area of the polygonal region.
5. The circumscribed circle of the polygon in problem 4 has a radius measurement of approximately 7.73 in. Using 3.14 as an approximation to π, find the circumference of this circle. Compare with the perimeter of the polygon in problem 4.

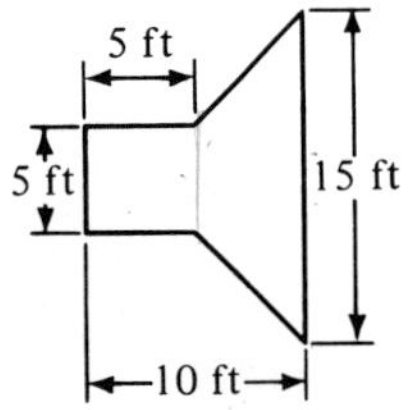

Figure 8-44

8-13 AREA OF A CIRCULAR REGION

Our treatment of the basic formula for the area of a circular region must be more intuitive than for any of the other figures analyzed so far, since the actual derivation of the formula depends on the limit concept and the calculus, which are not typical topics in elementary mathematics.

However, this need not prevent our getting a firm intuitive understanding of the relationship.

Figure 8-45 shows a picture of a circle with center at O. A set of n points on the circle can be selected so that the chords determined by successive pairs of these points are congruent. In Fig. 8-45(b), for which $n = 12$, the congruent chords are $\overline{AB}$, $\overline{BC}$, $\overline{CD}$, . . . , and $\overline{LA}$. What kind of polygon is $ABCDEFGHIJKL$? Besides being a dodecagon, it is a regular polygon. Note that the area of the circular region is approximately the same as the area of the polygonal region. We know from the previous section that the area of the polygonal region can be found using the formula $A = \frac{1}{2}ap$, where p is the perimeter and a is the apothem. We get an approximation of the area of the circular region by using the formula for the polygonal region. If we greatly increase the number of sides of the regular polygon by selecting a greater number of points on the circle, then the value of the perimeter of the polygon very nearly approximates the value of the circumference of the circle. That is, $p \approx 2\pi r$, where p is the perimeter of the polygon, r is the radius of the circle, and "$\approx$" is read "is approximately."

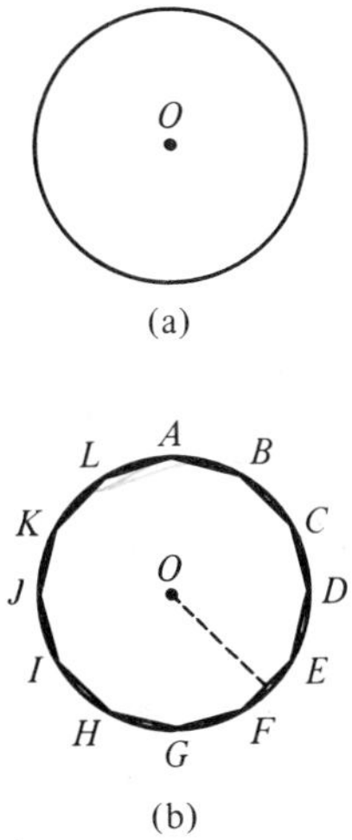

Figure 8-45

We also note that the apothem of the polygon is very nearly the same as the measure of the radius. That is, $a \approx r$. It also seems that the area of the circular region is very nearly approximated by the area of the polygonal region. If A is the area of the circular region, $A \approx \frac{1}{2}pa$. Since $p \approx 2\pi r$ and since $a \approx r$, we have $A \approx \frac{1}{2}(2\pi r)r$. Simplifying this, we have $A \approx \pi r^2$, where r is the radius of the circle. Using the techniques of more advanced mathematics, we can show that the area of any circular region is indeed equal to the product of π and the square of the radius. That is, for any circular region we have

$$\mathbf{A} = \pi \mathbf{r}^2.$$

Example: What is the area of the circular region whose radius has a measure of 1.5? We shall use 3.14 as the value of π and assume here for the purposes of computation that it is exact.

$$A = \pi r^2,$$
$$A = (3.14)(1.5)^2,$$
$$A = 70.650.$$

The area measurement is 70.650 square units.

Two closed regions related to circular regions are of interest because of their common occurrence and because a determination of formulas for their area illustrates the interrelationship of several geometric ideas.

The first figure is called an **annulus**. (See Fig. 8-46.) The boundary of an annulus is describable as two **concentric circles**. Circles are concentric if they are coplanar and have the same point as center. The area of an

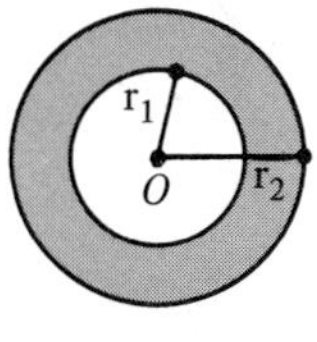

Figure 8-46

annulus can be obtained by subtracting the area of the circular region defined by the inner circle from the area of the circular region defined by the outer circle. If the radius of the inner circle is r_1, its area is $\pi r_1{}^2$. If the radius of the outer circle is r_2, its area is $\pi r_2{}^2$. The area, A, of the annulus is then given by $A = \pi r_2{}^2 - \pi r_1{}^2$. Using the distributive property, we have $A = \pi(r_2{}^2 - r_1{}^2)$.

The second figure is called a **sector** of a circle. Sometimes we are not looking for the area of the entire circular region. Instead we want the area of a piece of the region that has the shape of a "pie wedge." Such a subset of the circular region is called a sector. A sector is more formally defined as the closed region determined by two radii of a circle and the arc determined by the endpoints of the radii of the circle. (See Fig. 8-47.) The area of a sector is related to the degree measure of the defining arc. If m represents the degree measure of this arc, then the portion of the area of the circle that is also in the sector is

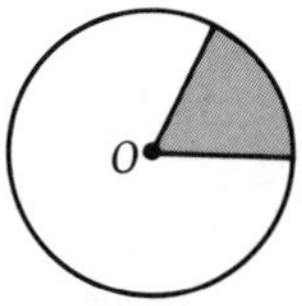

Figure 8-47

$$\frac{m}{360}.$$

Where does the 360 come from? What is the relationship of the degree measure of the arc to the measure of the angle determined by the two radii?

The area of a sector is then given by the formula

$$A = \left(\frac{m}{360}\right)\pi r^2,$$

where m is the degree measure of the arc and r is the radius of the circle.

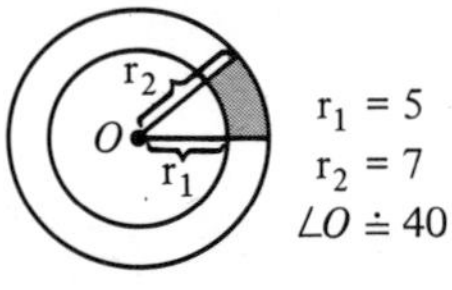

Figure 8-48

EXERCISE 8-13

1. Find the area of the following circular regions:
   (a) Radius has measurement of 2.51 m. (Use 3.14 as approximation of π.)
   (b) Radius has measurement of $12\frac{8}{11}$ in. (Use $\frac{22}{7}$ as approximation of π.)
2. Find the area of the shaded region in Fig. 8-48.
3. The bases of a circular cylinder have radius of 2 in. Find the sum of the areas of the two bases.
4. Find the area of the shaded portion of Fig. 8-49. The polygon is a square whose sides measure 1 in. The four arcs are "quarter-circles."
5. Will the formula for the area of an annulus be applicable if the two circles that define the "annulus" are not concentric?

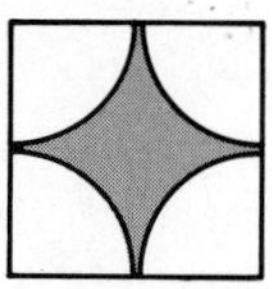

Figure 8-49

8-14 SURFACE AREA

Up to this point we have considered only areas of closed regions that lie in a single plane. You will recall that we have also considered some noncoplanar closed surfaces. (See Sec. 4-3.) It is possible to consider each of these surfaces as having an area—some number associated with the size of the closed surface. In the case of polyhedral surfaces, this **surface area** is easily determined by adding the areas of the polygonal regions that are the faces of the polyhedron. These areas we can usually determine, since we already know formulas for many of the closed planar regions. Describe the surface areas in Fig. 8-50.

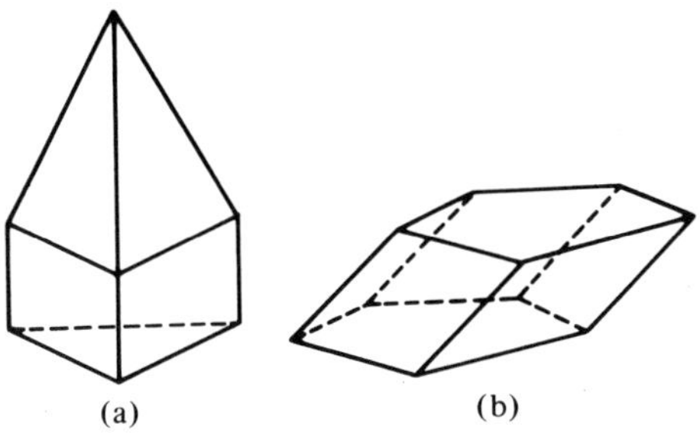

Figure 8-50

Some of the three-dimensional closed surfaces contain surfaces that are not plane regions. Common examples are the cylinder, cone, and sphere. (See Fig. 8-51.) However, it is possible to find a relationship between the area of these surfaces and the area of certain plane figures. This we shall investigate in Exercise 8-14.

For closed three-dimensional surfaces that have bases, the surface area is usually analyzed as the sum of two areas—the area of the bases and the area of the lateral surface. In Sec. 4-7 we described figures called prisms, cylinders, pyramids, and cones. In doing this we considered a pair of parallel planes and a set of lines called elements that intersected these planes. The portion of the cylinder (cone, or whatever) that is contained in the plane is called the base (or bases) of the figure. The area of these closed regions is called the **base area** of the figure. The area of one of the base regions is usually represented by B. The area of the part of the surface between the two planes is called the **lateral surface area.** (Remember that "lateral" means "side.")

Recalling the convention discussed in Sec. 4-7 concerning names for prisms, pyramids, cylinders, and cones, what would you call a prism whose base is a square region? A cylinder whose base is a circular region? A cone whose base is a circular region?

Those prisms and cylinders whose elements are perpendicular to the plane of the base are called **right prisms** or **right cylinders.** What would you call a cylinder whose base is a circular region and whose elements are perpendicular to the plane of the base? The measure of one of the perpen-

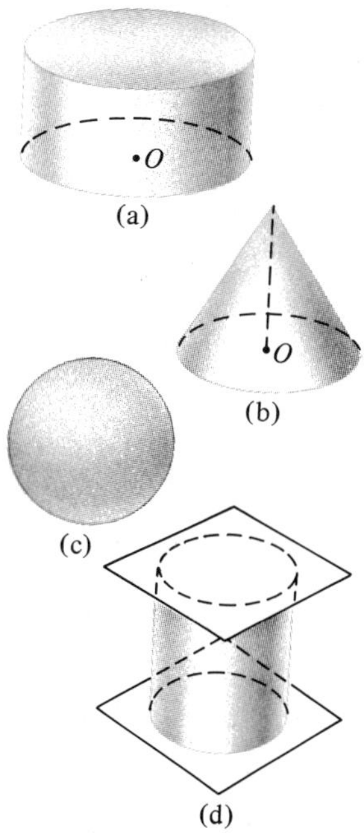

Figure 8-51

dicular elements is called the **altitude** or **height** *of the prism* or *cylinder*.

Circular cones and pyramids with regular polygonal regions as bases are called "right" or "regular" if the segment joining the vertex of the figure with the center of the base is perpendicular to the base. The measure of this segment (and often the segment itself) is called the **altitude** *of the cone* or *pyramid*.

For reference purposes we present a table of formulas for the surface area of some of the common closed three-dimensional surfaces. Those with an asterisk are discussed further in Exercise 8-14. The following symbolism is used: S, surface area; B, area of one base; r, radius; h, altitude of the three-dimensional figure; t, distance from vertex of the figure to edge of the base; p, perimeter of the base polygon.

Figure	*Formula*
Right prism*	$\mathbf{S = 2B + ph}$
"Right" pyramid* (regular polygonal region base)	$\mathbf{S = B + \frac{1}{2}pt}$
Right circular cylinder*	$\mathbf{S = 2\pi r^2 + 2\pi rh}$ or $\mathbf{S = 2\pi r(r + h)}$
Right circular cone*	$\mathbf{S = \pi r^2 + \pi tr}$ or $\mathbf{S = \pi r(r + t)}$
Sphere	$\mathbf{S = 4\pi r^2}$

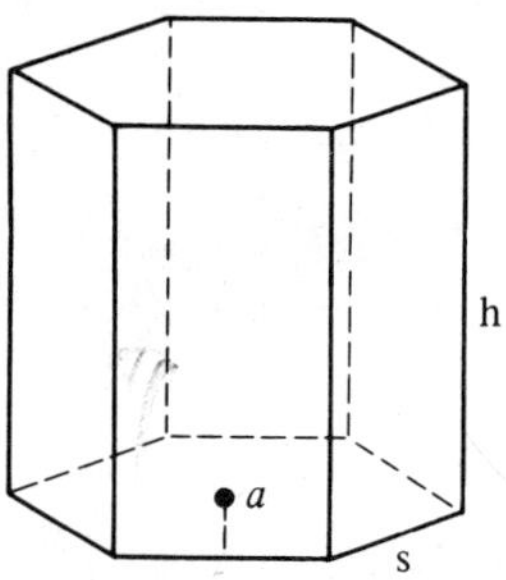

Figure 8-52

EXERCISE 8-14

1. Find the surface area of the right rectangular prism whose altitude measurement is 20 in. and whose base polygon has measurements of 10 in. and 5 in.

2. Find the following areas associated with the right prism shown in Fig. 8-52. The base is a regular polygonal region of n sides whose apothem has measure a and whose sides have measure s. The height of the prism is h.
   (a) Area of one base. (b) Sum of the areas of the bases.
   (c) Area of one lateral face. (d) Sum of the areas of the faces.
   (e) Surface area of the prism.
   In (e), if you let $B = \frac{1}{2}pa$ and $p = ns$, you will have the formula for the surface area of the right prism with a regular polygonal base.

3. Find the surface area of the square pyramid whose base has side measurements of 15 cm and for which the distance from the vertex to each edge of the base is 22 cm. (This distance is often called the **slant height** of the pyramid.)

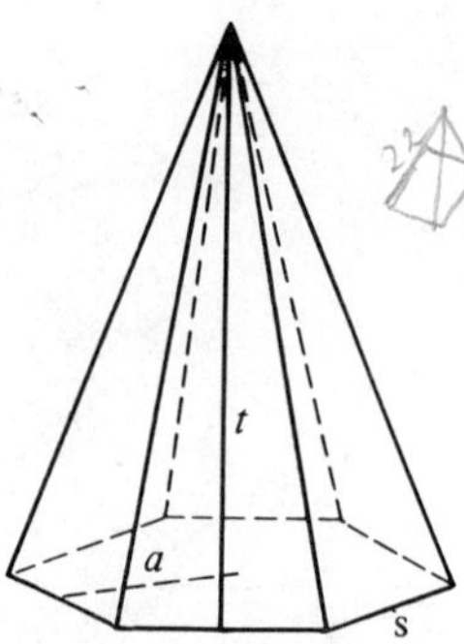

Figure 8-53

4. **A regular pyramid** is one in which the base polygon is regular and for which the segment joining the vertex of the pyramid and the center of the base is an altitude (Fig. 8-53). Find the following areas associated with a regular pyramid whose base polygon has n sides. The length of a side of the base is s; the apothem of the base polygon is a; the distance (slant height) from the vertex of the pyramid to an edge of the base is t.
   (a) Area of the base. (b) Area of one face.

(c) Sum of the areas of the faces. (d) Surface area of the pyramid.

In (d), if you let $B = \frac{1}{2}pa$ and $p = ns$, you will have the formula for the surface area of a regular pyramid.

5. (a) Take the label from a cylindrical soup or vegetable can. Make the cut in the paper perpendicular to the base of the can. Lay the paper out flat. What kind of polygonal region is formed? This suggests that the lateral area of this cylinder is equivalent to the area of this polygonal region.
(b) What is the height of this polygonal region? (Refer to Fig. 8-54.)
(c) What is the length of this polygonal region?
(d) Write a formula for finding the lateral area of a right circular cylinder if r is the radius of the base and h is the height of the cylinder.
(e) How do you find the area of the bases?
(f) Write a formula for the surface area of a right circular cylinder.

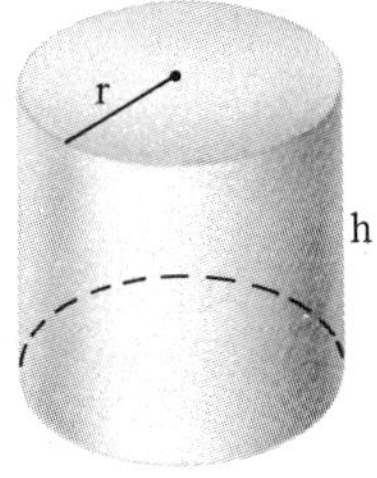

Figure 8-54

6. (a) Obtain a conical drinking cup. The cup corresponds to a right circular cone if it has a cover.
(b) How do you find the distance around the rim of the cup? Let r be the radius. [Refer to Fig. 8-55(a).]
(c) How do you find the area of the cover or base of the cone?
(d) Slit the cone along one element of the lateral surface and make the surface lie flat as in Fig. 8-55(b). Why is this a sector of a circle?
(e) If t is the measure of an element of the cone, what is the radius of the circle of which the lateral surface is a sector?
(f) Write a formula for the area of this sector.
(g) Write a formula for the surface area of the cone, where r is the radius of the base and t is the measure of an element of the lateral surface.

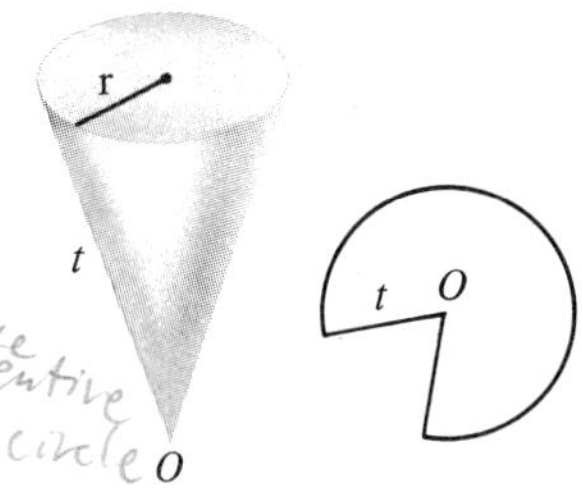

(a) (b)

Figure 8-55

8-15 AREA RATIOS

In the *Meno* of Plato we find Socrates quizzing a boy about how long the side of a square must be in order to yield an area double that of a square region whose side has measurement of 2 units. The boy says that you must double the side of the square. Is he correct?

The area of the square region whose sides have measurements of 2 units is 4. Double this area would be 8. The area of the square region whose sides have measurements that are the double of 2 units is 16. (The double of 2 is 4 and $4 \times 4 = 16$.) Apparently the boy is mistaken. Socrates' question is but a particular case of a more general principle concerning the relationship of the areas of two similar figures. You will investigate this in Exercise 8-15.

For the present, let us investigate the relationships of the areas of closed regions where we know the ratios of the measures of certain important segments in the figure. Suppose, first, that we have a square, as shown in Fig. 8-56(a). The measure of the side is s. Now suppose we have a second square such that the ratio of the length of the side of the first square to the

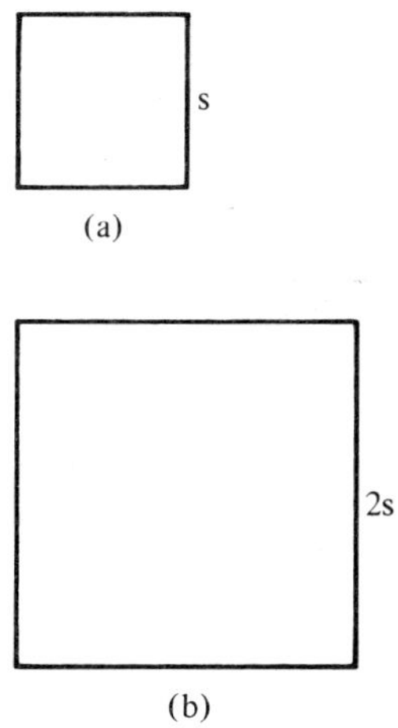

Figure 8-56

length of the side of the second square is 1: 2. What is the ratio of the area of the first square region to the area of the second square region? The length of each side of the second square in Fig. 8-56(b) must be $2s$. Thus, the area is $(2s)(2s)$ or $4s^2$. Since s^2 is the area of the first region, the ratio of the two areas is $s^2 : 4s^2$, or simply 1: 4.

For a second example, suppose we have a triangular region whose base has measure b and whose altitude has measure h; thus, its area is $\frac{1}{2}bh$. Now consider a second triangle with the following properties: the ratio of the base of the first triangle to the base of the second triangle is 1: 3; the ratio of the altitude of the first triangle to the altitude of the second triangle is 1: 2. What is the ratio of the area of the first triangular region to the area of the second triangular region? The base of the second triangle has measure $3b$ and the altitude has measure $2h$. Why? The area of the second region is $\frac{1}{2}(3b)(2h)$ or $\frac{1}{2}(6)bh$. The ratio of the two areas is $\frac{1}{2}bh : \frac{1}{2}(6)bh$, or simply 1: 6.

What is the ratio of the areas of two parallelogram regions if the ratios of the base of the first to the base of the second and of the altitude of the first to the altitude of the second are both $1 : n$?

EXERCISE 8-15

1. Complete the following table by finding the area of each rectangular region.

b	*h*	*Area*
10	0.5	
10	1	
10	5	
10	30	

If the base of a rectangular region remains constant, the area of the region (increases, stays the same, decreases) as the height increases.

2. Complete the following table by finding the height of each rectangle.

b	*h*	*Area*
0.5		72
1		72
3		72
8		72
12		72
72		72

If the area of a rectangular region remains constant, the height of the rectangle (increases, decreases, stays the same) as the base increases.

3. Complete the following tables. $\triangle ABC \sim \triangle XYZ$.

r.s.	$\triangle ABC$ b	$\triangle ABC$ h	$\triangle XYZ$ b'	$\triangle XYZ$ h'	*Area* $\triangle ABC$	*Area* $\triangle XYZ$	*Area ratio*
1 : 1	2	3					
1 : 2	2	3					
1 : 3	2	3					
1 : 4	2	3					
1 : n	2	3					

4. Complete the following table. $\triangle ABC \sim \triangle XYZ$.

r.s.	$\triangle ABC$ b	$\triangle ABC$ h	$\triangle XYZ$ b'	$\triangle XYZ$ h'	*Area* $\triangle ABC$	*Area* $\triangle XYZ$	*Area ratio*
2 : 3			3	9			
3 : 5			5	10			
5 : 7			14	21			
5 : 9			4.5	3.6			
m : n			b'	h'			

5. Read the section of the *Meno* from which the problem in the opening paragraph of this section was taken. See Ref. 39, pp. 38–43.

8-16 AREAS AND ERRORS OF MEASUREMENT

To close this chapter we reintroduce the problem of errors of measurement. You have undoubtedly noticed that all the area formulas involve products and/or sums in which the measures are those associated with certain linear measurements—base, height, radius. In the practical situation these measures are obtained by some sort of measuring process. If an instrument is used, design limitations introduce error, as discussed in Sec. 5-10. We have already seen that when linear measures are added, the greatest possible error in the sum is the sum of the greatest possible errors for each of the measures. In the area formulas we find products of measures. What is the nature of the error in these cases?

Figure 8-57 illustrates an analysis of errors of linear measurement and their relation to the area of a rectangular region. We think you will be surprised at the magnitude of the possible error.

The measurements of a rectangular house lot were made precise to the nearest foot. The base measurement is 95 ft and the altitude measurement is 75 ft. Using these measurements, the area measurement would be 7125 sq ft.

Since the precision of the measurement was 1 ft, the greatest possible error in the base or altitude is 0.5 ft. If both measurements are overestimated by 0.5 ft, then the true measurements for the base and altitude are 94.5 ft and 74.5 ft, respectively. Thus, the true measurement of the area would be 7040.25 sq ft. This differs from 7125 by 84.75. This error is certainly more than the sum of the greatest possible errors in the two dimensions of the rectangle.

If both measurements are underestimated by 0.5 ft, then the true measurements for the base and altitude are 95.5 ft and 75.5 ft, respectively. In this case the true area would be 7210.25. This differs from 7125 by 85.25.

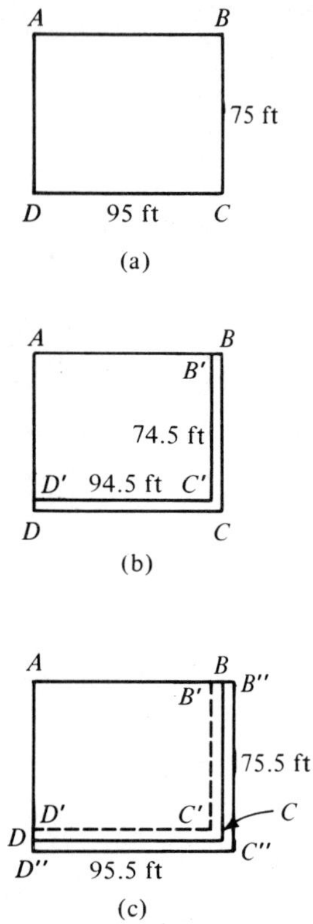

Figure 8-57

Thus, with the measurements and precision given, the true area of the rectangular plot falls within certain limits. If A represents the area, the interval of tolerance can be expressed as: $7040.25 \leq A \leq 7210.25$.

The value obtained by multiplying the given base and height falls somewhat near the midpoint of this interval, providing a fairly good estimate of the area. In practice, this value is often rounded so that the reported area measurement contains the same number of significant digits as the least accurate of the linear measurements. Thus, 7125 sq ft would be rounded to 7100 sq ft.

EXERCISE 8-16

1. Find the interval of tolerance for the areas of each of the following regions. Report the area measurement as suggested in the last paragraph of Sec. 8-16.
   (a) Rectangle: b, 15 in.; h, 24 in. (b) Triangle: b, 5 cm; h, 12 cm.
2. Express the answers to the *Examples* in this chapter using the convention on significant digits given in the last paragraph of Sec. 8-16.

SUMMARY OF PRINCIPLES

Every closed region in a plane has a unique area.

All congruent regions in a plane have equal areas.

One closed region has a greater area than a second closed region if it contains the second region or a closed region equal in measure to the second region as a proper subset.

Area measurement is additive.

Area of the region is determined as follows:

rectangle	$A = bh$
triangle	$A = \frac{1}{2}bh$

parallelogram $A = bh$
trapezoid $A = \frac{1}{2}h(b_1 + b_2)$
regular n-gon $A = \frac{1}{2}ap$

If the ratio of similitude of two plane regions is $m:n$, then the ratio of the areas of the two regions is $m^2:n^2$.

EXERCISES

1. Encircle A, S, or N (always, sometimes, or never true) for each statement below.
   (a) All congruent regions have the same area. A S N
   (b) Regions with the same area are congruent. A S N
   (c) Units for area measurement are square regions. A S N
   (d) A rectangular region whose base is 10 and height is 5 has a greater area than a parallelogram region whose base is 10 and whose height is 5. A S N
   (e) The ratio of the areas of two plane regions is the same as the ratio of similitude for the two figures. A S N
2. Find the area of the region determined by each of the following:
   (a) Rectangle: b, 10″; h, 25″.
   (b) Rectangle: b, 3 ft; h, 14 in.
   (c) Triangle: b, 10 m; h, 3.5 m.
   (d) Parallelogram: b, $2\frac{1}{2}$ yd; h, $\frac{3}{4}$ yd.
   (e) Trapezoid: b_1, 20 in.; b_2, 6 in.; h, $\frac{1}{2}$ in.
   (f) Regular octagon: s, 2′; a, 2.4′.
   (g) Circle: r, 35 cm.
   (h) Annulus: r_2, 17 m; r_1, 20 cm.
   (i) Sector: r, 13.4 ft; central angle, 60°.
3. Find the following surface areas:
   (a) Right hexagonal prism: h, 10 m; s, 2 m; a, 1.7 m.
   (b) Right circular cylinder: h, 5 cm; r, 0.5 cm.
   (c) Regular square pyramid: t (slant height), 150′; s, 100′.
   (d) Right circular cone: t, 0.003 cm; r, 0.001 cm.
4. For suggestions on possible activities related to teaching area concepts in the elementary schools, read Ref. 56, pp. 233–243.
5. We have used the square region as the basic unit for area measurement. The square is a regular polygon of four sides.
   (a) What other regular polygonal regions can be used to "cover the plane without overlapping and without leaving empty spaces"?
   (b) For an interesting article relating the problem of tiling (tessellation) to the concept of area, read Ref. 110, pp. 114–117.
6. In algebra the distributive property is used to show that $(a + b)^2 = a^2 + 2ab + b^2$. Read Ref. 75, pp. 40–44, to see how this algebraic sentence is related to the concept of area.

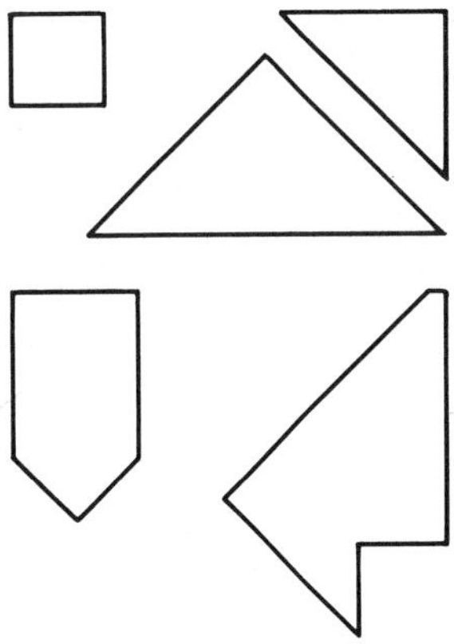

Figure 8-58

7. Find the surface area of several of the polyhedrons whose patterns are found in Appendix C.

8. Cut out copies of the regions shown in Fig. 8-58. Arrange the pieces to form a square region.

9. It is possible to find approximations of the areas of some closed regions for which we have no formulas.
   (a) In Fig. 8-59, describe how you might approximate the area of closed region *ABCDEF* if each of the "slender" regions is rectangular.
   (b) In Fig. 8-60, describe how you might find the area of the closed region *WXYZ* if each of the "slender" regions is trapezoidal.

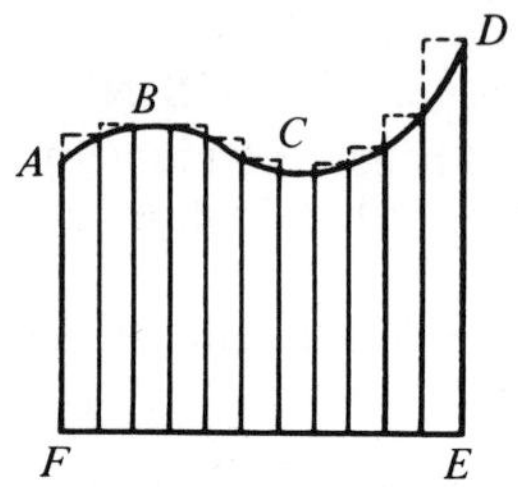

Figure 8-59

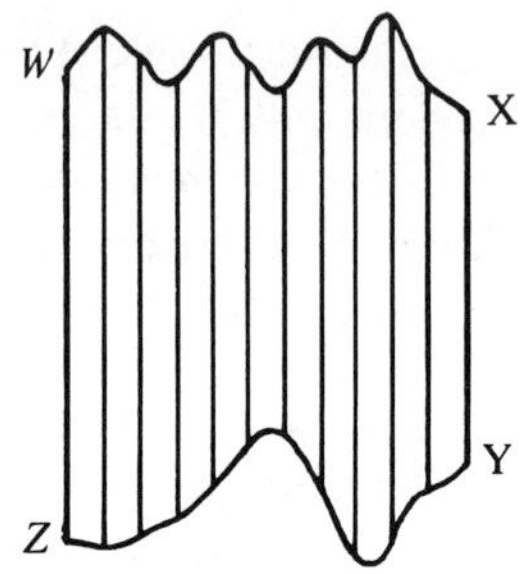

Figure 8-60

10. Find the ratio of the areas of two similar rectangular regions if the ratio of similitude is 2:5.

11. Find the ratio of similitude of two similar triangular regions if the ratio of their areas is 25:49.

12. (a) Draw a large circle on heavy paper.
   (b) Partition the circular region into 16 congruent sectors. (Use a protractor to draw the central angles.)
   (c) Cut out the circular region and each of the sectors.
   (d) Arrange the 16 sectors as in Fig. 8-61.

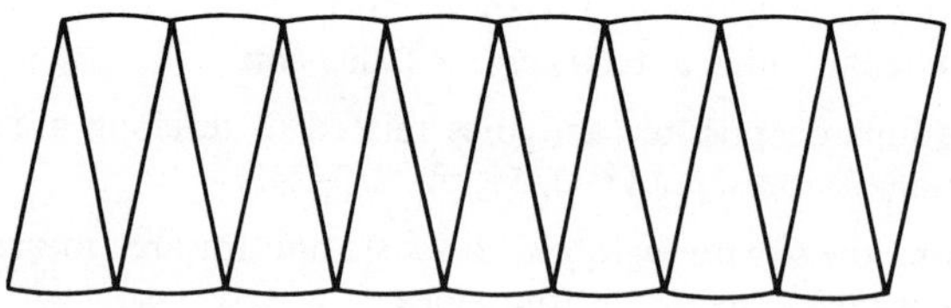

Figure 8-61

   (e) What polygonal region would be formed or suggested if the number of such sectors was greatly increased?
   (f) If *r* is the radius of the circle, what would the height of the polygon be?
   (g) If *r* is the radius of the circle, what would the length of the base of the polygon be? [*Hint:* Semicircle.]

(h) What is the area of the polygonal region? Compare this formula with the formula given in the text for the area of a circular region.

13. Draw a picture of a square, $WXYZ$, each of whose sides has measurement of 8 in.

(a) Draw $\overline{AD}$, $\overline{CB}$, and $\overline{BZ}$ such that the measurements are as in Fig. 8-62(a). What is the area of region $WXYZ$?

(b) Cut out the square region and cut along $\overline{AD}$, $\overline{CB}$, and $\overline{BZ}$. Rearrange the four parts to form the rectangle in Fig. 8-62(b). What is the area of the rectangular region?

(c) Compare the answers for (a) and (b). The additive property of area should imply that the two areas are identical. Explain the obtained results.

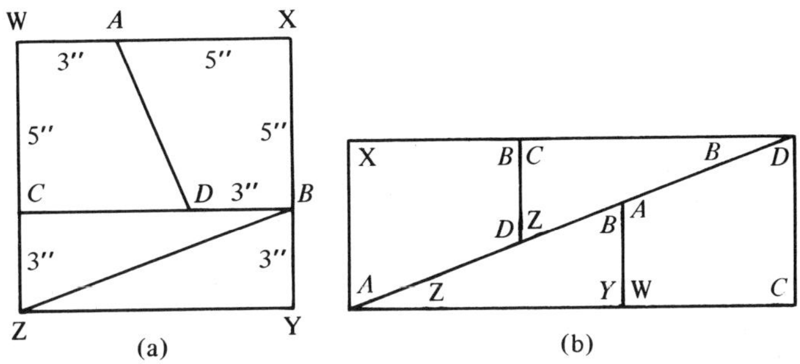

Figure 8-62

CHAPTER 9

Volume

9-1 INTRODUCTION

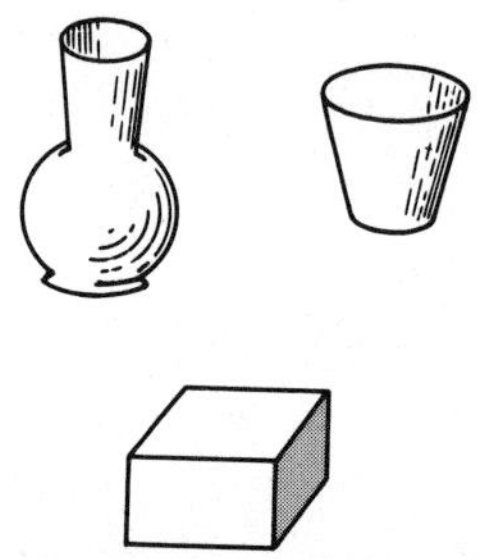

Figure 9-1

In this chapter we extend the concept of measurement to **volume**—the number associated with the size of a region in space. We find volume when we want to determine the capacity of a container or the amount of space occupied by a given figure. (See Fig. 9-1.)

Since the development of ideas about volume so closely parallels that for area, we may be more brief in our commentary. Essentially, we shall investigate the nature of volume measurement, some formulas for finding volumes of common solids, ratios of volumes of certain solids, and the question of errors of measurement as it relates to volume.

9-2 COMPARISON OF VOLUMES

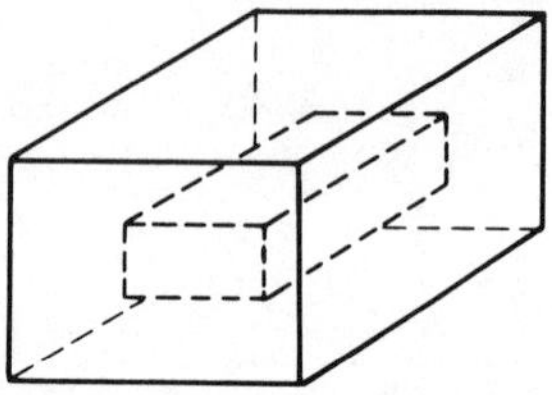

Figure 9-2

What fundamental notions do we need to understand in order to make comparisons of volumes? For example, what do we mean when we say that one space region has greater volume than another? We could start by noting that one solid has a greater volume than a second solid if the second solid is a proper subset of the first. Physically, this would mean that one solid is "inside" a second solid. (See Fig. 9-2.)

If one solid is not already in the interior of a second solid, we need to assume that there are exact copies of the solids in such positions that a direct comparison can be made. That is, there are space regions congruent to other space regions. Again, we are using the assumption that congruent figures have equal measures—in this case, congruent solids have equal volumes.

Finally, since one region may not be a proper subset of a second region and have no regions congruent to it that are proper subsets of the second region, we may need to subdivide the given solid. Here we seem to be using the property that if a solid is partitioned, then the volume of the original

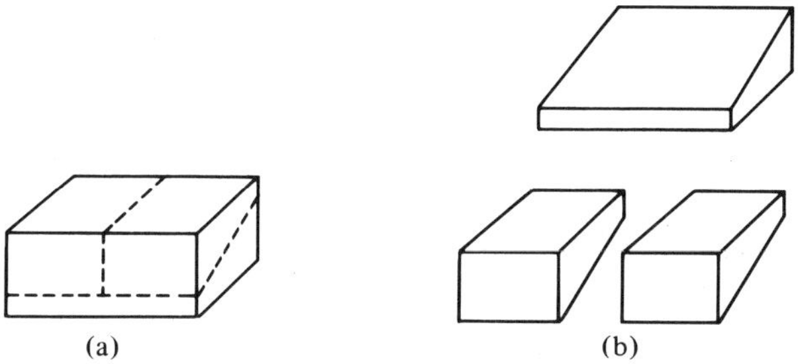

Figure 9-3

solid is equal to the sum of the volumes of each solid in the partition. Part (a) of Fig. 9-3 shows a partition of a solid. Part (b) shows the separate parts of the partition.

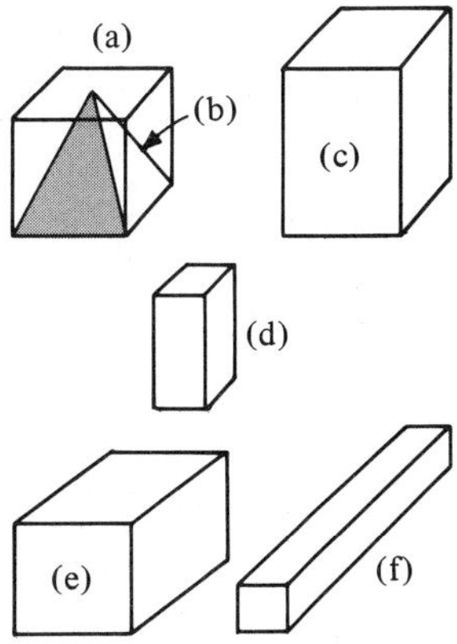

Figure 9-4

EXERCISE 9-2

1. Referring to Fig. 9-4, explain why (a) has greater volume than (b); why (c) has greater volume than (d); why (e) has greater volume than (f).
2. Suppose you have two containers and a quantity of sand. How could you determine which container has the greater volume?
3. Obviously, it is physically impossible to put a solid block of wood inside a second solid block of wood. Yet, in the text we speak of a solid "inside" a solid. What are we referring to?

9-3 VOLUME AND COUNTING

In order to answer the question, "How much greater volume?" we need a way to assign numbers to the size of space regions. We can begin by counting the number of congruent solids it takes to "fill" a given solid. For example, we could try to fill two containers with marbles. (See Fig.

9-5.) The container that held the greater number of marbles would have the greater volume. (Should all the marbles be of the same size?) What problems do you encounter when trying to fill containers with marbles? What geometric figure corresponds to a marble?

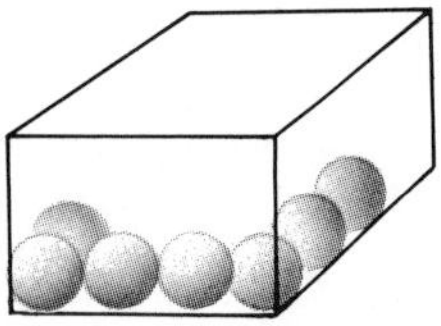

Figure 9-5

If the container is a box, it would probably be more convenient to use other boxes as the counting unit (Fig. 9-6). They fit into the corners better than do spheres. What problems do you encounter using boxes as units in a circular cylindrical solid? Again, we encounter situations in which it is desirable to use smaller units or to subdivide the given unit. Like linear and area measurement, volume measurement deals with a continuous set, so that we need to have the real-number system at our disposal if every solid is to have a unique volume and if congruent regions are to have equal volume.

EXERCISE 9-3

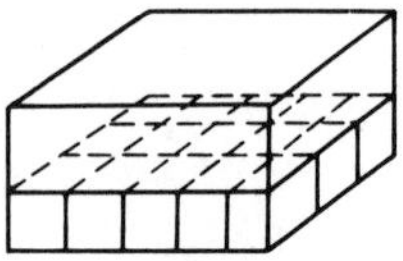

Figure 9-6

1. Obtain a box of lump sugar. What is the volume of the box if the unit of volume is one lump of sugar?
2. Using a soup can as the unit of volume, find the volume of a cardboard carton.
3. Find the volume of a handkerchief box using a bar of soap as the unit of volume. Make a new unit of volume by cutting the bar of soap into four congruent pieces. What is the volume of the same box using one of the pieces as the unit?
4. (a) Congruent space regions have the same volume. A S N
   (b) Space regions with the same volume are congruent. A S N

9-4 BASE AND ALTITUDE—REVIEW

We need to recall what is meant by the base and altitude of various solids. For any cylinder or prism, the bases of the surface are the closed regions determined by the two parallel planes mentioned in the definition of the figure. (See Sec. 4-7.) The altitude of the figure is the distance between the two parallel bases; and it is the length of a segment that has an end-point in each base and is perpendicular to one of the bases. Figure 9-7 shows several cylindrical and prismatic solids with equal altitudes.

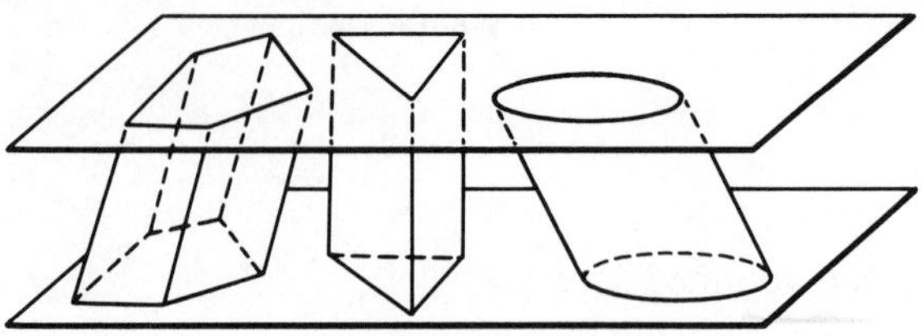

Figure 9-7

The base of a cone or pyramid is the closed plane region determined by the defining plane. The height or altitude of the cone or pyramid is the distance from the vertex of the figure to the plane of the base. (See Sec. 8-14 and Fig. 9-8.)

Further, those cylinders and prisms whose elements are perpendicular to the defining plane are called **right cylinders** or **right prisms**. What is the relationship between the length of a lateral edge of a right prism and the altitude of the prism? (See Figs. 9-9 and 9-10.)

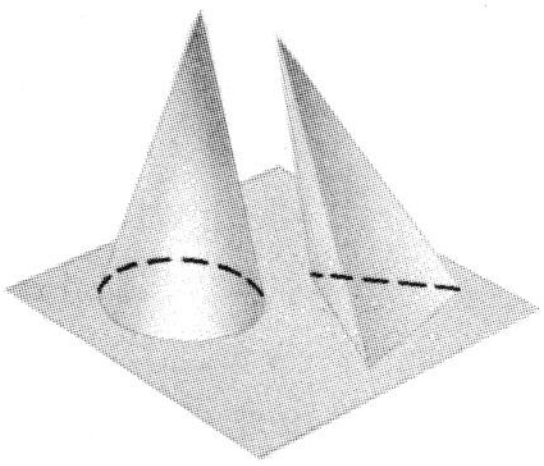

Figure 9-8

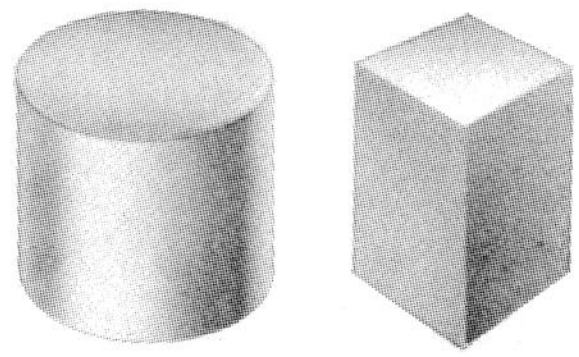

Figures 9-9 and 9-10

A pyramid with a regular polygonal base is called **regular** and a circular cone is called **right** if the altitude segment drawn from the vertex to the base intersects the base at the center of the base polygon or the center of the circle (Fig. 9-11).

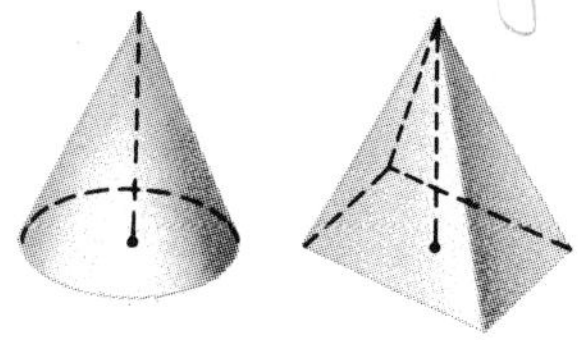

Figure 9-11

EXERCISE 9-4

1. Describe and draw a sketch of a right rectangular prism.

9-5 STANDARD UNITS

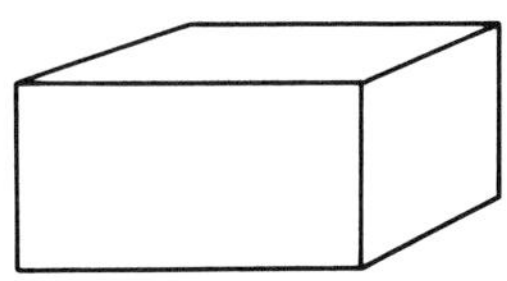

Figure 9-12

The basis of much work in volume measurement is the solid called a right rectangular prismatic solid. This is also known as a solid rectangular parallelepiped or a **rectangular solid**. (See Fig. 9-12.) In determining the volume of this solid, the unit of volume measurement that seems to work best is the cubical solid. A **cube** is a right rectangular prism in which any two adjacent edges have the same linear measure. Figure 9-13 shows a cube, each of whose edges has measurement of 1 in. Obviously, cubical solids can be used to "fill up" rectangular solids very nicely.

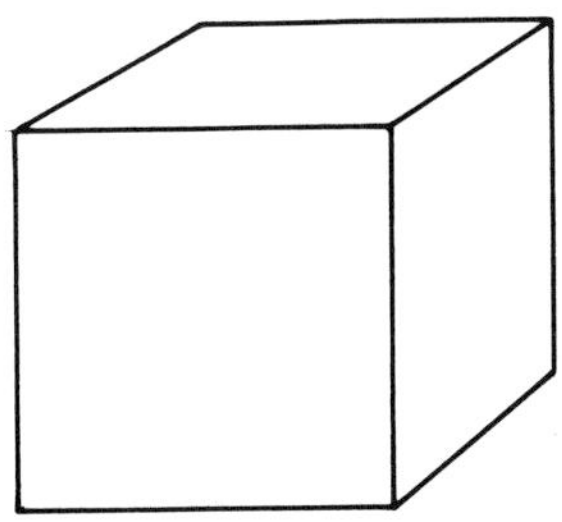

Figure 9-13

The standard units of volume measurement are cubical solids, each of whose edges measures one linear unit. The volume measurement of these cubical solids is said to be one cubic unit. In the English system the common standard units of volume measurement are the cubic inch (volume equal to the volume of a cube each of whose edges has measurement of

1 in.), the cubic foot, and the cubic yard. The standard abbreviations for "one cubic inch" are "1 cu in." and "1 in.³" What do you think the standard abbreviations are for "one cubic foot" and "one cubic yard"?

In the metric system are units of volume measurement called the cubic meter, the cubic decimeter (also commonly called the liter), the cubic centimeter (the familiar "cc" of the doctor's needle), and the cubic millimeter. (See Fig. 9-14.)

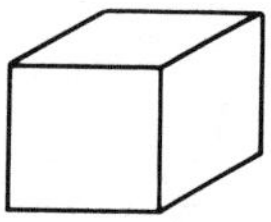

Figure 9-14

Notice that all of the units above are based on a corresponding standard unit of linear measurement. The reason will be obvious if you stop to think how many "one-inch cubes" or blocks it will take to fill a rectangular solid whose base dimensions are 4 in. and 5 in. and whose altitude measurement is 3 in. How many 1-in. blocks fit along the bottom edge whose measure is 4? (See Fig. 9-15.) How many rows of the blocks fit on the base? This means that twenty blocks cover the base of the solid. (Did you notice that this is the same number as the area of the base in square inches?) Now, how many layers of blocks will fit in the solid? Obviously, 3. How many blocks in all? $4 \times 5 \times 3$ or 60 blocks. So the volume measurement of the rectangular solid is 60 cu in. Perhaps you now see why cubical solids whose edges have a measurement of one unit are convenient units.

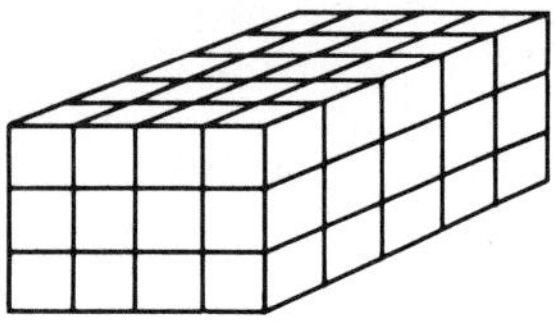

Figure 9-15

Because of the importance of volume in determining amounts of fluids, special standard units are used for gas and liquid volume measurement. You will often find these referred to as capacity measurements. (What other methods might be used to measure amounts of liquids?) In the English system the common units of fluid measurement are the gallon, quart, pint, and ounce. In the metric system the basic capacity unit is the liter. The liter was originally defined as the volume of 1000 grams of water at its maximum density. This turned out to be 1000 cc, since 1 cc of water weighs 1 gram under these conditions. (Since then, the liter has been redefined in terms of the kilogram using a metal standard, so that now 1 liter is equivalent to 1000.028 cc. But for most practical purposes it is convenient to assume that 1 liter is equivalent to 1000 cc.) The root word "liter" is combined with the familiar prefixes of the metric system to form the names of the other units, such as kiloliter, hectoliter, deciliter, centiliter, and milliliter (whose abbreviation is the fairly well-known "ml").

For comparison purposes, the liter is 1.0567 liquid quarts.

EXERCISE 9-5

1. What units other than those suggested in the text are used to measure volume? [*Hint:* What are the common cooking units?]

2. (a) How many cubic inches are in a quart?
   (b) In part (a) does it make a difference if you have a "liquid" quart or a "dry" quart?
   (c) How many ounces in a pint? In a quart? In a gallon?
3. How many cubic centimeters are in a liter?
4. Make a model of a 1-in. cube.
5. Make a model of a cube whose volume is $\frac{1}{8}$ that of a 1-in. cube. What are the measurements of the edges of this cube?
6. What is the area of one face of a 1-in. cube?
7. Using 1-in. cubical blocks, make models of cubes whose edges have the following measures: 1, 2, 3, 4, 5, 6.

9-6 STRATEGY

In devising formulas for finding the number associated with the volumes of various solids, we shall make an assumption about finding the volume of one basic solid. This is similar to the assumption we made about finding the area of a rectangular region and its relation to areas of other plane regions. We will then want to show how formulas for other solids are related to the basic one. To do this we shall need to make a second assumption—known as Cavalieri's principle—which tells how the volumes of certain solids are related to each other. We shall then show, quite informally, how the volumes of the solids determined by a prism, a pyramid, a cylinder, a cone, and a sphere are related to the volume of the basic figure—the right rectangular prismatic solid.

9-7 VOLUME OF A RIGHT RECTANGULAR PRISMATIC SOLID

The rectangular solid is taken as the basic solid in volume analysis because of its frequent occurrence in practical situations and because of the simplicity of determining its volume when the unit of volume measurement is a cubical solid.

We shall make the assumption that the volume of a rectangular solid is given by the product of the area of one of its bases and its height. If V represents volume, B the area of the base, and h the altitude, we have the formula:

Area of base (l×w)

$$\mathbf{V = Bh.}$$

We are willing to accept this assumption because it conforms to our experiences with counting solid cubical blocks, and it helps us devise formulas for the volume of other solids.

Consider the rectangular solid whose base has linear measurements of 3 in. and 2 in. and whose height measurement is 4 in. How many 1-in. square regions "fit" on the base? Is this not the area of the base? What is the area of one face of a cube, each of whose edges has measurement of 1 in.? Then, how many such cubical solids fit on the base? How many layers of cubical solids are needed to fill the solid? Thus, the volume measurement of the solid is 24 cu in. Do you now see why the volume is given by the product of the base area and the altitude?

EXERCISE 9-7

1. The volume of a rectangular solid is often obtained by using the formula $V = \ell wh$. What do ℓ, w, and h represent? What is ℓw?
2. Find the volume measurement of each of the rectangular solids whose measurements are given below:

Solid	ℓ	w	h	*Volume*
(a)	5′	7′	10′	
(b)	3 cm	8 cm	9 cm	
(c)	11 in.	11 in.	11 in.	
(d)	2 yd	4 ft	6 in.	
(e)	3.5 m	0.8 m	20.1 m	
(f)	$7\frac{1}{2}''$	$\frac{1}{4}''$	$3\frac{1}{2}''$	

3. Find the volume of a rectangular solid whose measurements are 4 ft, 8 ft, and 9 ft if the unit of volume is a rectangular solid whose measurements are 1 ft, 2 ft, and 3 ft.
4. If e represents the measures of one edge of a cube, write a formula for finding the volume of a cubical solid in terms of e.

9-8 CAVALIERI'S PRINCIPLE

Having the formula for the volume of a rectangular solid, we now seek formulas or techniques for finding the volume of other solids. There are many interesting intuitive approaches to this problem, some of which we consider in later sections of this chapter. More formal analysis relies on a relationship known as the principle of Cavalieri. (Cavalieri was a seventeenth-century Italian mathematician.) In this text we simply state the essence of the principle. In Exercise 9-8 you will find some experiments that further illuminate the implications of this principle.

Cavalieri's principle: Given two solids with equal altitudes, as in Fig. 9-16, if the cross sections at equal heights from the base have equal areas, then the two solids have equal volumes.

EXERCISE 9-8

1. Complete each of the following:
   (a) Every cross section of a circular cylindrical solid is a ______ ______ congruent to the base of the solid.
   (b) Every cross section of a prismatic solid is a ______ ______ congruent to the ______ of the solid.
   (c) The cross sections of a circular conical solid are either ______ ______ or a ______.
   (d) The cross sections of a pyramidal solid are either ______ ______ or a ______.
2. (a) Make 20 cardboard copies of a rectangular region whose measurements are 1 in. and $1\frac{1}{2}$ in.
   (b) Make two stacks of 10 cards each. Each stack should be a model of a right rectangular prismatic solid. What is the relationship between the base areas of each of the prisms? What is the relationship between the altitudes of each of the prisms? What is the relationship between the volumes of each of the prismatic solids?
   (c) Each card in the stack represents a cross section of the prism. What is the relationship among the areas of the carboard sections?
   (d) Make the two stacks look like the picture in Fig. 9-16.
   (e) Has the altitude of either solid changed? Has the area of any cross section changed? Has the volume of either solid changed?
3. (a) Get 20 washers all with the same diameter and thickness.
   (b) Make models of two cylindrical solids by stacking the washers in two piles of 10 each. Each stack should have the same height and the same volume as the other, but a different shape.
   (c) If each washer represents a cross section, what relationship exists between areas of cross sections at equal heights from the bases?
4. (a) Make two copies each, using cardboard, of square regions with the following side measurements in inches: $\frac{1}{4}$, $\frac{1}{2}$, $\frac{3}{4}$, 1, $1\frac{1}{4}$, $1\frac{1}{2}$, $1\frac{3}{4}$, 2.
   (b) Make two models of regular pyramidal solids by stacking the cardboard regions as in Fig. 9-17(a).
   (c) How do the volumes of the two pyramidal solids compare?
   (d) Make the two stacks look like the picture in Fig. 9-17(b).
   (e) Do the two solids have the same altitude? Do the cross sections at equal altitudes from the bases have equal areas?
   (f) Do the pyramidal solids have the same volume?
5. Repeat the steps of problem 4 using circular regions whose diameters have the measurements given.

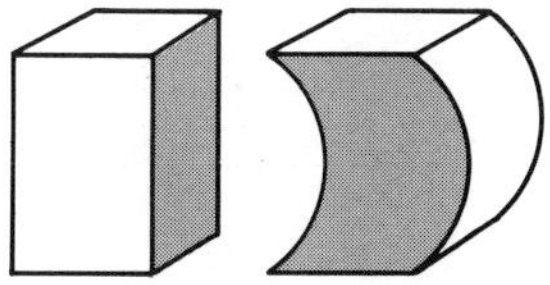

Figure 9-16

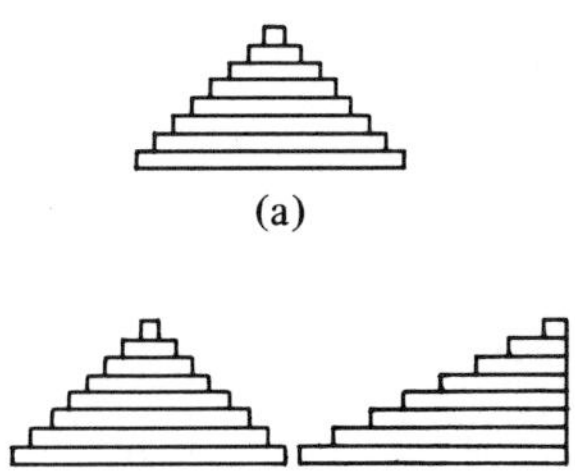

(a)

(b)

Figure 9-17

6. In problems 2-5 above we have used pairs of solids whose corresponding cross sections have the same shape as well as the same area. Does it seem reasonable that the equal-volume relationship holds if the cross section shapes differ but the areas remain equal?

9-9 VOLUME OF A PRISMATIC SOLID

The relationship between the volume of any prismatic solid and the volume of a rectangular solid is not difficult to accept. The formula is the same:

$$\mathbf{V = Bh.}$$

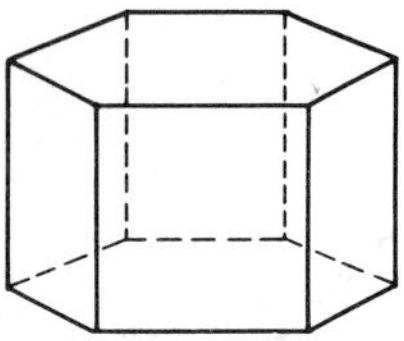

Figure 9-18

Using Fig. 9-18 as an example, it seems apparent that cubical solids will not fit on the base without being subdivided, but it also seems fairly obvious that the area of the base region still indicates the number of cubical solids needed to cover the bottom even if they are subdivided. The height of the prism tells the number of layers required. The product of the base area and the altitude gives the volume.

Figure 9-18 shows a right prismatic solid. Now, think of the stack of cards pictured in Fig. 9-19(a) as representing that right prismatic solid. Part (b) of the figure represents the stack of cards pushed to one side. The stack is pushed without tilting the bases, giving a "leaning-tower" effect. The height of the prism in (a) is the same as the height of the prism in (b). Any card in the stacks represents a cross section of the solid. All cross section areas are equal. Cavalieri's principle assures us that the volume of the "leaning" solid is the same as the volume of the right solid. (Note that the leaning solid and the right solid have the same base area and the same altitude.) Then, for any solid prism the formula for finding the volume of the solid is

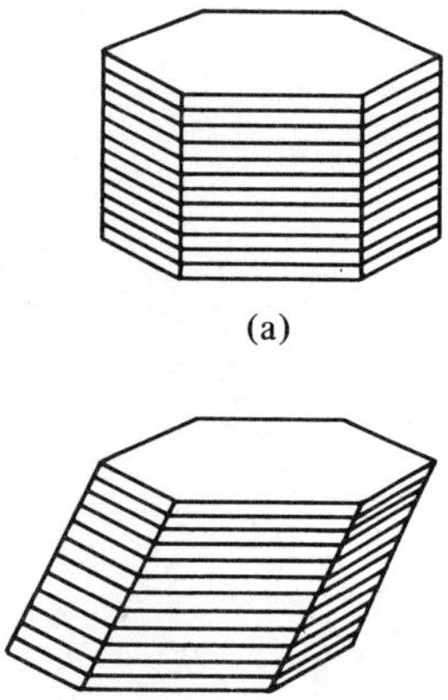

(a)

(b)

Figure 9-19

$$\mathbf{V = Bh,}$$

where B is the area of the base and h is the altitude of the prism.

EXERCISE 9-9

1. Find the volume of each of the following prismatic solids:
   (a) A right regular hexagonal prism whose altitude is 10 and whose base has apothem of 17.3 and side of 10.
   (b) A triangular prism whose altitude is 3 and whose base has area of 7.
   (c) A square prism whose altitude is 13, whose lateral edges have measures of 15, and whose base edges measure 3.

2. Make a cardboard model of a right rectangular prism whose altitude measurement is 3 in. and whose base has measurements of 1 in. and 2 in. Make a cardboard model of a right triangular prism whose altitude measurement is 3 in. and whose base is a right triangle with leg measurements of 1 in. and 2 in.
   (a) What should be true of the volumes of the two prismatic solids?
   (b) Fill the triangular prism with sand and pour it into the rectangular prism. Is the belief about the relationship of the two volumes confirmed in experience?

9-10 VOLUME OF A PYRAMIDAL SOLID

Figure 9-20

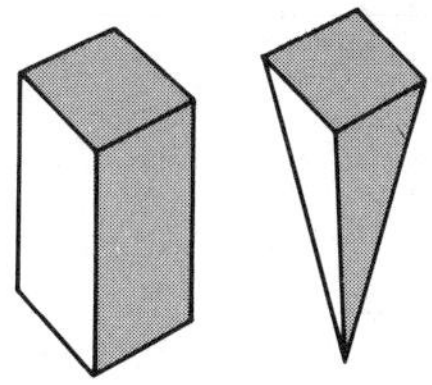

Figure 9-21

With our intuitive understanding of Cavalieri's principle, it seems evident that two pyramidal solids with congruent bases and equal altitudes have the same volume. But how does the volume of a pyramidal solid relate to the volume of a rectangular solid? (See Fig. 9-20.) To find out, perform the following experiment. In Appendix C you will find patterns for models of a right square prism and a square pyramid that have congruent bases and altitude segments. (See Fig. 9-21.) Make each model without fastening the "covers." Fill the pyramid with some granular substance (such as sand, salt, sugar) and then pour this into the prism. Continue until the prism is full. How many times did you have to fill the pyramid in order to fill the prism?

This simple experiment provides an intuitive appreciation of the relation between the volume of a pyramidal solid and the volume of a prismatic solid. Given any pyramidal solid, its volume is equal to one third the volume of a prismatic solid whose base and altitude are congruent to the base and altitude of the pyramid. Thus, we have

$$\mathbf{V} = \tfrac{1}{3}\mathbf{Bh}.$$

More formal discussions of the derivation of this formula are a bit more complicated and often use the Cavalieri principle in the proof. (The interested reader may wish to consult Ref. 22, pp. 154–155, for a derivation of the formula for the volume of a triangular pyramidal solid from the formula for the volume of a triangular prismatic solid.)

EXERCISE 9-10

1. Find the volume of a pyramidal solid whose altitude is 15.2 and whose base area is 75.1.

2. Find the volume of a pyramidal solid whose altitude is $10\frac{1}{2}$ and whose base rectangle has sides of $3\frac{1}{2}$ and $4\frac{1}{2}$.

3. Make models of a right hexagonal prism and a regular hexagonal pyramid with congruent bases and altitudes. Illustrate the relationship between the volumes of the two solids determined by the models by filling the pyramid with sand and pouring it into the prism.

4. Figure 9-22 shows the intersection of a plane and a pyramidal solid. The intersection is a cross section. What is the relationship of the plane and the base? That part of the pyramid between and including the polygonal regions in the base and intersecting plane is called a **frustum** of the pyramid. Find the volume of the solid determined by the frustum in the figure if the measurements are as indicated.

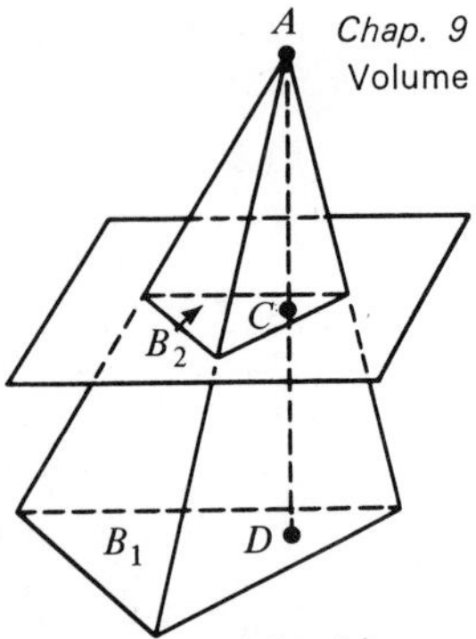

AC = 5
AD = 10
B_1 : 4 sq units
B_2 : 1 sq unit

Figure 9-22

5. Figure 9-23(a) shows a triangular pyramid *A-DEF*. Figure 9-23(b) shows a triangular prism related to the pyramid. In this prism find triangular pyramids *A-DEF*, *F-ABC*, and *F-ABE*. These three pyramids partition the prismatic solid into three pyramidal solids. Why do the solids *A-DEF* and *F-ABC* have the same volume? Notice that *A-DEF* and *F-EDA* are the same solids. Why do solids *F-ABE* and *F-EDA* have the same volume? What then is true about the volumes of the three solids *A-DEF*, *F-ABC*, and *F-ABE*? Then, the volume of each must be one third the volume of the prismatic solid.

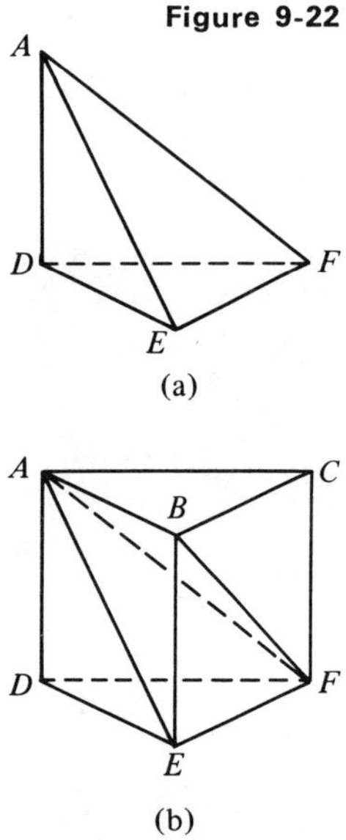

Figure 9-23

9-11 VOLUME OF A CYLINDRICAL SOLID

Once we accept the formula for the volume of a prismatic solid, the formula for the volume of a cylindrical solid seems to be highly related to it, since every prism is a cylinder. For the special case of the circular cylindrical solid (Fig. 9-24), B in the formula for the volume of a prismatic solid is simply the area of the circular base of the cylinder. That is, we can determine the number of blocks that fit on the bottom layer of the solid by finding the area of the base. Again, the height of the cylinder tells us the number of layers. If B is the area of the base, then $B = \pi r^2$, where r is the radius of the base. Thus, the formula for the volume of a circular cylindrical solid becomes

$$\mathbf{V = \pi r^2 h},$$

where r is the radius of the base and h is the height of the cylinder.

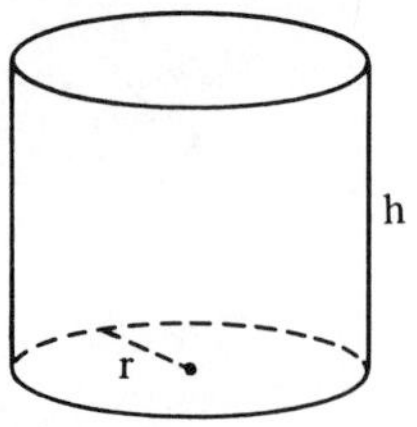

Figure 9-24

EXERCISE 9-11

1. Find the volume of a cylindrical solid whose radius is 170 and whose height is 30.

2. Find the volume measurement of a cylindrical solid whose radius measurement is 3 m, whose height measurement is 10 m, and in which the measurement of an element of the surface is 15 m.
3. Design a rectangular prism whose altitude is the same as the altitude of an orange juice can and whose base area is equal to the base area of the can. What should be the relationship between the capacities of the two containers? Check this by filling one with sand and pouring it into the other.

9-12 VOLUME OF A CONICAL SOLID

The relationship between the volume of a circular conical solid and the volume of a circular cylindrical solid with base and altitude congruent to those of the conical solid is the same as that between the pyramidal solid and the prismatic solid. For the conical solid, then

$$\mathbf{V} = \tfrac{1}{3}\pi \mathbf{r}^2\mathbf{h},$$

where r is the radius of the base and h is the altitude.

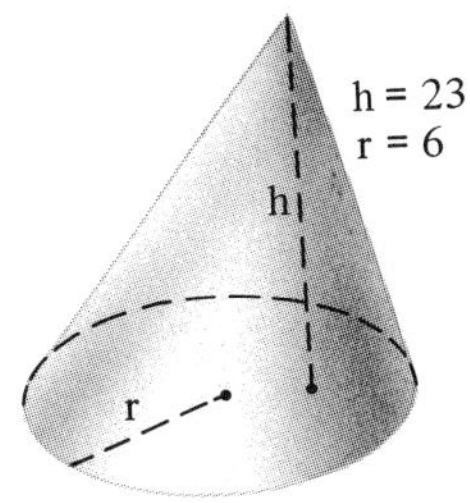

Figure 9-25

EXERCISE 9-12

1. Find the volume of the conical solid shown in Fig. 9-25.
2. Describe a frustum of a cone and a way to compute the volume of the solid determined by the frustum.

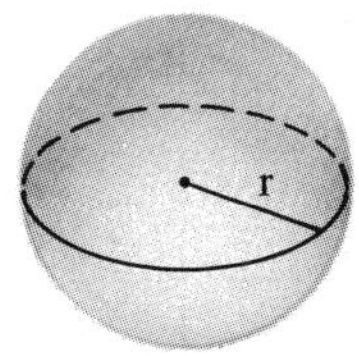

Figure 9-26

9-13 VOLUME OF A SPHERICAL SOLID

The volume of a spherical solid (Fig. 9-26) can be found using the formula:

$$\mathbf{V} = \tfrac{4}{3}\pi \mathbf{r}^3,$$

where r is the radius of the sphere.

There is an interesting relationship between the volumes of the solids determined by a cylinder, a cone, and a hemisphere. (A hemisphere is one of the portions of a sphere determined by a plane that contains the center of the sphere.) As in Fig. 9-27, suppose that you have a hemisphere with radius r and a cone and a cylinder each of whose heights is r and whose bases have radius r. The volumes of the related solids are:

cylinder	$V_{cyl} = \pi r^2 \cdot r$	or πr^3,
cone	$V_{co} = \frac{1}{3}\pi r^3$,	
hemisphere	$V_h = \frac{1}{2}(\frac{4}{3}\pi r^3)$	or $\frac{2}{3}\pi r^3$.

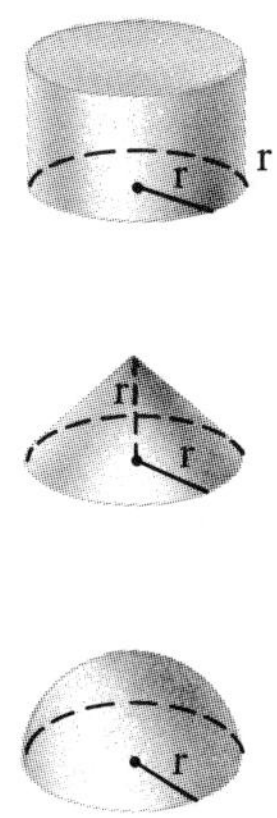

Figure 9-27

Thus, the volume of the hemispherical solid is equal to the difference between the volume of the cylindrical solid and the volume of the conical solid.

EXERCISE 9-13

1. Find the volume measurement of each of the following spherical solids. Use appropriate values as approximations of π.
   (a) Radius has measurement of 10 ft.
   (b) Radius has measurement of $3\frac{1}{2}$ yd.
   (c) Diameter has measurement of 24 m.
2. Models that approximate the situation in Fig. 9-27 can be made from one half of a rubber ball (hollow) and cardboard models of a cone and a cylinder. Make these models or buy a set of ready-made models (they are available from several distributors of mathematics teaching aids). Verify the relationship between the volumes of the solids determined by the hemisphere, cone, and cylinder by filling the containers with sand.
3. Read Ref. 66, p. 268.

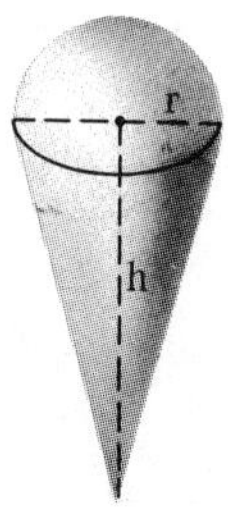

Figure 9-28

9-14 USING BASIC FORMULAS

Now that we have formulas for finding the volume of some of the basic solids, we can use them and the additive property of volume measurement to find the volume of composite figures. We present two examples.

Example: Find the volume of the solid shown in Fig. 9-28. The ice cream cone can be partitioned into two solids: a cone and a hemisphere. If r is the radius of the hemisphere and h is the height of the cone, the volume of the ice cream cone is

$$\tfrac{1}{3}\pi r^2 h + \tfrac{2}{3}\pi r^3.$$

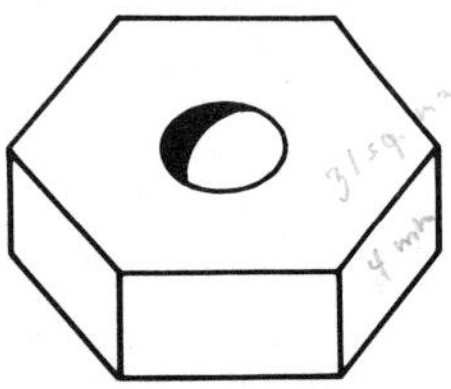
Figure 9-29

Example: Find the volume of the metal nut shown in Fig. 9-29. The nut is 4 mm thick. The area measurement of the hexagonal base of the prism is about 31 sq mm; the area of the circular base region of the hole is about 3 sq mm.

Volume of the prismatic solid:	$(31)(4) = 124.$
Volume of the cylindrical solid:	$(3)(4) = 12.$
Volume of the nut:	$124 - 12 = 112.$

The volume measurement of the nut is 112 cu mm.

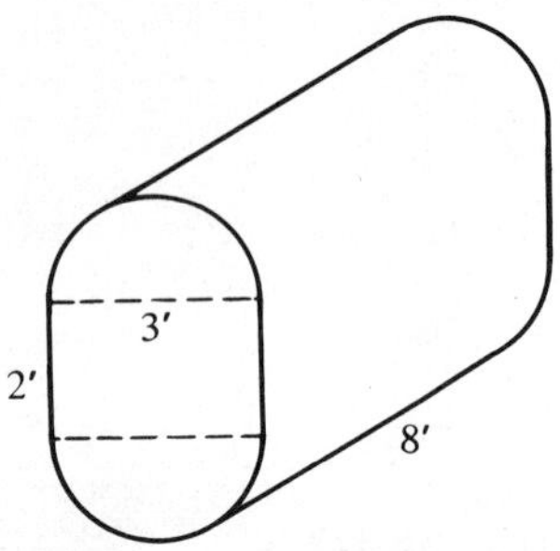

Figure 9-30

EXERCISE 9-14

1. Find the volume in cubic inches of a half-gallon milk carton.
2. Find the volume of the oil storage tank shown in Fig. 9-30. Use the measurements as shown.

Chap. 9
Exercise 9-4

3. Find the volume of the cake pan pictured in Fig. 9-31(a). Figure 9-31(b) shows that the two parts of the pan are frustums of circular cones. To find the altitude of the cones you will need to use the principle concerning the relationship of the sides of a pair of similar triangles.

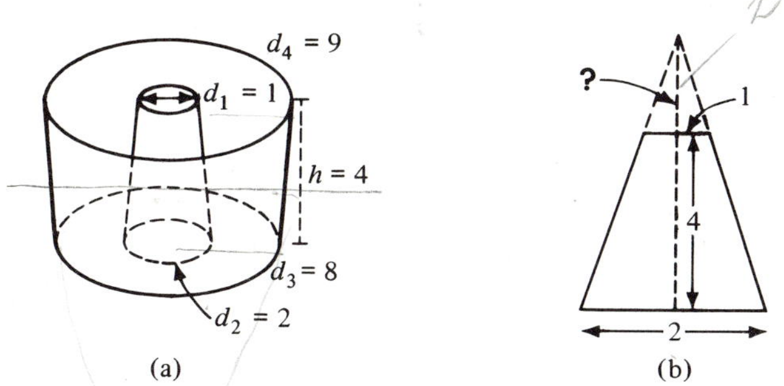

Figure 9-31

9-15 VOLUME RATIOS

In Sec. 8-15 we discussed what happens to the area of a rectangular region if the measure of each side of the polygon is doubled. We noted that the area measure is multiplied by 4. What happens to the volume of a rectangular solid if the measure of each edge is doubled?

Using the formula $V_1 = \ell wh$ for the volume of the solid in Fig. 9-32(a), the volume of the solid in (b) is $V_2 = (2\ell)(2w)(2h)$. Simplifying, we have $V_2 = 8\ell wh$. Doubling the measures of the edges multiplies the volume by 8! What would happen to the volume if the measures of each of the edges were tripled? What would happen to the volume if you doubled the length, tripled the width, and quadrupled the height?

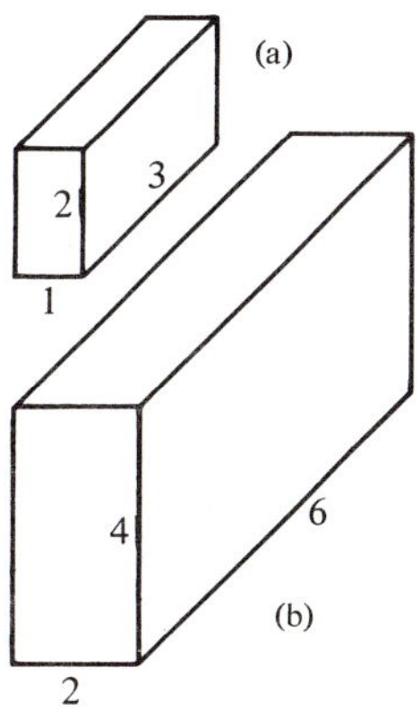

Figure 9-32

EXERCISE 9-15

1. Complete the following table:

Dimensions of rectangular solid			*Volume*
ℓ	w	h	
2ℓ	$2w$	$2h$	
3ℓ	$3w$	$3h$	
4ℓ	$4w$	$4h$	
5ℓ	$5w$	$5h$	
6ℓ	$6w$	$6h$	
$n\ell$	nw	nh	

2. Make two models of rectangular solids, one of which has dimensions twice those of the other.

3. Try to design two cubical solids, one of which has twice the volume of the other. This is known as the problem of the duplication of the cube.

9-16 VOLUMES AND ERRORS OF MEASUREMENT

Our analysis of the effect of the approximate nature of physical measurement would not be complete if we did not consider the case of volume measurement. You recall that the problem is to find the interval of tolerance for the volume of a given solid if the precision used to obtain the associated linear measurements is known.

Suppose measurements of a box were made using an instrument whose precision is 1 in. The measurements are 7 in. and 8 in. for the base. The altitude measurement is 9 in. What is the maximum value of the true measure of each of the measurements? The minimum value?

The maximum value for the true volume is (7.5)(8.5)(9.5) or 605.625. The minimum value for the true volume is (6.5)(7.5)(8.5) or 414.375. The value obtained using the reported measurements is 504. This differs from the upper limit by 101.625 and from the lower limit by 89.625. A fairly good approximation of the volume measurement is 500 cu in. Notice that using 500 is in accord with the agreement to use the same number of significant digits in a product as occurs in the least accurate of the factors.

EXERCISE 9-16

1. Find the interval of tolerance for the volume of a rectangular solid whose measurements are 15 m, 11 m, and 37 m.
2. For problem 1, express the greatest possible error in the volume as a percent of the computed volume.

SUMMARY OF PRINCIPLES

Every closed region in space has a unique volume.

All congruent regions in space have equal volumes.

One closed region in space has a greater volume than a second closed region if it contains the second region or a closed region equal in measure to the second region as a proper subset.

Volume measurement is additive.

Volume of the solid is determined by:

right rectangular prism	$V = Bh$,
prism	$V = Bh$,
circular cylinder	$V = \pi r^2 h$,
pyramid	$V = \frac{1}{3}Bh$,
circular cone	$V = \frac{1}{3}\pi r^2 h$,
sphere	$V = \frac{4}{3}\pi r^3$.

If the ratio of similitude of two space regions is $1 : n$, then the ratio of the volumes of the two regions is $1 : n^3$.

EXERCISES

1. Find the volume of each of the solids pictured in Fig. 9-33.

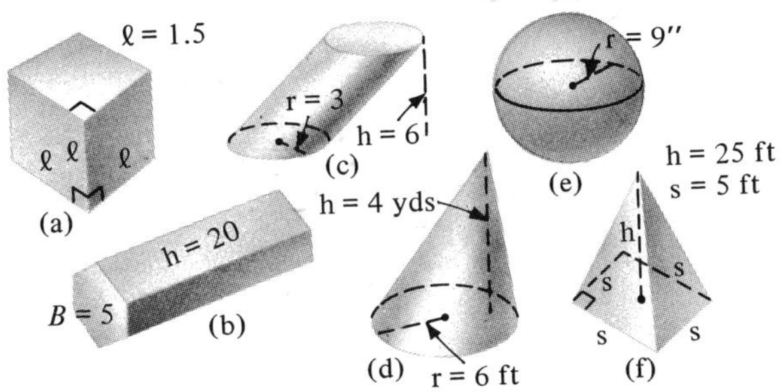

Figure 9-33

2. (a) Congruent space regions have the same volume. A S N
(b) Space regions with the same volume are congruent. A S N
(c) Units for volume measurement are square regions. A S N
(d) If two conical solids have the same volume, they have congruent bases and altitudes. A S N

3. What is the ratio of the volumes of two solid prisms whose bases are congruent but for which the altitude of one is twice the altitude of the other?

4. Find the volume of a fish tank whose measurements are 12 in., 10 in., and 8 in. How many gallons of water will the tank hold?

5. Find the volume of a refrigerator whose inside measurements are 14 in., 24 in., and 36 in.

6. (a) 1 cu ft is equivalent to ________ cu in.
(b) 1 cu yd is equivalent to ________ cu ft.
(c) 1 cu m is equivalent to ________ cu cc.

7. Express the answer to problem 5 above in cubic feet.

8. Find the volume of the Earth, assuming that the Earth is a spherical solid whose radius measurement is 4000 mi.

9. The radius of Saturn is approximately nine times the radius of Earth. Without computing the radius of Saturn, find its volume.

10. What is the normal capacity of the lungs of an adult?

11. What volume of water will a washing machine hold if the tub is a cylinder whose radius has measurement of 9 in. and whose height has measurement of 12 in.?

12. Not all textbooks use the terms "prism" and "prismatic region" as this text does. Look in various sources for alternative definitions of "prism," "prismatic region," "prismoid," and "prismatoid."

13. Some authors use the phrases "volume of a pyramid," "volume of a cylinder," or "volume of a prism." Apparently, what agreement do they use?

14. Read Ref. 59, pp. 44–45, and Ref. 108, pp. 257–263.

CHAPTER 10

Indirect Measure

10-1 INTRODUCTION

You have been introduced to the basic concepts of measurement in Chapter 5. A measure obtained by direct comparison with some other like object (for example, by comparing the length of a foot ruler to the length of a table) is called a **direct measure**. A measure obtained as the result of computations with other measures is called an **indirect measure.** Thus, if the side of a square is measured to be 3.2 in inches to the nearest tenth of an inch, the perimeter can be computed to be 12.8 in inches with the same precision. The 3.2 in. is a direct measurement, the 12.8 in. is an indirect measurement. (Note that the perimeter might also be obtained by direct application of the foot ruler.)

In this chapter you will encounter three specific methods of obtaining indirect measures, each of which depends upon the concept of similar triangles. In some cases the method will present a more convenient way to obtain measures that could be obtained directly. In other cases the method will allow us to find measures which would otherwise be unobtainable.

10-2 SCALE DRAWINGS

From the concept of a similarity transformation developed in Sec. 7-4 we know that if two figures are similar, any two corresponding linear measures (image distances) have the same ratio as r.s., the ratio of similitude. Also, in Exercise 7-4, 5, you noted that any two triangles are similar if the corresponding angles are respectively congruent, or if two sides of one are proportional to two sides of the other and the included angles are congruent.

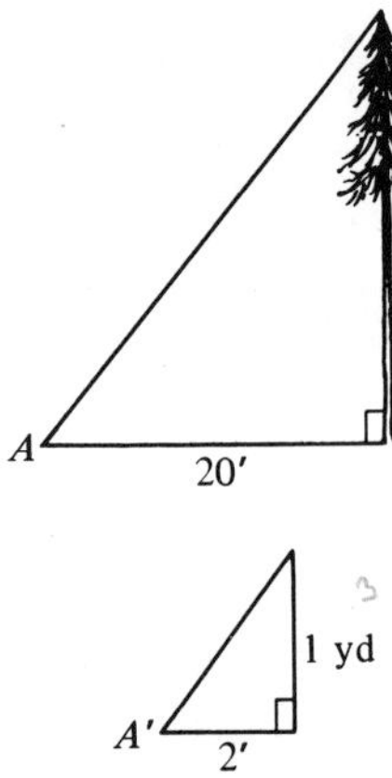

Figure 10-1

The tree shown in Fig. 10-1 has a height represented by h and casts a shadow of 20 ft. At the same time a yardstick casts a shadow of 2 ft. The usual assumptions are: (1) the ground is a level plane (horizontal), (2) the tree and yardstick are parallel (vertical), and (3) the rays of the sun are parallel. Rarely, if ever, would all three of these hold. (When would the rays of the sun be parallel?) However, the small errors that these approximations entail do not greatly affect the result.

If we assume these conditions hold, we can see that we have similar triangles. The angles A and A' are congruent since they are determined by the tops of the tree and stick, the tips of the shadows of the tree and stick, and the horizontal ground. (An angle determined by the horizontal plane and a line of sight raised from the horizontal is called an **angle of elevation**.) Which of the three assumptions justifies this congruence? The right angles that the tree and stick make with the ground are also congruent. Explain. This gives two pairs of congruent angles in these two triangles. Why are the angles in the third pair congruent? Thus the two triangles of Fig. 10-1 are similar, the corresponding sides are in proportion, and one way to find h is to write:

$$\frac{20}{2} = \frac{h}{3} \qquad \text{(where did the "3" come from?);}$$

then

$$2h = (3)(20),$$
$$2h = 60,$$

and

$$h = 30.$$

Thus the height of the tree is estimated (computed) to be 30 ft. The precision and accuracy of the result are subject to all of the conditions above in addition to the usual errors from the original measures. In textbook problems of this sort it is usual to assume that the numbers given are exact and that the above conditions hold; the reader must be aware of these assumptions. We shall follow this convention unless we clearly state otherwise.

A second way to obtain h is to compare measures in one triangle with the corresponding measures in the other; thus:

$$\frac{h}{20} = \frac{3}{2},$$

then

$$2h = 60,$$

and

$$h = 30, \qquad \text{as before.}$$

The first method makes use of the statement of Sec. 7-4 that if you compare measures of corresponding distances between the two figures, the ratio is the same as the ratio of similitude. But, if you compare measures between two sides within each figure, the ratios are the same for each figure. (This second method can also be verified by elementary operations on the proportion used for the first method.) Thus, within each triangle of Fig. 10-1 the ratio of height to shadow is 3 : 2, but r.s. between the two figures is 20 : 2, or 10 : 1.

Other common examples of scale drawings using the concept of similarity are the enlarging of a photograph, microfilming, map reading, and drafting (blueprints). For example, if a picture is to be enlarged in a linear ratio of 2 : 5 and the given dimensions are 14 in. and 22 in., then the enlargement would have what dimensions?

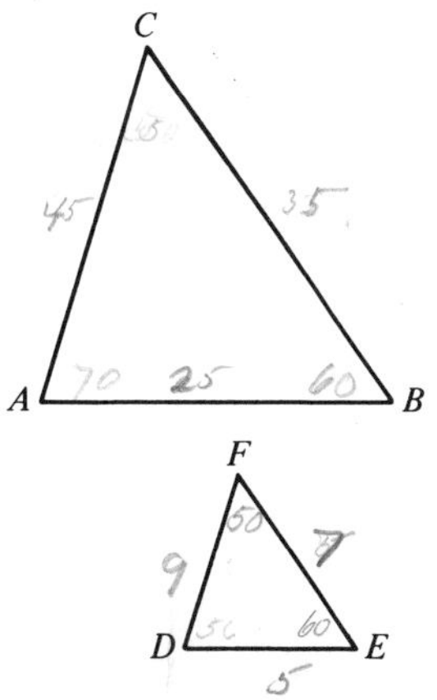

Figure 10-2

EXERCISE 10-2

1. (a) In Fig. 10-2, $m(\angle A) \overset{\circ}{=} 70$, $m(\angle B) \overset{\circ}{=} 60$, $m(\angle F) \overset{\circ}{=} 50$, and $m(\angle E) \overset{\circ}{=} 60$. Are the triangles similar? Explain. If $a = 45$, $c = 35$, $f = 7$, and $e = 5$, find the measures of b and d.

(b) Suppose a translation were carried out that resulted in $\triangle DEF$ being placed as $\triangle D'E'F'$ in Fig. 10-3. Would all that was concluded in part (a) now be true? What about the ratios of AD' to $D'E'$ and CF' to $F'E'$? What can you conclude about the ratio of AC to $D'F'$? What relationship exists between $\overleftrightarrow{AC}$ and $\overleftrightarrow{D'F'}$?

(c) In Fig. 10-3 suppose $BC = 12$, $BF' = 8$, $D'F' = 10$, and $\overleftrightarrow{D'F'} \parallel \overleftrightarrow{AC}$. Find AC. Can you find AD'? Justify each step. [Note: For this part, use only the facts listed in this part (c); use none from parts (a) or (b).]

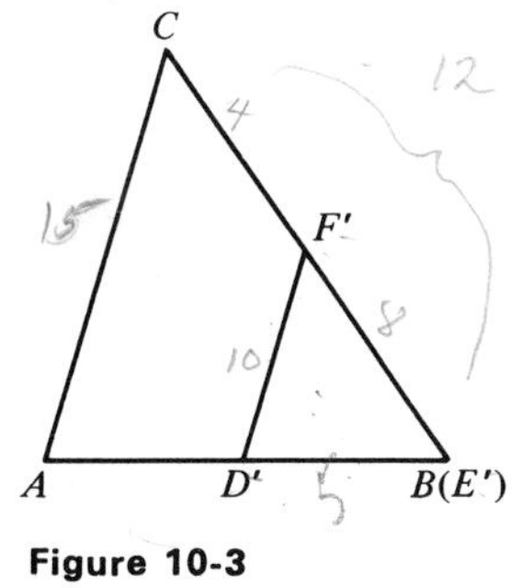

Figure 10-3

2. (a) In $\triangle PQR$ (no figure pictured) $m(\angle P) \overset{\circ}{=} 20$, $m(\angle Q) \overset{\circ}{=} 60$, $PQ = 32$, and $PR = 14$. In $\triangle ABC$ $m(\angle B) \overset{\circ}{=} 100$, $m(\angle C) \overset{\circ}{=} 60$, and $AC = 96$. From this information, which of the following is true:

(1) $AB = 7$. (2) $AB = 42$.
(3) $AB = 28$. (4) AB cannot be determined.

(b) With the same information listed in part (a), is it possible to determine BC? If so, find it; if not, what additional information is necessary?

3. (a) To find the distance across the swamp in Fig. 10-4(a), $\overline{RT}$ and $\overline{ST}$ are extended as in Fig. 10-4(b), and $\overline{PQ} \parallel \overline{RS}$. If $PT = 3$ and $PQ = 4$, find RS. Is $\triangle RST \sim \triangle QPT$? Explain. (Note that nothing was said as to how $\overline{PQ}$ was made parallel to $\overline{RS}$; we are assuming that someone can do so.) Did you need to know RT to find RS?

   (b) Figure 10-4(c) is formed by again extending $\overline{RT}$ and $\overline{ST}$ through T so that D is on $\overleftrightarrow{ST}$ and $\overline{TS} \cong \overline{TD}$; E is on $\overleftrightarrow{RT}$ and $\overline{TR} \cong \overline{TE}$. With points D and E determined, $\overline{DE}$ is drawn. From this, if $\overline{DE}$ is measured to be 40 ft, can RS be determined? How? Is $\overline{DE} \parallel \overline{RS}$? Explain.

   (c) Suggest a way in which $\overline{PQ}$ could be determined parallel to $\overline{RS}$ other than by making $\triangle TDE$ as in part (b).

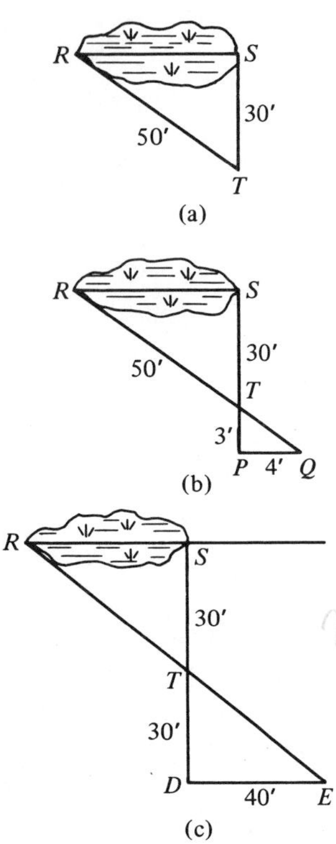

Figure 10-4

4. (a) A certain map is being constructed to a scale of 1 in.: 20 mi. Then a city block that has measurement of 1320 ft would be represented by a segment with what measurement?

   (b) On this same map a segment that is $2\frac{1}{2}$ in. long represents what measurement?

   (c) On this same map how do angles between images of segments compare with angles between the original segments? (By angles between images we mean the angles determined by images.)

   (d) Suggest a scale that might be more appropriate for showing the city block mentioned in part (a).

5. A certain picture frame has an "inside" area of 96. If the picture contained in this frame is being enlarged using a linear ratio of 1:4, what must be the inside area of the frame for the enlargement?

6. In Fig. 10-5 altitude $\overline{CD}$ is drawn in the right triangle ABC, which has its right angle at C.

   (a) Notice that $\angle A$ is a complement of $\angle 1$; $\angle 2$ is also a complement of $\angle 1$. Then what can be concluded about $\angle A$ and $\angle 2$? What relationship exists between $\angle 3$ and $\angle 4$? Explain. Then what can you conclude of $\angle 1$ and $\angle B$? This then shows that $\triangle ACD$ and $\triangle CBD$ are related how?

   In $\triangle ACD$, $CD = h$ and corresponds to ________ in $\triangle CBD$.
   In $\triangle ACD$, $AC = b$ and corresponds to ________ in $\triangle CBD$.
   In $\triangle ACD$, $AD = x$ and corresponds to ________ in $\triangle CBD$.

   (b) In a proportion such as $a:b = b:c$, b is called the **mean proportional** between a and c. (Sometimes written $\frac{a}{b} = \frac{b}{c}$.) Then a is the first term, b is the second term, and c the third. In a proportion such as $a:b = c:d$, or $\frac{a}{b} = \frac{c}{d}$, a, b, c, and d are called the first, second, third, and fourth terms, respectively. Then a and d are the **extremes**, b and c are the **means**. Recall that in any proportion the product of the means is equal to the product of the extremes. Then in part (a) we can write: $h:x =$ ___$:h$, or ___$:h = h:x$. Then $h^2 =$ (___)(x), using this last rule. Using the new term "mean proportional," we can say that if the altitude is drawn to the hypotenuse of a right triangle, this altitude is the mean proportional between . . . (complete this).

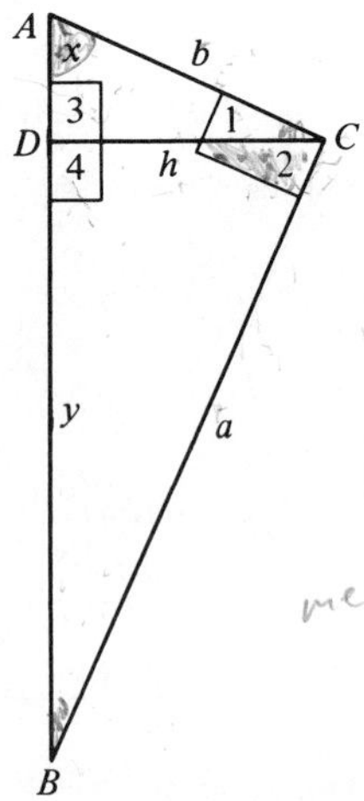

Figure 10-5

   (c) In like manner show that $\triangle ACD \sim \triangle ABC$. Write a proportion relating b, c, and x. Then $b^2 =$ ____. Write a verbal statement, similar to that for part (a), summarizing this part.

(d) In a similar way show that $\triangle BCD \sim \triangle BAC$ and write a proportion relating a, c, and y. Write a summary statement (verbal) combining all four parts of this exercise.

7. In Fig. 10-6 a picture of the Earth (considered as a sphere) is shown with an exaggeration of the situation of Fig. 10-1. Refer to Fig. 10-1, but answer the following questions as they relate to Fig. 10-6.

(a) In Fig. 10-6 is the tree parallel to the yardstick?

(b) Is the yardstick vertical? The tree? Where is the horizontal ground?

The answer to several of these questions depends upon your understanding of the word "vertical." In a plane, vertical lines are usually parallel, since they are perpendicular to the "horizontal" lines. "Horizontal" is in quotes since this is not what we really mean by horizontal lines. For a sphere, a **vertical line** is a line determined by a point on the sphere and the center of the sphere. If the Earth is considered as a model of a sphere, vertical lines are determined by a plumb bob, which is usually a piece of string to which is attached a heavy pointed piece of metal. Thus, for the Earth a vertical line is one that passes through the gravitational center of the Earth. Now consider the question: Are vertical lines parallel? A **horizontal line** is a line that is perpendicular to a vertical line. Are horizontal lines parallel? When? Discuss the error involved in the assumptions we made in solving the problem of Fig. 10-1. Name a vertical line and a horizontal line in Fig. 10-6.

(c) The World Trade Center in Manhattan contains two skyscraper towers. Use the discussion above to explain why the two towers are not parallel to each other.

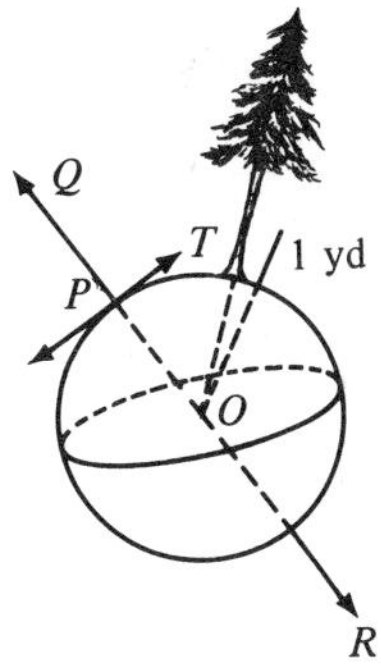

Figure 10-6

10-3 PYTHAGOREAN THEOREM

The ancient Egyptians were skillful builders. It was most important to them that their pyramids be "squared up"—that is, that the corners be very accurate right angles. They knew how to use a twelve-unit rope to determine a "good" right angle. Eleven knots (or thirteen if they had one at the beginning and one at the end) were tied at equal intervals. If three people (perhaps four) held the rope, one at the third knot, one at the seventh knot, the other(s) holding the ends, then when the ropes were drawn tight, a right angle was formed at the third knot, as in Fig. 10-7. The Babylonians also knew of this relationship and used it. The Greek mathematician Pythagoras (circa 500 B.C.) is given credit for proving it. Since his time several hundred proofs have been given, two of which are explored later.

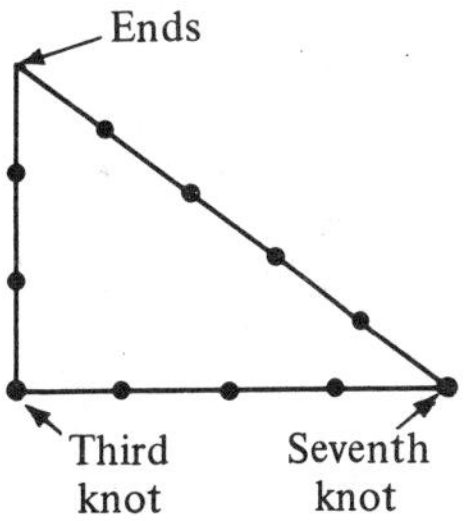

Figure 10-7

The **Pythagorean theorem**, sometimes called the rule of the right triangle, says that in a right triangle the square of the hypotenuse is equal to the sum of the squares of the legs. This is illustrated in Fig. 10-8. Actually the Egyptians and Babylonians were using the converse form of this theorem. (Converse forms are discussed in Sec. 13-4.) If the sum of the squares of

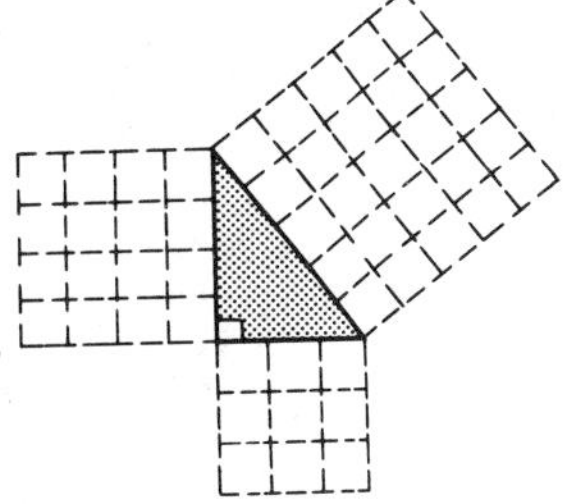

Figure 10-8

two sides of a triangle is equal to the square of the third side, this third side is the hypotenuse of a right triangle. That is, their knots showed that since $(3)^2 + (4)^2 = (5)^2$, the right angle is at the third knot of Fig. 10-7.

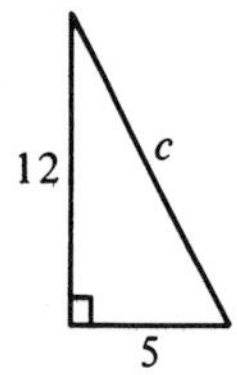

Figure 10-9

Further, if any two sides of a right triangle are known, the theorem gives us a method of finding the third side (assuming we know elementary operations with number sentences). Thus, in Fig. 10-9 with legs of 5 and 12 in the given right triangle, the hypotenuse can be found by

$$\begin{aligned}(c)^2 &= (5)^2 + (12)^2,\\ c^2 &= 25 + 144,\\ c^2 &= 169,\\ c &= \sqrt{169},\end{aligned}$$

hence

$$c = 13,$$

since $(13)^2 = 169$. And if we return to Fig. 10-4(a) we have [if we know $m(\angle S) \stackrel{\circ}{=} 90$]

$$\begin{aligned}(50)^2 &= (30)^2 + (RS)^2,\\ 2500 &= 900 + (RS)^2,\\ (RS)^2 &= 2500 - 900,\end{aligned}$$

hence

$$RS = \sqrt{1600},$$

thus

$$RS = 40.$$

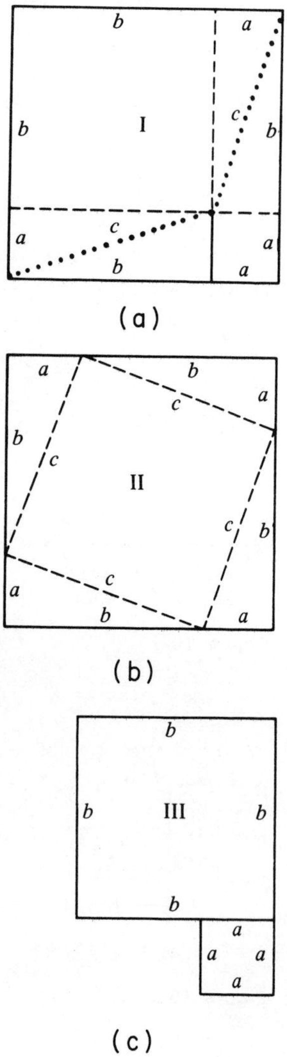

Figure 10-10

A demonstration of the validity of the Pythagorean theorem uses the following model shown in Fig. 10-10, which should be made of some substantial material. Let $a = 2\frac{5}{8}$, $b = 8\frac{1}{8}$, and $c = 8\frac{1}{2}$. Cut out four right triangles ABC, two rectangles $a \times b$, and three squares, one $a \times a$, one $b \times b$, and one $c \times c$. Arrange as in Fig. 10-10, I and II, with tape on one side, perhaps the back, where there are dashed lines; cut, or leave untaped, at the solid lines. (The dotted lines merely indicate the diagonals of the rectangles and represent c for those triangles.) Fold back the four triangles of figure II. Referring to figure I, fold back the lower left rectangle, then the right rectangle and the $a \times a$ square. The result is figure III. Since the original $(a + b)$ by $(a + b)$ squares were equal in area, and we removed the same area from each (four congruent right triangles), then the resulting areas have the same measure; thus, $b^2 + a^2 = c^2$.

The discerning reader might notice that the dimensions listed for a, b, and c are not sides of a right triangle; however, the error is but $\frac{3}{4}$ sq in. out of $\frac{289}{4}$ sq in., which is less than 0.7 percent. This is surely less area than could be detected after all the measuring, taping, cutting, and so on. And this gives us one more opportunity to point out the limitations of physical models. Recall problem 13 of the Exercises at the end of Chapter 8. Would this same limitation apply to the demonstration of Fig. 10-11 in problem 1 below?

EXERCISE 10-3

1. In Fig. 10-11 we have drawn the altitude $\overline{CD} \perp \overline{AB}$ and selected x to represent AD. Then $BD = c - x$. Explain. From Exercise 10-2, Fig. 10-5, we conclude that $x{:}b = b{:}c$, and $(c - x){:}\ a = a{:}c$. Explain. Then $b^2 =$ _____; and $a^2 =$ _____. Then $a^2 + b^2 =$ _____. Complete this demonstration of the truth of the Pythagorean theorem.
2. In Fig. 10-12, find c: (a) in radical form, (b) in decimal form to the nearest tenth.
3. In Fig. 10-13, find a: (a) in radical form, (b) in decimal form to the nearest tenth.
4. In Fig. 10-14 find c in radical form. To the nearest tenth. To the nearest hundredth. Do you think there is a rational number that represents this value? A group called the Pythagoreans devoted considerable time to this problem. Until being confronted by problems of this sort, they had believed that all measurable quantities could be expressed by integers or rational numbers. There is no such number of the form $\frac{M}{N}$, where M and N are integers, that names the answer to this exercise. The proof of this appears in most college algebra books—for example, Ref. 9, pp. 38–40. An alternate proof appears in Ref. 50, pp. 56–59.
5. Given rectangular solid $ABCD$-$EFGH$ in Fig. 10-15.
   (a) If $AE = 3$, $FG = 4$, and $EF = 12$, find EC. [*Hint:* Draw $\overline{EC}$ and $\overline{EG}$. Now is there any right triangle containing the sides we know as well as the segment whose length we seek? What do you know about $\triangle EGC$? About $\triangle EFG$?]
   (b) If the length, width, and height of a rectangular solid are ℓ, w, and h, respectively, as indicated in Fig. 10-15, show that d, the measure of a diagonal of the solid, perhaps EC, is given by $d = \sqrt{\ell^2 + h^2 + w^2}$.

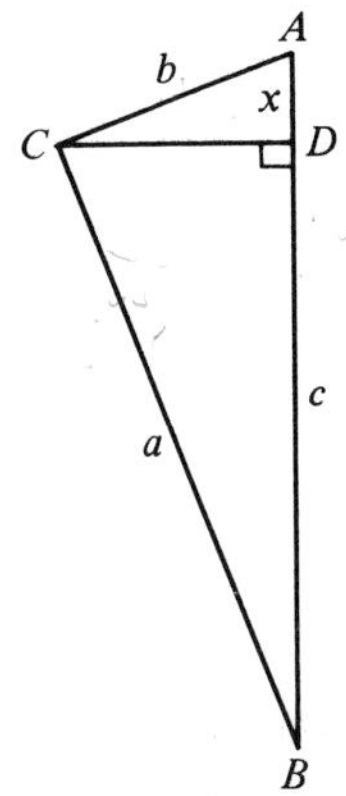

Figure 10-11

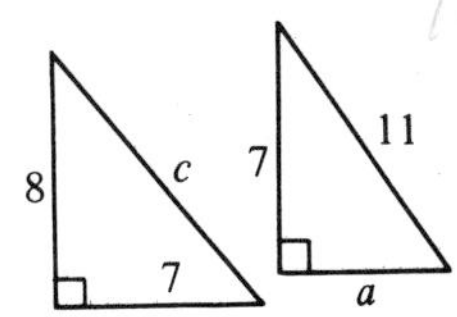

Figures 10-12 and 10-13

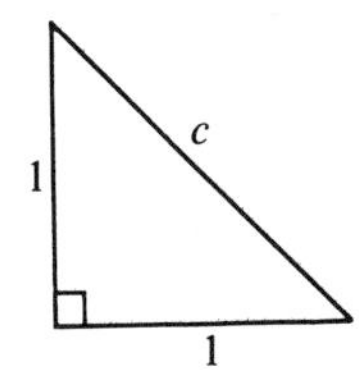

Figure 10-14

10-4 TRIGONOMETRY—TANGENT RATIO

Literally, "trigonometry" means three-angle measure. However, the part of mathematics that is included as trigonometry is not limited to triangles. As a matter of fact, a complete trigonometry can be developed without any consideration of an angle concept. But in its historical development, and, indeed, in many of its fundamental applications, the triangle plays a most prominent role. The relationships among similar triangles provide the basis for our development. Of special convenience is the right triangle. If you know that two right triangles have an acute angle of one

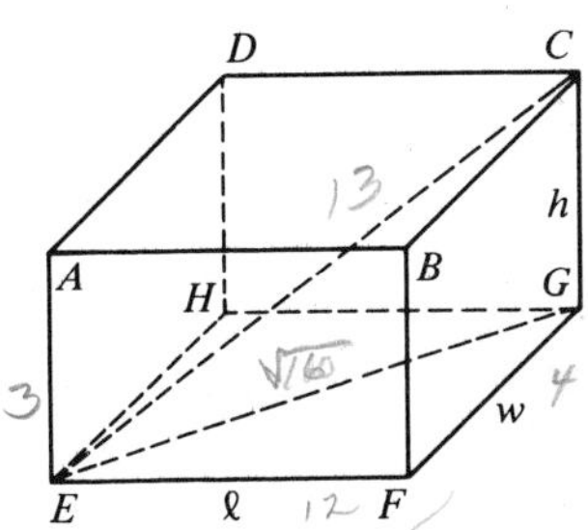

Figure 10-15

congruent to an acute angle of the other, what can you conclude about the triangles?

Now consider Fig. 10-16. Since $\triangle AED$ and $\triangle ACB$ each contain $\angle A$, and each has a right angle, at E and C, respectively, they are similar. How many triangles similar to $\triangle ACB$ can be formed by drawing lines perpendicular to $\overline{AC}$? Using the scale indicated, what is the ratio of DE to AE? Of BC to AC? Let us describe $\overline{BC}$ and $\overline{DE}$ as being the *legs opposite* the acute angle A, and describe $\overline{AE}$ and $\overline{AC}$ as *legs adjacent* to the acute angle A. What is the ratio of the measure of the opposite leg to the measure of the adjacent leg in each of the other similar triangles just mentioned? Suppose we now consider a right triangle in which there is an acute angle congruent to $\angle A$, and in which the leg adjacent to this angle has a measurement of 20 ft. What is the measurement of the leg opposite this angle? (You should have said, "10 ft.") This ratio, in a right triangle, of the leg opposite an acute angle and the leg adjacent to the angle is so useful that it is given a name: the **tangent ratio**. (There is no obvious connection between the name of this ratio and the name of a line that contains only one point of a circle.) Thus the tangent of an acute angle of a right triangle is equal to

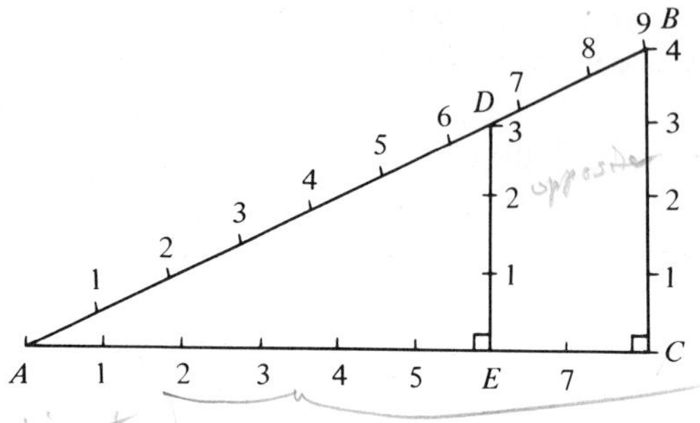

Figure 10-16

$$\frac{\text{measure of the opposite leg}}{\text{measure of the adjacent leg}}.$$

In practice the statement for $\triangle ACB$ of Fig. 10-16 is

$$\tan A = \frac{BC}{AC},$$

which is read, "The tangent of angle A is equal to the ratio of BC to AC in triangle ACB." What is the symbolic statement for the tangent ratio of angle A in $\triangle AED$? Write the verbal statement also. (A note of caution: The terms "adjacent" and "opposite" never apply to the hypotenuse when we use this ratio; recall that the legs of a right triangle are the sides not opposite the right angle.)

Suppose the angle of Fig. 10-16 is made smaller but AC is kept constant, as suggested by Fig. 10-17. Using the scale on the diagram, what is $\frac{DE}{AE}$? $\frac{BC}{AC}$? Although the ratio for this diagram has a value different from that of Fig. 10-16, notice that in this diagram it is a constant. That is, if we draw lines perpendicular to $\overline{AC}$ and compute the ratio in the right triangles thus formed, we will get the same

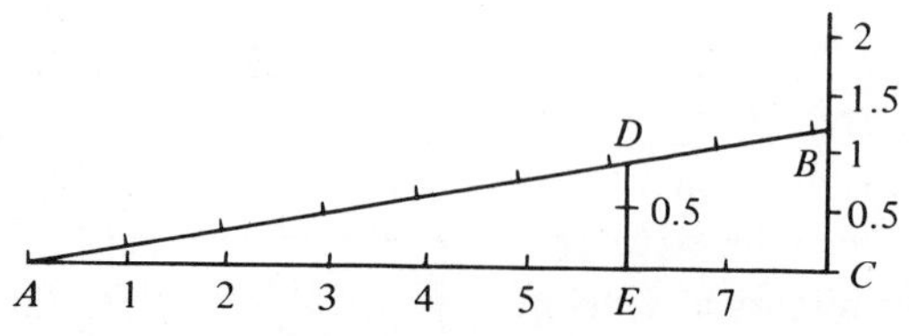

Figure 10-17

result. Thus we are led to believe that the value of the ratio depends only on the size of the angle. Perhaps for this reason this relationship is often called the tangent function. Thus if we draw many of these triangles, using different values for the measure of $\angle A$, and keep track of the values for the ratio that correspond to a particular measure of the angle, then any time we have a right triangle in which we know the measures of an acute angle and one leg we can look up this ratio and use it to compute with, much as we did with the triangle with the leg measurement of 20 ft. Fortunately for us, someone has performed this tabulation, and Table 10-1 lists values for some acute angles. Use these in calculating the answers for the problems in the next exercise. First, we illustrate the method.

TABLE 10-1
SOME TANGENT VALUES

Angle	*Tangent*
30°	0.577
35°	0.700
40°	0.839
50°	1.192
60°	1.732

A tree and its shadow are illustrated in Fig. 10-18. The measurement of the shadow is 42 ft and the angle of elevation of the top of the tree has a measure of 35 in degrees. We wish to know the height of the tree. Since the shadow is represented by the leg of the triangle that is adjacent to the angle, and the height is represented by the leg opposite this angle, we write:

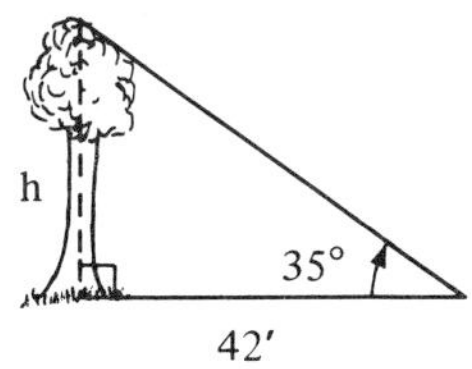

Figure 10-18

$$\tan 35° = \frac{h}{42}.$$

From Table 10-1 we see that $\tan 35° = 0.700$; therefore,

$$0.700 = \frac{h}{42};$$

then

$$h = (0.700)(42),$$

and

$$h = 29.400.$$

Ordinarily we would list this answer as 29.4 ft, since "0.577" implies three significant digits. Since the given length indicates two significant digits, how should the answer be listed?

In this example we have considered the angle measure to be exact, which is the usual practice. If serious consideration must be given to the precision or accuracy in the result, then the error introduced by the approximation in the angle measure must be taken into consideration. Almost any high school trigonometry text contains a discussion of this. Not only do we consider our angle measures to be exact, but, unless we clearly state otherwise, we consider given linear measures exact, and the precision or accuracy of the answer is determined by the trigonometric ratio value used. Thus, in the absence of any additional statement in the problem above, we would list the answer as 29.4 ft.

EXERCISE 10-4

Figure 10-19

1. Suppose that in Fig. 10-19 the angle of elevation of the top of the tree had a measure in degrees of 50. What would be the height of the tree?

2. A certain ladder is designed in such a way that its "safe angle" is 60° (as pictured in Fig. 10-20) when it reaches a height of 20 ft. To the nearest tenth of a foot how far from the base of the building should the bottom of the ladder be placed in order that these conditions be met?

3. If the safe angle of a different ladder is 40°, how far up a building will it reach safely if the bottom of the ladder is placed 20 ft from the base of the building?

4. A new thrill ride is being designed for a carnival. The surface is to rise at an angle of 30°. With this design the surface must be braced every 25 ft of horizontal distance (see Fig. 10-21). What, then, is the difference in the lengths of successive braces?

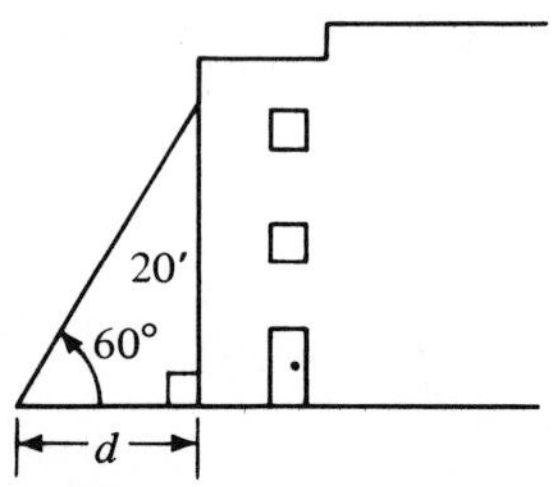

Figure 10-20

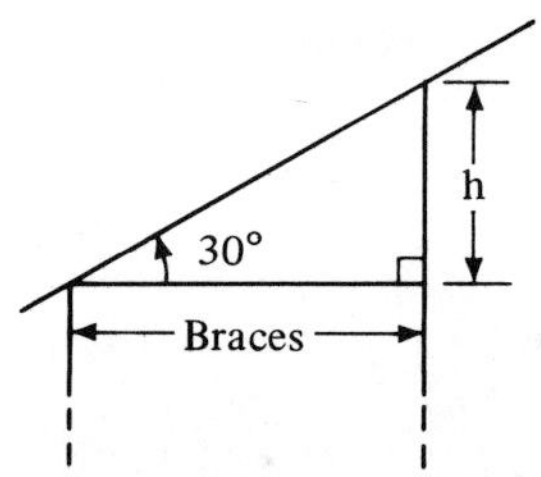

Figure 10-21

5. Suppose a certain ladder has been classified safest if, when it reaches a height of 30 ft, the foot of the ladder is 25 ft from the base of the building. Can we determine from this the safe angle for this ladder? In Fig. 10-20, using the facts stated, the leg opposite the angle would have what measure? What would d be? Thus the tangent of our safe angle would be represented by a rational number in what fractional form? Express this as a decimal numeral showing three decimal places. In Table 10-1 is there an angle having a value for its tangent that is close to the value you obtained? Then we are able to conclude that the safe angle for this ladder has approximately what degree measure?

10-5 TRIGONOMETRY—OTHER RATIOS

Figure 10-22 is a reproduction of Fig. 10-16. In Sec. 10-4 we considered the ratio of the measure of the leg opposite the acute angle to the measure of the leg adjacent to the acute angle. In similar triangles the ratio of any two corresponding sides is a constant for that pair. We have six arrangements of pairs of sides—hence, six possible ratios we might have selected. Each has a name as listed below. We shall be using only the first three;

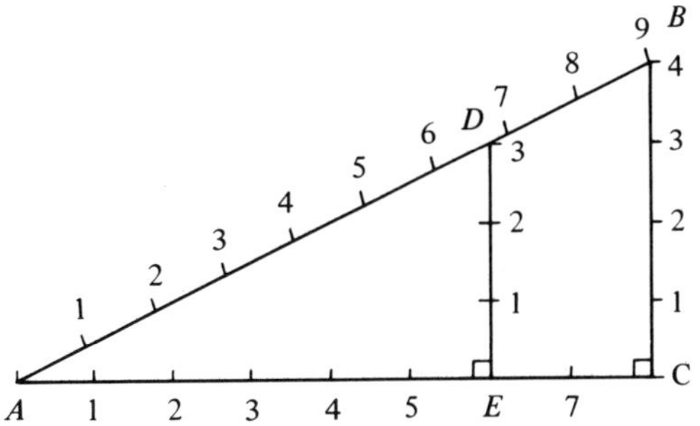

Figure 10-22

the others are included for your information. You should identify a partitioning of this list into three pairs such that each member of a pair is the reciprocal of the other member of that pair. Using the scale on the diagram, compute to three decimal places the value of each of the ratios listed.

Ratio name (Function)	*Abbreviation*	*Definition*	*In △ABC (Fig. 10-22) Value*
tangent	tan	$\frac{\text{opposite leg}}{\text{adjacent leg}}$	$\tan A = \frac{BC}{AC} = \frac{4}{8} = 0.500$
sine	sin	$\frac{\text{opposite leg}}{\text{hypotenuse}}$	$\sin A = \frac{BC}{AB} = \frac{4}{9} = 0.444$
cosine	cos	$\frac{\text{adjacent leg}}{\text{hypotenuse}}$	$\cos A = \frac{AC}{AB} = \quad =$
cotangent	cot ctn	$\frac{\text{adjacent leg}}{\text{opposite leg}}$	$\cot A = \frac{AC}{BC} = \quad =$
cosecant	csc	$\frac{\text{hypotenuse}}{\text{opposite leg}}$	$\csc A = \frac{AB}{BC} = \quad =$
secant	sec	$\frac{\text{hypotenuse}}{\text{adjacent leg}}$	$\sec A = \frac{AB}{AC} = \quad =$

Table 10-2 lists values of the sines and cosines for some angles. Use these in calculating the answers for the problems in the next exercise. First we illustrate the method using the sine and cosine ratios.

TABLE 10-2
SOME RATIO VALUES

Angle	*Sine*	*Cosine*
30°	.500	.866
35°	.574	.819
40°	.643	.766
50°	.766	.643
60°	.866	.500

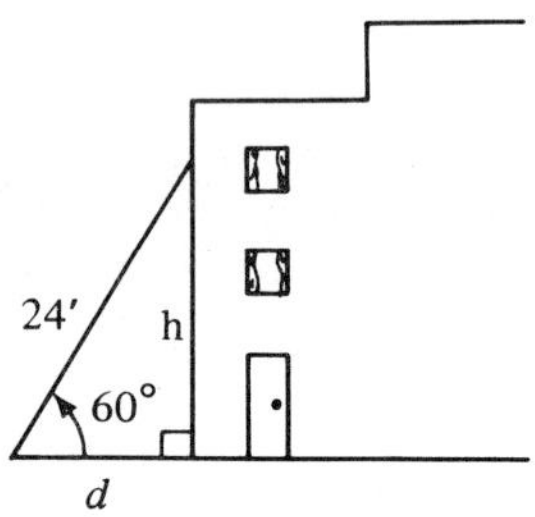

Figure 10-23

Figure 10-23 illustrates a ladder leaning against a building. Suppose this is a 24-ft ladder that requires a safe angle of 60°. How far up the side of the building will it reach, and how far from the base of the building should the bottom of the ladder be placed?

Method: Of h, d, and the 24-ft side, identify the hypotenuse, the adjacent leg, and the opposite leg. Select the function on the basis of this and proceed as before. Thus:

$$\sin 60° = \frac{h}{24}, \qquad \cos 60° = \frac{d}{24}.$$

Then from Table 10-2:

$$0.866 = \frac{h}{24}, \qquad 0.500 = \frac{d}{24},$$
$$h = (0.866)(24), \qquad d = (0.500)(24),$$
$$h = 20.784, \qquad d = 12.000,$$

and we list

$$h = 20.8, \qquad d = 12.0.$$

Notice that after we found h to be 20.8, we could have used this with the tangent of 60° to find d, rather than using the cosine of 60°. Do this. In practice d would be found as we did above, because in calculating it in this way, we use only the given information. In general we do not use indirect measures when it is possible to compute from the given information. However, this convention is not always followed, and if rules of operation with approximate measures are followed, the answers are usually satisfactory by either method.

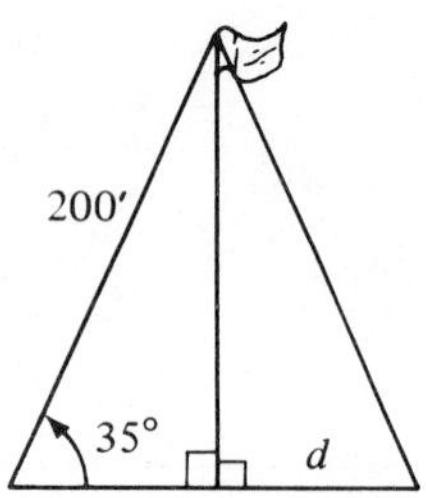

Figure 10-24

EXERCISE 10-5

1. In Fig. 10-24, guy wires 200 ft long are to be attached to support the flagpole. The angle of elevation of the top of the pole is to be 35°. How far from the base of the pole should the wires be attached?
2. In right triangle ABC with right angle at C, angle B has a degree measure of 50 and $AB = 45$. Find AC.

3. In right triangle *ABC* with right angle at *C*, $BC = 10$, $AB = 20$.
   (a) Express *AC* in radical form. You might find the remainder of this question easier if this answer is expressed in simplest form.
   (b) Express sin *A* as a rational number in fractional form.
   (c) Using your answer to part (a), express the following in fractional form (some will contain radicals): (1) cos *A*, (2) sin *B*, (3) tan *A*, (4) tan *B*.
   (d) Express sin *A* in decimal form, showing three decimal places.
   (e) What angle in Table 10-2 has the same value for its sine as that expressed in part (d)? Then what is the degree measure of angle *A*?

4. From a window 40 ft above ground level a cable is to be attached, making an angle of 40° with the ground. (See Fig. 10-25.) What is the length of the cable?

5. For the carnival ride of Exercise 10.4, 4, what is the distance along the surface between braces?

6. In right triangle *ABC* of problem 2, what is the degree measure of $\angle A$? What is cos *A* in terms of *a*, *b*, and *c*? Likewise, what is sin *B*? From Table 10-2 what is sin *B*? What is cos *B* from this table? What is cos *A*? What is sin *A*? Now suppose that $m(\angle A) \overset{\circ}{=} 30$. Answer each of these questions again. Complete the following statement summarizing your findings: In these two cases the sine ratio of an acute angle of a right triangle has the same value as the __________ ratio of the __________ angle.

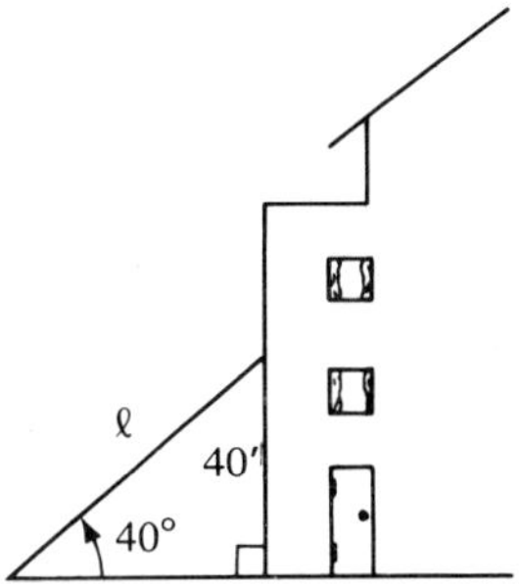

Figure 10-25

10-6 APPLICATIONS

In Exercises 10-4 and 10-5 you have worked some problems that illustrate uses of the trigonometric ratios, or functions. Table 10-3 gives values of the sine, cosine, and tangent functions for angles of each degree measure between 0 and 90. The Exercises at the end of this chapter illustrate other general applications. First, however, let us investigate three other situations of special interest.

First, consider $\triangle ABC$ in Fig. 10-26. Assume that $\angle A$ is acute and *b* and *c* are known. The question is: Can we express the area of any such triangular region in terms of *A*, *b*, and *c*? Consider altitude h_b as indicated. Then in $\triangle ABD$ what angle is a right angle? What is sin *A* in this triangle? Then

$$h_b = (c)(\sin A).$$

But the area of $\triangle ABC$ is $(\frac{1}{2})(b)(h_b)$; hence,

$$\text{area of } \triangle ABC = (\tfrac{1}{2})bc \sin A.$$

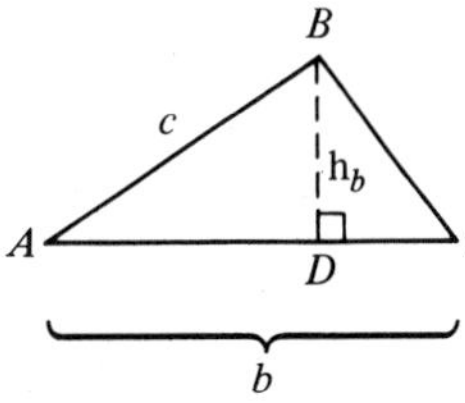

Figure 10-26

Would this relationship hold if $\angle A$ were obtuse? Draw a picture in which $\angle A$ is obtuse and show that this formula is valid. [You may have to be informed, or reminded, that sin $(180 - A)$ has the same value as sin *A*.

TABLE 10-3
VALUES OF TRIGONOMETRIC RATIOS

Angle	Sine	Cosine	Tangent	Angle	Sine	Cosine	Tangent
1°	.017	1.000	.017	46°	.719	.695	1.036
2°	.035	.999	.035	47°	.731	.682	1.072
3°	.052	.999	.052	48°	.743	.669	1.111
4°	.070	.998	.070	49°	.755	.656	1.150
5°	.087	.996	.087	50°	.766	.643	1.192
6°	.105	.995	.105	51°	.777	.629	1.235
7°	.122	.993	.123	52°	.788	.616	1.280
8°	.139	.990	.141	53°	.799	.602	1.327
9°	.156	.988	.158	54°	.809	.588	1.376
10°	.174	.985	.176	55°	.819	.574	1.428
11°	.191	.982	.194	56°	.829	.559	1.483
12°	.208	.978	.213	57°	.839	.545	1.540
13°	.225	.974	.231	58°	.848	.530	1.600
14°	.242	.970	.249	59°	.857	.515	1.664
15°	.259	.966	.268	60°	.866	.500	1.732
16°	.276	.961	.287	61°	.875	.485	1.804
17°	.292	.956	.306	62°	.883	.469	1.881
18°	.309	.951	.325	63°	.891	.454	1.963
19°	.326	.946	.344	64°	.899	.438	2.050
20°	.342	.940	.364	65°	.906	.423	2.145
21°	.358	.934	.384	66°	.914	.407	2.246
22°	.375	.927	.404	67°	.921	.391	2.356
23°	.391	.921	.424	68°	.927	.375	2.475
24°	.407	.914	.445	69°	.934	.358	2.605
25°	.423	.906	.466	70°	.940	.342	2.747
26°	.438	.899	.488	71°	.946	.326	2.904
27°	.454	.891	.510	72°	.951	.309	3.078
28°	.469	.883	.532	73°	.956	.292	3.271
29°	.485	.875	.554	74°	.961	.276	3.487
30°	.500	.866	.577	75°	.966	.259	3.732
31°	.515	.857	.601	76°	.970	.242	4.011
32°	.530	.848	.625	77°	.974	.225	4.331
33°	.545	.839	.649	78°	.978	.208	4.705
34°	.559	.829	.675	79°	.982	.191	5.145
35°	.574	.819	.700	80°	.985	.174	5.671
36°	.588	.809	.727	81°	.988	.156	6.314
37°	.602	.799	.754	82°	.990	.139	7.115
38°	.616	.788	.781	83°	.993	.122	8.144
39°	.629	.777	.810	84°	.995	.105	9.514
40°	.643	.766	.839	85°	.996	.087	11.430
41°	.656	.755	.869	86°	.998	.070	14.301
42°	.669	.743	.900	87°	.999	.052	19.081
43°	.682	.731	.933	88°	.999	.035	28.636
44°	.695	.719	.966	89°	1.000	.017	57.290
45°	.707	.707	1.000				

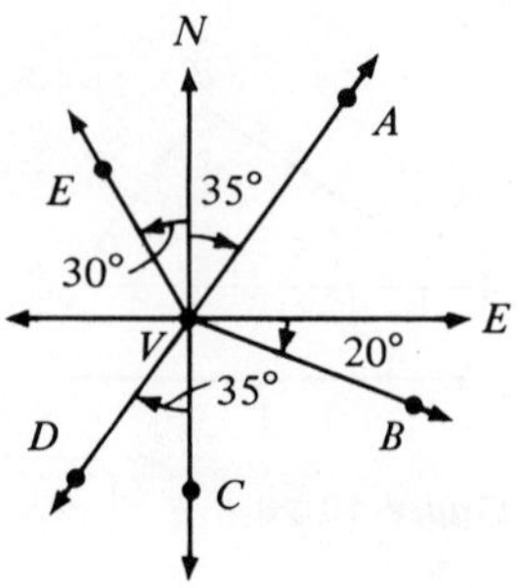

Figure 10-27

(Illustrating this requires defining the sine function for other than acute angles. See Exercise 11-4, 14.) Likewise, this area formula holds if $\angle A$ is a right angle.] Thus, the area of any triangular region is equal to one half the product of the measures of any two sides of the triangle and the sine of the included angle.

Second, consider Fig. 10-27, which shows the positions of five boats, indicated as *A*, *B*, *C*, *D*, and *E*, with respect to a north-south line through

V. In this situation we say the **bearing** of A from V is 35°. The bearing of B from V is 110°. The bearing of C from V is 180°. What do you think the bearings of D and E from V are? [Answers: 215°, 330°.] By this method we can locate any point of this diagram on a ray through V by stating its bearing from V as determined by an angle formed by $\overrightarrow{VN}$ and the ray through V and the point. Notice that B is completely located by the introduction of a distance measure on $\overrightarrow{VB}$. Thus we might say that B is 100 mi on a bearing of 110° from V. Another method would say that B is E 20° S of V, or S 70° E of V.

Suppose we must find RT in Fig. 10-28, knowing that R is 30 mi on a bearing of 60° from V, and T is 70 mi on a bearing of 330° from V. Draw $\overline{RT}$. What is the degree measure of $\angle NVT$? Of $\angle RVT$? Consider two methods of finding RT. Find RT by the shorter method.

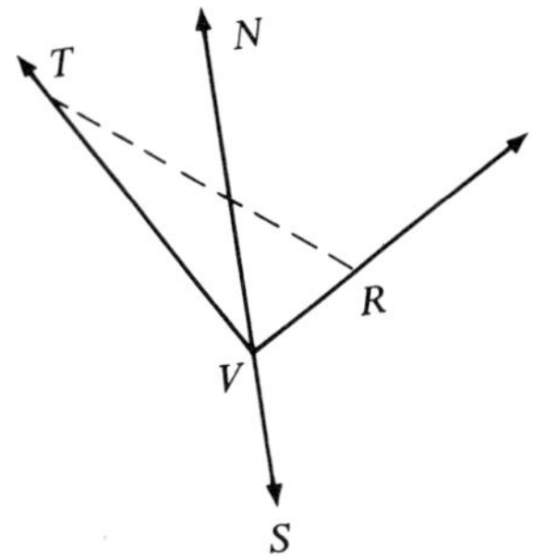

Figure 10-28

In Fig. 10-29, B is 100 mi on a bearing of 120° from A. How far east of A is B? What is the degree measure of $\angle BAC$? Which trigonometric function involves AC and AB with respect to $\angle BAC$? Find AC. This measure represents the distance that B is east of A. How far west of B is A? What is the bearing of A from B?

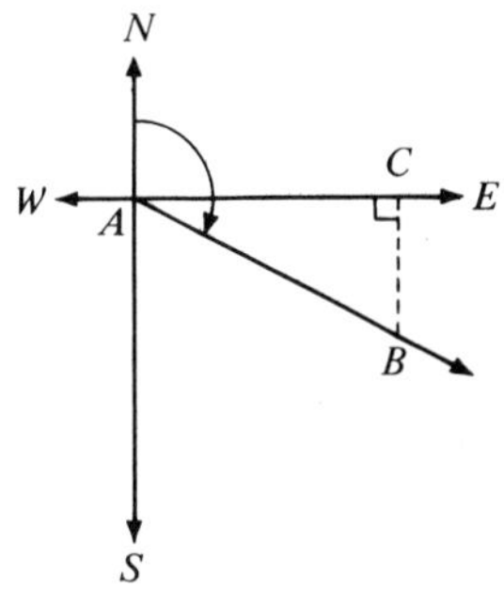

Figure 10-29

Third, consider the instrument pictured in Fig. 10-30(a). It is comprised of a yardstick, the protractor from Appendix C, and a piece of string with a weight attached. Figure 10-31 illustrates how the instrument might be employed to calculate the height of a tree. Suppose the string falls along the protractor, giving a reading of 50°. What, then, is the angle of elevation of the top of the tree? Figure 10-30(b) shows a detailed view of the string across the protractor. Line HVH' indicates the horizontal; $\angle GVP$ shows the reading of 50. What is the degree measure of $\angle HVG$? Of $\angle TVH'$ How does $m(\angle TGP)$ compare with $m(\angle TVH')$? Explain. Is not $\angle TGP$

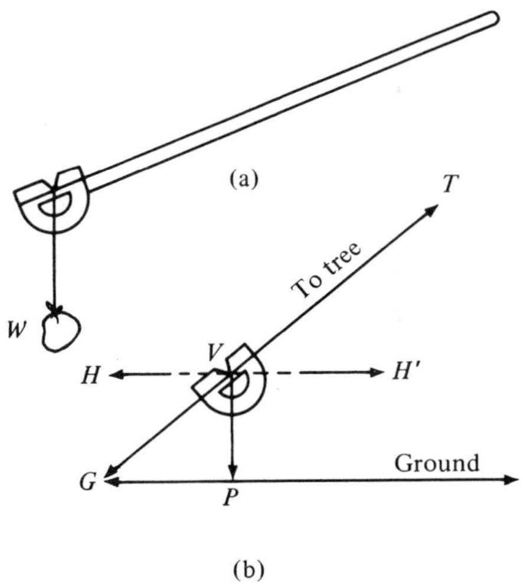

Figure 10-30

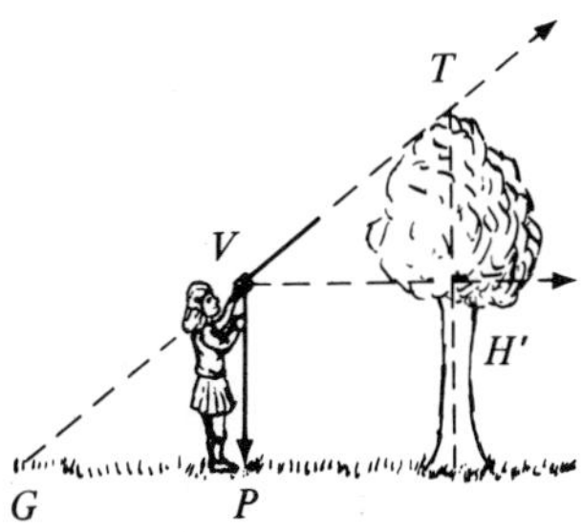

Figure 10-31

the angle of elevation of the top of the tree? Suppose the eye of the individual pictured is 5 ft off the ground and the distance from her eye to the tree is 20 ft. What is the height of the tree? Since $m(\angle TVH') \overset{\circ}{=} 40$ and $VH' = 20$,

$$\tan 40^\circ = \frac{h}{20},$$
$$h = (20)(0.839),$$
$$h = 16.780 \qquad \text{or approximately } 16.8.$$

Then the height of the tree is $(16.8 + 5)$ ft, or 21.8 ft.

Thus we see that the trigonometric ratios, or functions, have widely diversified applications in the real world. In the exercises we explore their use in other than right triangles.

EXERCISE 10-6

1. In $\triangle RST$, $RS = 12$, $ST = 17$, and $m(\angle RST) = 75$.
(a) Find the area of $\triangle RST$.
(b) Using this area, find h_r and h_t. (Recall that "h_r" names the measure of the altitude from vertex R.)

2. Suppose a boat is 100 mi on a bearing of 220° from a port and is sailing due north. How far must a ship sail due west from this port to cross the path of the boat?

3. Suppose you are holding the instrument in Fig. 10-30(a), the string yields a reading on the protractor of 35°, and you are standing 30 ft from the tree. To the nearest integer what is the height of the tree?

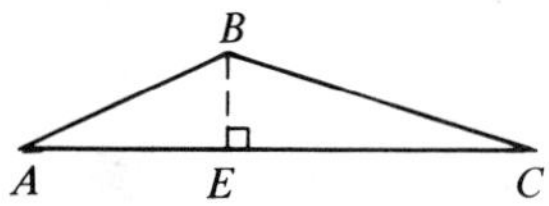

Figure 10-32

4. In Fig. 10-32, $AB = 12$, $BC = 18$, and $m(\angle A) \overset{\circ}{=} 25$. Draw $\overline{BE} \perp \overline{AC}$ with E on $\overline{AC}$.
(a) Find BE.
(b) Find AE.
(c) In right triangle BEC and using BE and BC, express as a fraction the value of $\cos(\angle EBC)$. Change this to a decimal numeral showing three places. In Table 10-3 what angles have a value close to this for their cosine? Then to the nearest degree what is the measure of $\angle EBC$?
(d) Use the result of part (c) and compute EC.
(e) What is the size of $\angle ABE$? Of $\angle C$?
We see that, knowing only three parts, we were able to find the measures of all the parts of the triangle. In the Exercises at the end of this chapter the method is generalized.

If two angles of a triangle are congruent to two angles of another triangle, respectively, the triangles are similar.

If the altitude is drawn to the hypotenuse of a right triangle, (1) the altitude is the mean proportional between the segments of the hypotenuse formed by the altitude, and (2) either leg is the mean proportional between the (whole) hypotenuse and the segment of the hypotenuse adjacent to that leg.

All vertical lines associated with a given sphere intersect.

In a right triangle the square of the hypotenuse is equal to the sum of the squares of the legs.

If in a triangle the square of one side is equal to the sum of the squares of the other two sides, then the triangle is a right triangle with a right angle opposite this first side.

A diagonal of a rectangular solid has a measure equal to the square root of the sum of the squares of the measures of the solid.

The sine ratio of an acute angle has the same value as the cosine ratio of its complementary angle.

The area of any triangle is equal to one half the product of any two sides and the sine of the included angle; hence, $A = (\frac{1}{2})\, ab \sin C$.

EXERCISES

1. To the nearest degree what is the angle of elevation of the sun when a vertical stake with a measurement of 18 in. casts a 5-ft shadow?
2. Assuming the wind is strong enough to hold the kite string straight, how high is a kite above the ground level if the string makes an angle of 48° with the ground and there is 150 ft of string out?
3. In Chapter 7 we *agreed* that if two or more triangles are each similar to the same triangle or to similar triangles, all of the triangles are similar each to each. Essentially, this states that similarity displays the transitive property. Using the first statement in the Summary of Principles above and the fact that if two triangles are similar the corresponding angles are congruent, show that, indeed, if two triangles are similar to a third triangle, the two triangles are similar to each other. (This verifies the transitivity property for similarity.)
4. (a) In Fig. 10-33 a fish has "struck" a line in such a way that the line makes an angle of 12° where it enters the water. (The diagram has been drawn without showing the bending of the light rays as they enter the water—thus it shows the "true" nature of the problem.) If the string leaves the boat at a height of 6 ft above the water and there is 3 ft of line in the water, how much line is out over the edge of the boat?
   (b) For part (a), how far is the fish, horizontally, from the boat?

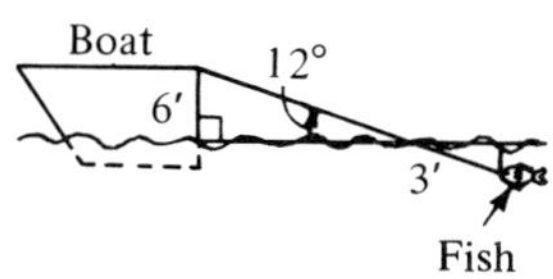

Figure 10-33

5. In equilateral triangle ABC of Fig. 10-34, $\overline{CD}$ is drawn perpendicular to $\overline{AB}$ with D on AB.

(a) Prove that $\triangle ACD \cong \triangle BCD$.

(b) List the degree measure of each angle in the figure.

(c) Let $AC = 2t$; then, without using trigonometric functions, express in terms of t, BC, AB, AD, DB, and CD.

(d) Using your results in part (c) and $\triangle ADC$, determine in fractional form sin 30°, sin 60°. Then write these values in decimal form and compare with the values in Table 10-3.

(e) Follow the instructions in part (d) and find cos 30°, cos 60°, tan 30°, tan 60°. Which results are "better"—the fractional forms or the tabular forms?

6. In square $ABCD$ of Fig. 10-35, $\overline{AC}$ is drawn.

(a) What is the word name for $\overline{AC}$?

(b) Is $\triangle ACD \cong \triangle CAB$? Explain.

(c) Is $\triangle ACD \cong \triangle ACB$? Explain.

(d) Then what can you conclude of $\angle 1$, $\angle 2$, $\angle 3$, $\angle 4$?

(e) Let $AB = s$. Then, without using trigonometric functions, find in terms of s, BC and AC.

(f) Use $\triangle ABC$ and your answers to part (e) to determine in fraction form sin 45°. Change this to a decimal form showing three decimal places and compare with the value in Table 10-3. Do likewise with cos 45°, tan 45°. Are these values dependent upon the size of s?

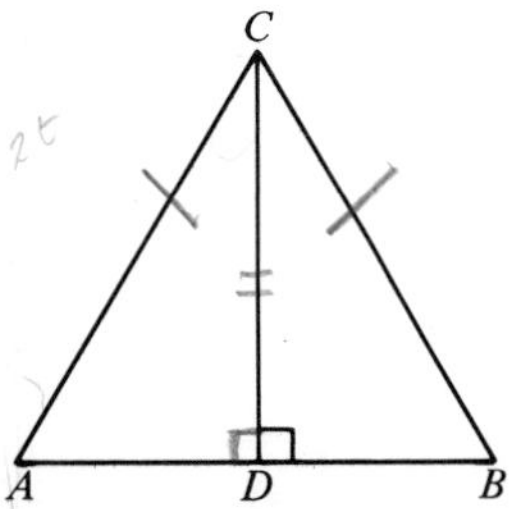

Figure 10-34

7. Given right triangle ABC with right angle at C. Let $BC = a$, $AC = b$, $AB = c$.

(a) Using a, b, and c, write expressions for each of the six trigonometric ratios of $\angle A$.

(b) As in (a), express the ratios of $\angle B$.

(c) Comparing your answers for (a) and (b), you should see some ratios that express the same value. List these by pairs and try to reach some general conclusions about the pairs. Specifically, see if you can devise a scheme whereby you can determine the sine and cosine values of any angle between 0 and 90 and use only the left column of Table 10-3.

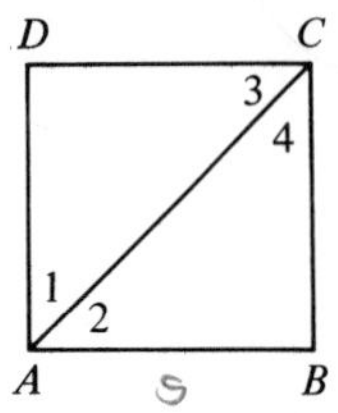

Figure 10-35

8. Build the instrument pictured in Fig. 10-30(a) by attaching your protractor to a yardstick and fastening a piece of string at the center point of your protractor. A weight should be tied at the other end of the string and permitted to hang loose. Then:

(a) Use this instrument as illustrated in Fig. 10-31 to approximate the height of some tall object, such as a building or tree.

(b) If the string is replaced by a stick, perhaps another yardstick, and attached in some fashion so that it can be sighted along what would be the path of the string, the instrument can be used in a horizontal position to determine such a measure as the distance across the swamp pictured in Fig. 10-36. If you use the instrument at the location suggested by S in the figure, with the sticks pointing along $\overline{SB}$ and $\overline{SA}$, respectively, you can determine a measure of $\angle S$. Arrange this setup in the following manner. Establish a line of sight across the swamp, indicated by $\overline{AB}$ in

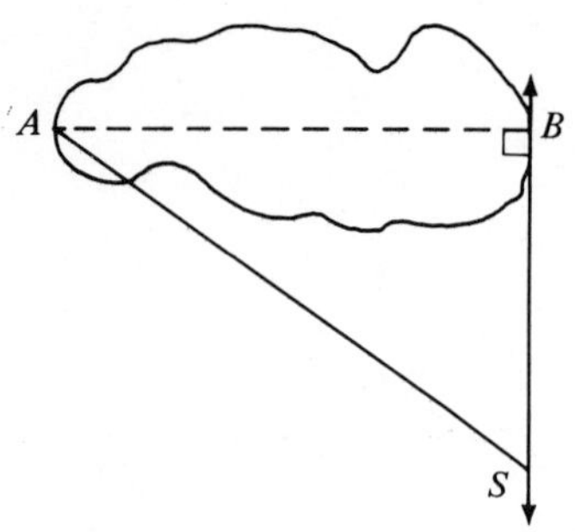

Figure 10-36

the figure. Then establish $\overleftrightarrow{BS} \perp \overline{AB}$ and measure $\overline{BS}$. Now with the measure of $\angle S$, where S is just some convenient point along the established $\overleftrightarrow{BS}$, and the measure of $\overline{BS}$, AB can be computed. Suppose $BS = 65$ and $m(\angle S) \stackrel{\circ}{=} 73$. Find AB.

(c) Use the method suggested in part (b) and find some otherwise unobtainable measurement. (You could even "make believe"—that is, find the distance across a park, pretending that the park is a body of water, or whatever.)

9. See Ref. 15 for further insight into the Pythagorean theorem and its historical development.

10. Associated with any angle of elevation is an angle of depression. Figure 10-37 indicates an airplane sighted at an angle of elevation of 62° from an airport. Thus $\angle GPA$ has a measure of 62, since this is the angle through which $\overrightarrow{PA}$ must be "elevated" from horizontal ray PG. And this is the angle through which someone at the airport has to raise his line of sight from the horizontal. The associated **angle of depression** is the angle through which a person in the plane must lower his line of sight from the horizontal. What angle in the figure represents this? What can you say of the measure of this angle? Explain. Is it true that an angle of depression always has the same measure as its associated angle of elevation? Is the reverse true?

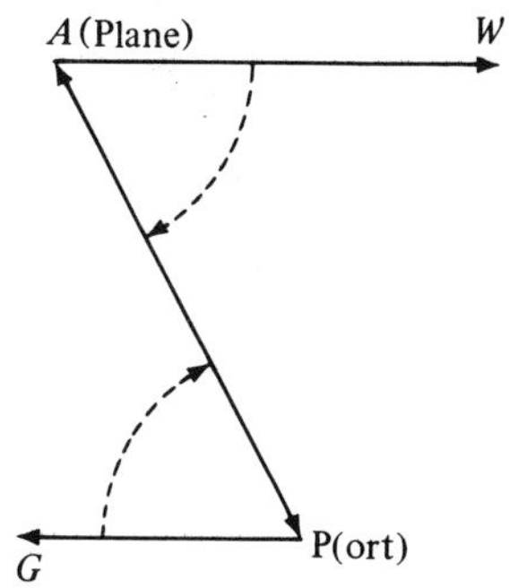

Figure 10-37

11. In $\triangle RST$ of Fig. 10-38, let $\overline{SD} \perp \overline{RT}$ with D on $\overline{RT}$, $RS = t$, $ST = r$, $SD = h_1$, and $RT = s$. Use these letters to express any measures or ratios called for.

(a) In this figure name any triangles that are right triangles.

(b) (1) In $\triangle RSD$ what is sin R?

(2) In $\triangle TSD$ what is sin T?

(3) Using your answers for parts (1) and (2), write two different expressions for h_1.

(4) Since both of these expressions for h_1 name the same quantity, they may be set equal to each other. Do this and then rearrange the expression you get so that letters, such as R and r, are on the same side of the equal sign.

(c) Now make a copy of $\triangle RST$ but do not show $\overline{SD}$. Draw $\overline{TV} \perp \overline{RS}$ with V on $\overline{RS}$. Let $TV = h_2$ and the other segments be as above.

(1) In right triangle SVT, what is sin S? What is sin R in $\triangle RVT$?

(2) As in part (b), solve for h_2, set the expressions equal and rearrange.

(d) By proper rearrangement—or, perhaps, simple observation—you should now have two fractional forms equal to another fractional form. The three fractional forms name the same rational number and show that in any triangle the sines of the angles are proportional to the measures of the opposite sides, respectively. This relationship is known as the *law of sines*. Have we really verified it for "any" triangle? What else would be required?

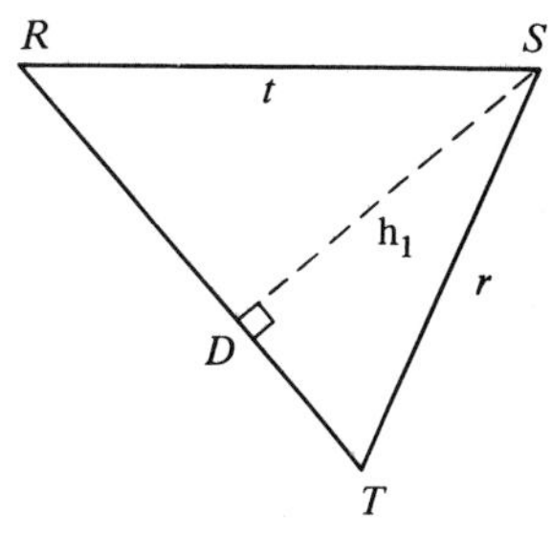

Figure 10-38

CHAPTER 11

Coordinate Geometry

11-1 INTRODUCTION

In Chapter 1 we introduced geometry as the "study of shapes and relative positions." We associated the fundamental concept of point with position. We even suggested ways of naming and thus, in a way, distinguishing between points. However, our concept of relative position has been limited to separation properties or to relations among distances between points. In this chapter we introduce several systems for specifying locations of points, given that we have an arbitrary reference system. Once we have seen how the location systems work, we shall use them as a means to introduce (1) the relationship of algebra to our geometry, and (2) a geometry that differs from the Euclidean geometry of the plane.

11-2 LOCATING POINTS IN ONE-, TWO-, AND THREE-DIMENSIONAL SPACE

The basic one-dimensional figure is the line. We have named individual points on a line (using capital letters), but how do we specify the location of a point on a line? The system we use comes from the measurement model. If three persons, A, B, and C, stand on one line, and you wish to describe their relative positions, you could take one person, such as A, as a reference point and say the others are to the left or right of him. You could also take one person, say A again, as reference point and measure the distance from him to each person. You would then say that B is so many linear units from A and C is so many linear units from A. Suppose that B is 5 ft from A (on the given line). On which side of A is B? Can you tell from the information given?

We have already seen (Sec. 5-8) that there exist one-to-one correspondences between the set of nonnegative real numbers and the set of points on a ray. The correspondence that had certain desirable properties (order, additivity, and so on) we used to coordinatize the ray for measurement purposes.

It is also true that there exist one-to-one correspondences between the set of points on a line and the complete set of real numbers (positive, zero, negative), which also have the properties used in Sec. 5-8. This correspondence we use as a basis for a system of locating points on a line. Every point on a line can be associated with a unique real number, and every real number corresponds to a unique point of the line. The unique number associated with a given point is called its **coordinate**.

Given a line m, as in Fig. 11-1, select any point on m and call it O. Point O is the endpoint of two rays. If P is on one side of O on the line, and N is on the other side of O on the line, name the two rays determined by O. If we select a unit, we can coordinatize $\overrightarrow{OP}$ using the set of nonnegative real numbers (Fig. 11-2). The coordinate of any point of m on the P side of O tells the distance of this point from O. Thus, the coordinate of a point locates it with reference to O. We could also coordinatize $\overrightarrow{ON}$ using the same unit system and the set of nonnegative real numbers, as in Fig. 11-3. The coordinate of any point on the N side of O would indicate its distance from O and thus its location on the ray. However, coordinatizing the line in this fashion would mean that two points of m would be associated with the same real number. This dilemma can be eliminated if we use the set of negative real numbers to coordinatize the ray on one side of O. The usual choice, as in Fig. 11-4, is to associate negative real numbers with the points to the left of (or below) the reference point. The points to the right of (or above) the reference point

Figure 11-1

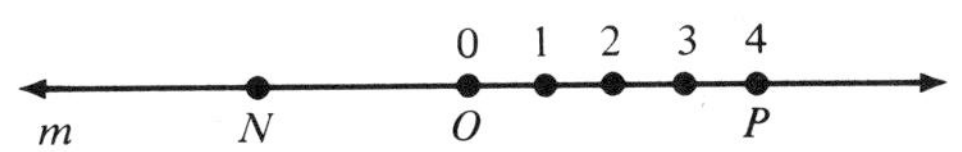

Figure 11-2

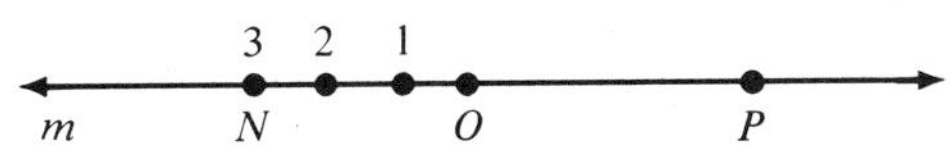

Figure 11-3

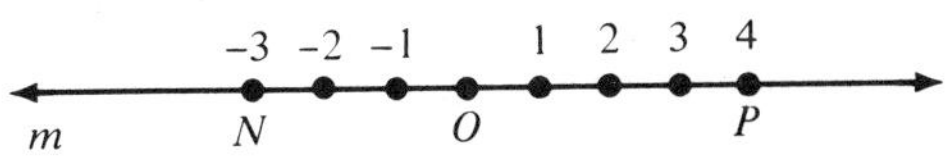

Figure 11-4

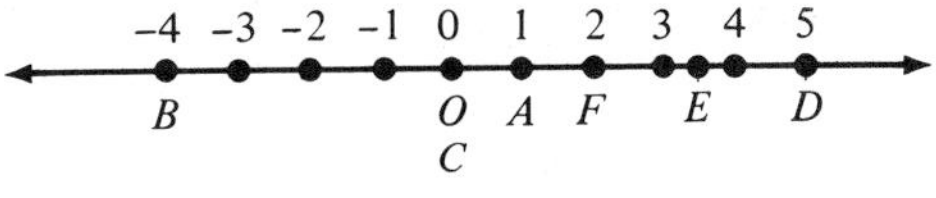

Figure 11-5

are associated with positive real numbers. The reference point, called the **origin,** is associated with zero. Notice that the negative or positive sign then merely indicates which of the two rays contains the point. The numerical value, sometimes called **absolute value**, of the coordinate of a point specifies the distance of the point from the origin.

In Fig. 11-5 what is the coordinate of *A*? Of *B*? Of *C*? Which point in the figure has 5 as its coordinate? Which has 2?

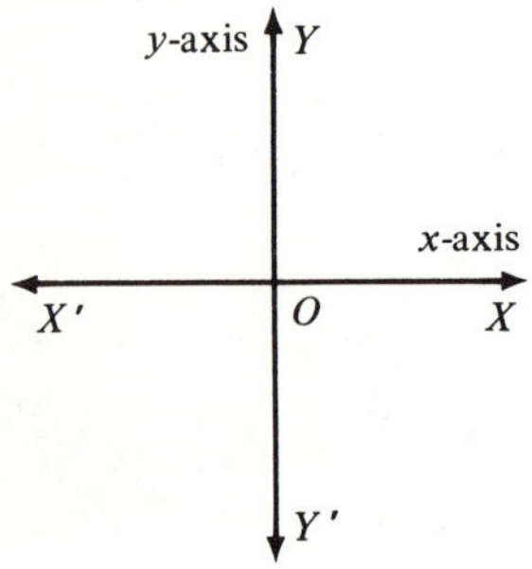

Figure 11-6

Positions in a plane may be located by use of a coordinate system based on two intersecting lines. Recall that a plane is determined by two intersecting lines. If the two reference lines are perpendicular, as in Fig. 11-6, the resulting coordinate system is called a **rectilinear** or **Cartesian** (after René Descartes) **coordinate system.** Each reference line is usually called an **axis** (plural: axes). They are often named the *x-axis* and the *y-axis.*

Every point, *P*, in the plane is the intersection of a pair of lines, as in Fig. 11-7. One of the lines is the *y*-axis or a line parallel to it and the other line is the *x*-axis or a line parallel to it. Notice that every line parallel to the *y*-axis must intersect the *x*-axis in a point; also, every line parallel to the *x*-axis must intersect the *y*-axis in a point. If the *x*- and *y*-axes are coordinatized, as in Fig. 11-8, every line parallel to the *y*-axis can be associated with the real number associated with the point on the *x*-axis where the line intersects the *x*-axis. Likewise, every line parallel to the *x*-axis can be associated with the real number associated with the point on the *y*-axis where the line intersects the *y*-axis. These numbers are called the ***x*-intercept** and ***y*-intercept**, respectively.

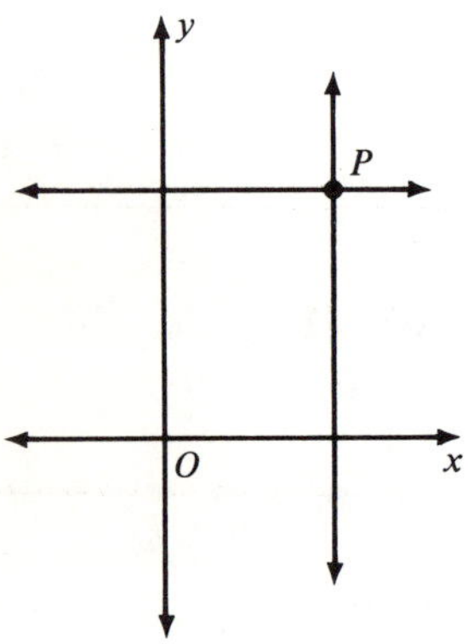

Figure 11-7

The two axes are usually coordinatized so that the intersection of the two axes, called the origin, is associated with zero for each line. Positive numbers are associated with points to the right of the origin on the *x*-axis and with points above the origin on the *y*-axis. The points to the left and below the origin on the axes are associated with negative numbers.

Then, every point on the plane can be associated with a pair of real numbers: one number associated with the lines parallel to the *y*-axis (including the *y*-axis itself) and one number associated with lines parallel to the *x*-axis (including the *x*-axis itself). A very convenient notation can be established if we agree to name the number associated with the parallel to the *y*-axis first and name the number associated with the parallel to the *x*-axis second. The numerals that name the two numbers are enclosed in parentheses and separated by a comma: (3, 2). This is called **ordered pair** notation, since the positions of the numerals in the pair are significant. In Fig. 11-8 the ordered pair associated with point *A* is (3, 4). The numbers mentioned in the pair are called the **coordinates** of the point. To name *A* and its coordinates we write: *A*(3, 4). What are the coordinates of *B*?

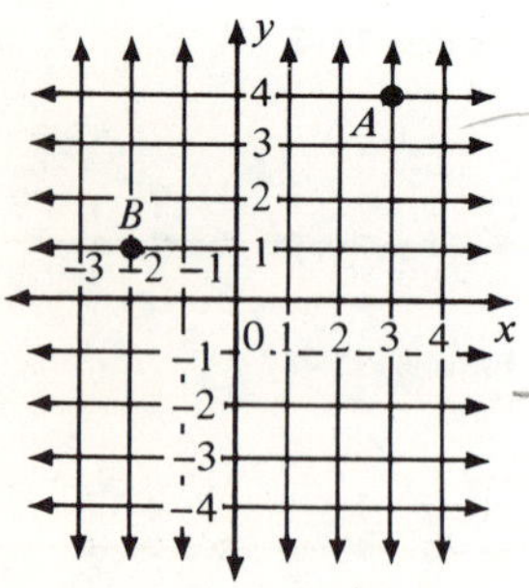

Figure 11-8

The first member of the ordered pair is called the ***x*-coordinate** (or **abscissa**) of the point, since we obtain the number by reading the scale on the x-axis. (Notice, however, that the number is associated with a line parallel to the y-axis. Don't let this confuse you.) The second member of the pair is called the ***y*-coordinate** (or **ordinate**) of the point, since we obtain the number by reading on the y-axis. (With what lines are the numbers on the y-axis associated?) In Fig. 11-9 with what point is the ordered pair $(1, 2\frac{1}{2})$ associated? The pair $(0, 2)$? $(-3\frac{1}{2}, 0)$? What are the coordinates of C? Of D?

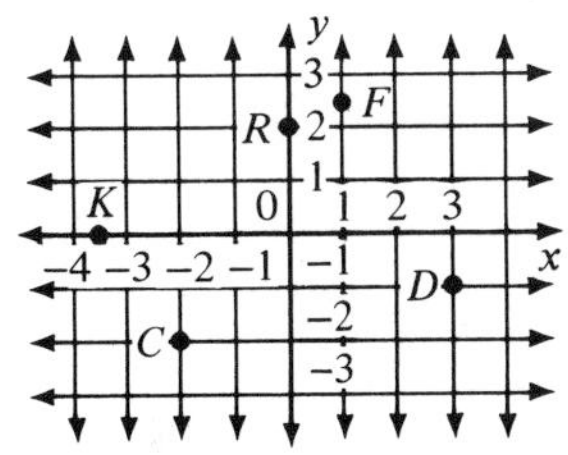

Figure 11-9

To locate points in three-dimensional space, three concurrent reference lines, again called axes, are used. The axes are often called the x-axis, the y-axis, and the z-axis. Figure 11-10 shows one possible labeling of the three lines. (There is not complete agreement on the best way to assign the names to the three lines. Thus, you will find other arrangements in other texts.) Any two of the axes, usually perpendicular lines, determine a plane. The third axis is usually perpendicular to each of the other lines and is thus perpendicular to the plane determined by them. This axis with either of the other two determines a plane. Thus, the three lines determine three planes in space. (See Fig. 11-11.)

Any point, P, in space can be considered as the intersection of three planes: one parallel to the plane determined by the y- and z-axes (this plane intersects the x-axis at some point); one parallel to the plane determined by the x- and z-axes (this plane intersects the y-axis at some point); and one parallel to the plane determined by the x- and y-axes (this plane intersects the z-axis at some point). (See Fig. 11-12.)

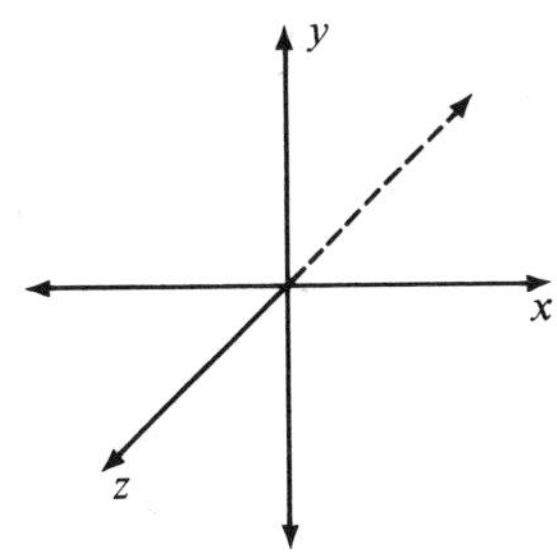

Figure 11-10

Each axis is coordinatized, the point corresponding to zero on each axis being the point of concurrence. Again this point is called the origin of the system. Each point in space is associated with an **ordered triple** of real numbers and usually has the form: (x, y, z). In this case, x is the number associated with the plane parallel to the plane determined by the y- and

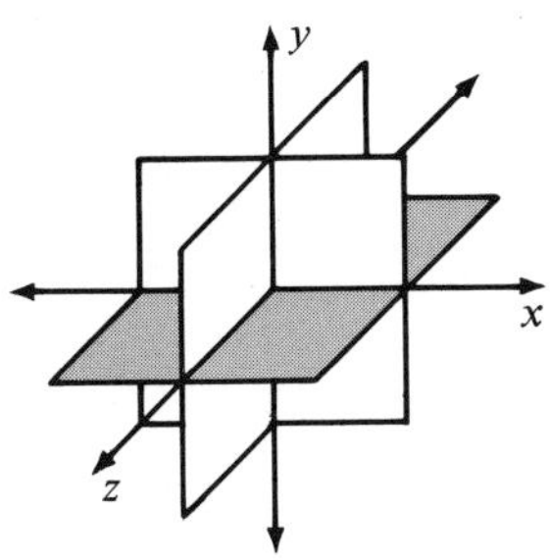

Figure 11-11

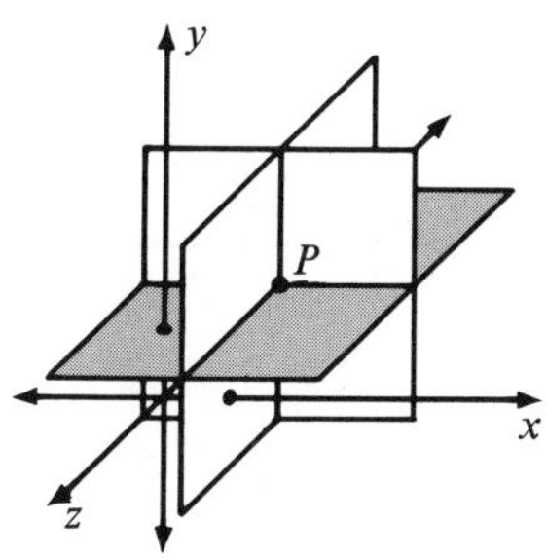

Figure 11-12

z-axes, y is the number associated with the plane determined by the x- and z-axes, and z is the number associated with the plane determined by the x- and y-axes. In Fig. 11-13, (2, 3, 1) is associated with A, and is indicated: $A(2, 3, 1)$. Thus, $(-1, -2, 0)$ is associated with B. (Note that drawings of three-dimensional coordinate systems can be difficult to read.) What point seems to be associated with $(-2, 1, 1)$? With $(2, -1, -\frac{1}{2})$?

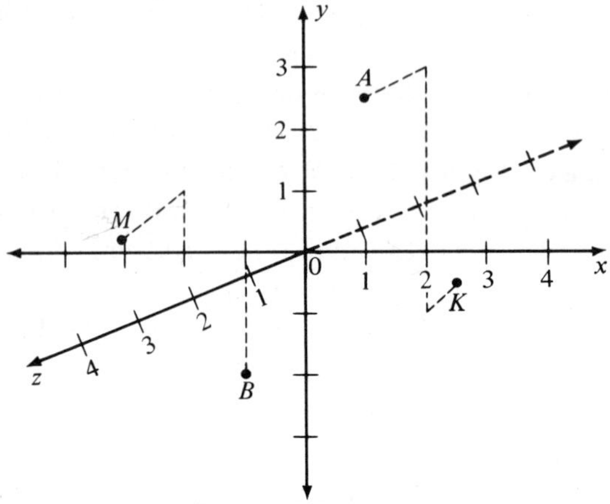

Figure 11-13

EXERCISE 11-2

1. What is the fundamental purpose of a coordinate system?
2. Look up the etymology of the word "rectilinear." Why was this word chosen to describe the Cartesian coordinate system? Why is this system commonly called the rectangular coordinate system?
3. (a) What are the line coordinates of the following points in Fig. 11-14? A, B, C, D, E.
   (b) Coordinatize a line and locate the following points. $A(3)$, $B(3\frac{1}{2})$, $C(-\frac{1}{2})$, $D(4.75)$, $E(0)$, $F(-1\frac{1}{4})$, $G(-15)$.

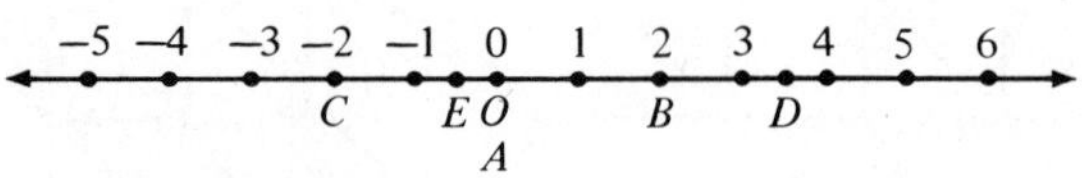

Figure 11-14

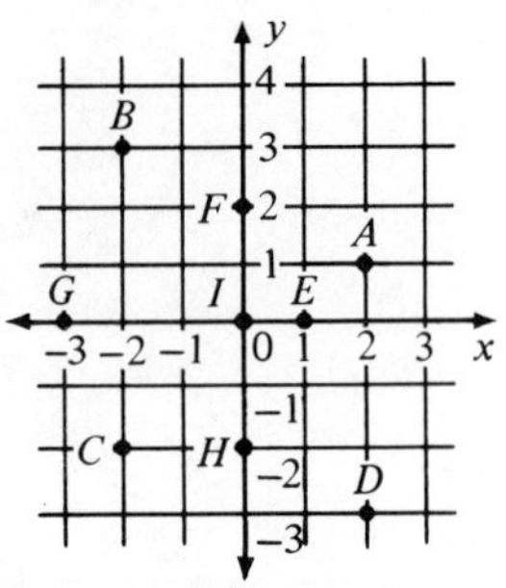

Figure 11-15

4. (a) What are the plane coordinates of the following points in Fig. 11-15? $A, B, C, D, E, F, G, H, I$.
   (b) Locate the following points on a plane. $A(1, 2)$; $B(-3, 4)$; $C(-3, -3)$; $D(1, -2)$; $E(3, 0)$; $F(0, 3)$; $G(-3, 0)$; $H(0, -3)$.

5. The two axes of the coordinate system for a plane determine four regions of the plane. Each region is called a **quadrant**. Figure 11-16 indicates the names given to the four quadrants. Quadrant I contains only those points whose x-coordinate is positive and whose y-coordinate is positive. In a similar manner describe the coordinates of points for each of the other three quadrants.

6. How many regions in space are determined by the three planes of the coordinate system for space? What might these regions be called?

7. (a) On a line what are the coordinates of the points one unit from $P(5)$? How many such points are there?
   (b) On a plane what are the coordinates of a point one unit to the right of $P(3, 4)$? One unit to the left of P? One unit above P? One unit below P?
   (c) In space what are the coordinates of a point one unit to the right of $P(3, 4, 5)$? [*Hint:* The x-coordinate is usually associated with right-left direction, the y-coordinate with up-down direction, and the z-coordinate with front-back direction.] One unit to the left of P? One unit above P? One unit below? One unit in front of? One unit in back of?

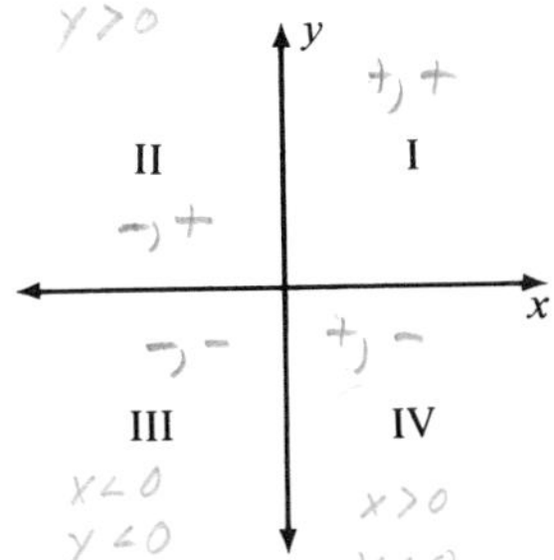

Figure 11-16

11-3 COORDINATE GEOMETRY OF A LINE

The language of algebra can be used to describe the sets of points that are subsets of a line. Figure 11-17 shows a coordinatization of line r. Consider $\overrightarrow{AB}$ in r. Every point in $\overrightarrow{AB}$ has a coordinate on r. Notice that these coordinates are either equal to 2, in the case of point A, or greater than 2, in the case of every other point on $\overrightarrow{AB}$. If we let the variable[1] m represent the coordinate of any point in the ray, it seems natural to say that $m \geq 2$ describes the coordinate of every point in the ray. The statement "$m \geq 2$" is an open sentence in one variable, m; and the solution set of this open sentence contains the coordinates of every point in $\overrightarrow{AB}$. The picture of $\overrightarrow{AB}$ is often called the graph of the open sentence. What open sentence describes $\overrightarrow{AC}$?

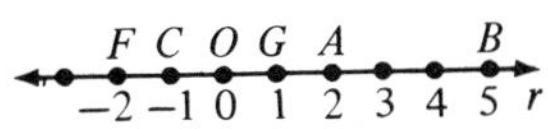

Figure 11-17

Does $\overset{\circ\!\!\rightarrow}{AB}$ have an endpoint? What open sentence describes $\overset{\circ\!\!\rightarrow}{AB}$?

Since a segment can be described as a set of points on a line between and including two given points, the coordinates of the points in a segment must be between or equal to the coordinates of the endpoints. Thus, if m represents the coordinates of the points in $\overline{FG}$ of Fig. 11-17, $\overline{FG}$ is described by $-2 \leq m \leq 1$.

[1] Unless otherwise specified, the replacement set for any variable used in this chapter is the set of real numbers.

EXERCISE 11-3

Questions 1–8 refer to Fig. 11-18.

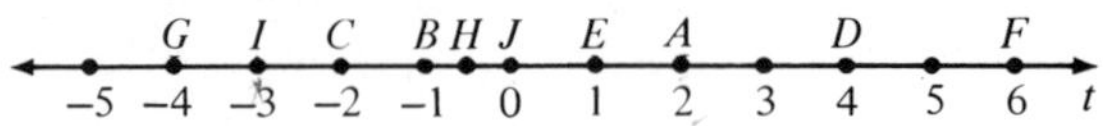

Figure 11-18

1. If m represents a coordinate, what kind of geometric figure is described by $m = 3$?

2. Name the segments that are described by the following open sentences:
(a) $2 \leq m \leq 6$. (b) $-4 \leq m \leq -1$.
(c) $-3 \leq m \leq 0$.

3. Describe each of the following with an open sentence: (a) $\overline{CD}$, (b) $\overline{EF}$, (c) $\overline{GH}$.

4. Name the ray described by $m \leq 6$.

5. Write an open sentence that describes $\overrightarrow{GH}$.

6. Name the half-line described by $m > 1$.

7. Write an open sentence that describes $\overset{\circ\!\!\rightarrow}{CD}$.

8. What set of points is described by $m^2 = 4$?

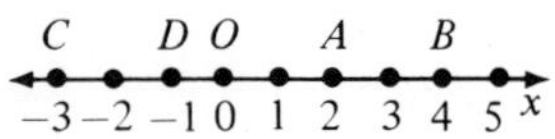

Figure 11-19

9. Complete the following table. Each segment listed is contained in line x of Fig. 11-19.

Segment	*Distance between endpoints*	*Coordinates of endpoints*	*Difference of coordinates*
$\overline{AB}$	2	4 and 2	2
$\overline{CD}$			
$\overline{CA}$			
$\overline{OA}$			
$\overline{DO}$			

What relationship seems to exist between the distance between the endpoints of the segment and the difference of the coordinates of the endpoints of the segment?

10. Can a triangle or a circle be represented on a line? Why?

11-4 COORDINATE GEOMETRY OF A PLANE

Perhaps you noticed that for open sentences corresponding to sets of points in a line only one variable was used. When open sentences are used to describe sets of points in a plane, two variables are usually, but not always, required. One variable corresponds to the x-coordinate of a point, the other to its y-coordinate. The next exercises will explore some of the relationships of algebra to plane geometry and the relationship of coordinate geometry to some of the concepts discussed earlier in the text.

EXERCISE 11-4

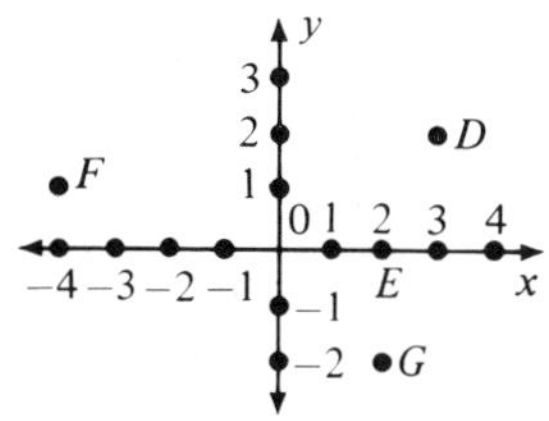

Figure 11-20

1. (a) Locate the following points in a plane: $A(2, 3)$; $B(-4, 5)$; $C(-6, 0)$.
(b) What are the coordinates of the following in Fig. 11-20: D, E, F, G?

2. Using a Cartesian coordinate system, locate the following points: $A(3, 5)$; $B(0, 2)$; $C(-3, -1)$; $D(\frac{1}{2}, 2\frac{1}{2})$; $E(-5, -3)$; $F(-2, 0)$.
(a) What geometric figure is suggested?
(b) What relationship exists between the y-coordinate and the x-coordinate of each point located?
(c) Write an open sentence that expresses this relationship using the variables x and y.

3. Using a Cartesian coordinate system, locate the following points: $A(1, -3)$; $B(0, 0)$; $C(-2, 6)$; $D(3, -9)$; $E(\frac{1}{2}, -1\frac{1}{2})$; $F(-1, 3)$.
(a) What geometric figure is suggested?
(b) What relationship exists between the y-coordinate and the x-coordinate of each point located?
(c) Write an open sentence that expresses this relationship using the variables x and y.

4. Using a Cartesian coordinate system, locate the points $A(0, -2)$ and $B(3, 1)$.
(a) Draw a picture of the line determined by points A and B.
(b) Select several other points on $\overleftrightarrow{AB}$ and write their coordinates in tabular form:

x-coordinate	*y-coordinate*

(c) What relationship seems to exist between the y-coordinate and the x-coordinate of each point?
(d) Write an open sentence that expresses this relationship.

5. (a) Given the open sentence $y = 2x$, find several ordered pairs that make the sentence true.
(b) Locate the points corresponding to the ordered pairs in part (a) using a Cartesian coordinate system.
(c) What geometric figure is suggested?

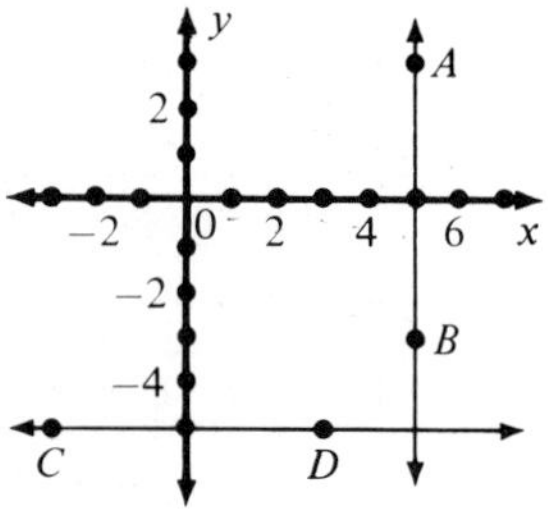

Figure 11-21

6. (a) What is the x-coordinate of every point in $\overleftrightarrow{AB}$ in Fig. 11-21? Does $x = 5$ describe this line?
   (b) What is the y-coordinate of every point in $\overleftrightarrow{CD}$ in Fig. 11-21? Does $y = -5$ describe this line?
7. What geometric figure in a plane corresponds to the open sentence $x > 5$?
8. (a) Using a Cartesian coordinate system, locate points corresponding to ordered pairs that make $y = x + 1$ a true sentence. What figure is suggested?
   (b) On the same set of axes, locate points corresponding to ordered pairs that make $y = x + 3$ a true sentence. What figure is suggested?
   (c) What is the relationship between the two figures suggested in parts (a) and (b)?
9. (a) Using a Cartesian coordinate system, locate points corresponding to ordered pairs that make $y = 2x$ a true sentence. What figure is suggested?
   (b) On the same set of axes, locate points corresponding to ordered pairs that make $y = -\frac{1}{2}x$ a true sentence. What figure is suggested?
   (c) What is the relationship between the two figures suggested in parts (a) and (b)?
10. Using a Cartesian coordinate system, locate the points, $A(3, 4)$, $B(-3, 1)$, and $C(2, -3)$. Draw $\overline{AB}$, $\overline{BC}$, and $\overline{CA}$. What figure is suggested? What do A, B, and C represent in the figure?
11. On a Cartesian coordinate system locate the points $A(-1, 2)$, $B(5, 2)$, $C(5, -3)$, and $D(-1, -3)$. Draw $\overline{AB}$, $\overline{BC}$, $\overline{DC}$, and $\overline{DA}$. What figure is suggested? What is its perimeter? What is its area?
12. In each part below, the coordinates of the vertices for a pair of triangles are given. Draw three separate Cartesian coordinate axes, one each for parts (a), (b), and (c). For each part locate the vertices and draw the sides of the triangles. Then decide which of the transformations—translation, reflection, rotation—is illustrated. (See Sec. 7-5.) If it is a reflection, specify the line of reflection.
    (a) $\triangle ABC$: $(-5, 5)$, $(-1, 1)$, $(-5, 1)$; $\triangle DEF$: $(5, 5)$, $(1, 1)$, $(5, 1)$.
    (b) $\triangle GHI$: $(-1, 4)$, $(-1, 2)$, $(-6, 2)$; $\triangle JKL$: $(1, -1)$, $(3, -1)$, $(3, -6)$.
    (c) $\triangle MNO$: $(-4, 7)$, $(-1, 2)$, $(-7, 3)$; $\triangle PQR$: $(6, 12)$, $(9, 7)$, $(3, 8)$.
13. Since the lines that are used to locate points in the Cartesian coordinate system are perpendicular, the Pythagorean theorem can be used to develop a formula for calculating the linear distance between two points in a plane if the coordinates of the points in that plane are known.
    (a) Locate on a rectangular coordinate axis system: $A(3, 1)$ and $B(6, 5)$. Draw $\overline{AB}$. Label C as the intersection of the line parallel to the x-axis that contains A and the line parallel to the y-axis that contains B. Draw $\overline{AC}$ and $\overline{BC}$. Notice that a right triangle is formed with the right angle at C. How do you know this? Subtract the x-coordinate of A from the x-coordinate of B. What is the difference? What is the measure of $\overline{AC}$? Subtract the y-coordinate of A from the y-coordinate of B. What is the difference? What is the measure of $\overline{BC}$? Use the Pythagorean theorem to find the measure of $\overline{AB}$.

(b) The following discussion refers to Fig. 11-22. $P_1(x_1, y_1)$ and $P_2(x_2, y_2)$ are two points in a plane. N is the intersection of the line through P_1 parallel to the x-axis and the line through P_2 parallel to the y-axis. Triangle P_1NP_2 is a right triangle. The measure of $\overline{P_1N}$ is $(x_2 - x_1)$ and the measure of $\overline{P_2N}$ is $(y_2 - y_1)$. Using the Pythagorean theorem, the measure d of $\overline{P_1P_2}$ is given by

$$d = \sqrt{(x_2 - x_1)^2 + (y_2 - y_1)^2}.$$

Find the distance between the two points whose coordinates are $(-2, 1)$ and $(5, 7)$.

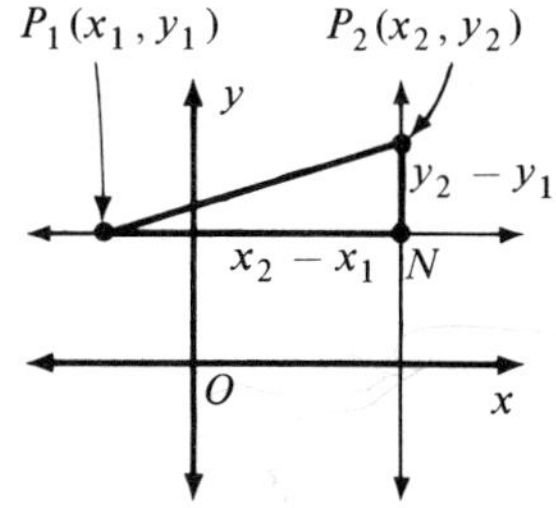

Figure 11-22

14. The trigonometric ratios were discussed for acute angles in Secs. 10-4 and 10-5. With the introduction of coordinate geometry the trigonometric ratios can be described for any angle. In this exercise we investigate only the case of the sine ratio. In Fig. 11-23, $\overrightarrow{OA}$ is drawn on a Cartesian coordinate axis system. Notice that the endpoint of the ray is the origin. Notice also that $\overrightarrow{OA}$ and the x-axis determine acute angle AOX. Point A is in quadrant I.

(a) If P is a point in quadrant II, what kind of angle is determined by $\overrightarrow{OP}$ and $\overrightarrow{OX}$? By $\overrightarrow{OP}$ and $\overrightarrow{OX'}$? What is the relationship of these two angles?

(b) If P is in quadrant III, what kind of angle is determined by $\overrightarrow{OP}$ and $\overrightarrow{OX}$? By $\overrightarrow{OP}$ and $\overrightarrow{OX'}$?

(c) If P is in quadrant IV, what kind of angle is determined by $\overrightarrow{OP}$ and $\overrightarrow{OX}$? By $\overrightarrow{OP}$ and $\overrightarrow{OX'}$?

(d) If P is in $\overrightarrow{OX}$, what is $m(\angle XOP)$?

(e) If P is in $\overrightarrow{OY}$, what is $m(\angle XOP)$?

(f) If P is in $\overrightarrow{OX'}$, what is $m(\angle XOP)$?

(g) If P is in $\overrightarrow{OY'}$, what is the measure of reflex angle XOP?

(h) If P_1 and P_2 are on the same ray in quadrant II, why is

$$\frac{AP_1}{OP_1} = \frac{BP_2}{OP_2}?$$

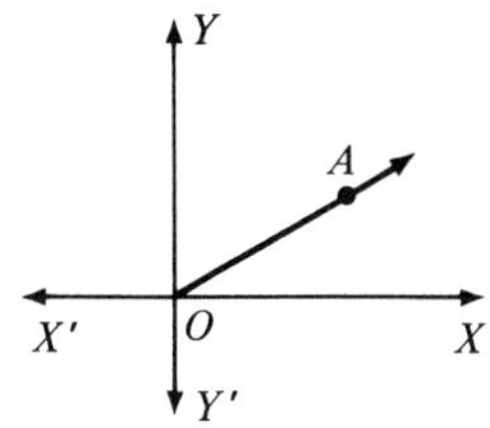

Figure 11-23

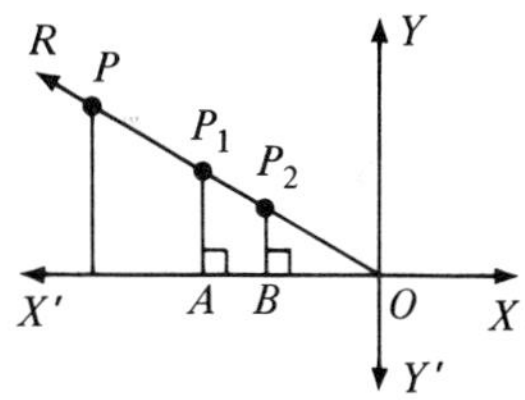

Figure 11-24

(See Fig. 11-24.) Notice that P_1A is also the y-coordinate of P_1, and P_2B is the y-coordinate of P_2. In general, for a given $\angle XOR$, where P is any point in $\overrightarrow{OR}$, the ratio of the y-coordinate of P to distance OP is a constant for that angle. This is true no matter what quadrant $\overrightarrow{OR}$ is in. This fact provides the basis for one definition of the sine of any angle: If a copy of an angle is made on a Cartesian coordinate axis system so that the vertex of the angle is the origin, the initial side of the angle is $\overrightarrow{OX}$, and any point P other than the vertex is selected on the terminal side, then the sine of the angle is given by the ratio of the y-coordinate of point P to OP. (Notice that in this discussion OP is always positive.)

(i) Using a Cartesian coordinate system, draw angles with initial sides $\overrightarrow{OX}$ and terminal sides such that the angles have the following measures in angular degrees: 25, 115, 200, 310, 0, 90, 180, 270. Select P in each of the terminal sides and determine the sine of the angle using the definition in (h). You will save yourself considerable effort if you select P so that $OP = 1$.

(j) Using the definition given in (h) and the technique suggested in (i), find the sines of the angles whose degree measures are: 30, 45, 60, 80. Then find the sine of the supplement of each angle. What relationship do you notice?

15. The following articles present some ideas for using coordinate geometry in the elementary school: Ref. 73, pp. 107–109; Ref. 78, pp. 549–552; Ref. 97, pp. 41–43; and Ref. 83, pp. 689–693.

16. Think of a moving point P that has the following restrictions placed on its motion: (1) motion is restricted to a plane, (2) positions are restricted to a fixed distance of 5 units from a fixed point, which we will call C.

(a) Describe the path of P in terms of a well-known geometric figure.
(b) In such a figure what is C called?
(c) What is the fixed distance called?
(d) The distance from P to C is always how many units?

17. Suppose for the situation described in problem 16 we locate $C(3, 4)$ using a Cartesian coordinate system. Since P has a great many different positions, let us use a variable name to specify any one of its positions—for example, (x, y).

(a) What must be the distance between the point named by (3, 4) and the point named by (x, y)?

Note that this situation is related to the distance formula given in problem 13 above: $d = 5$, $x_2 = x$, $x_1 = 3$, $y_2 = y$, $y_1 = 4$. Thus we can write

$$\sqrt{(x - 3)^2 + (y - 4)^2} = 5,$$

or

$$(x - 3)^2 + (y - 4)^2 = 5^2,$$

or

$$x^2 - 6x + y^2 - 8y = 0.$$

(b) What information about the path is given by the "3" and "4" in the equation?
(c) What information is given by the "5"?
(d) What do "x" and "y" represent?
(e) What does the equation represent?
(f) Write an equation of a circle whose center is at (5, 6) and whose radius is 10.
(g) Write an equation of a circle whose center is at (−2, 3) and whose radius is 11.
(h) Write an equation of a circle whose center is at (h, k) and whose radius is r.
(i) Devise a technique for finding the coordinates of the center and the radius of a circle whose equation is

$$x^2 - 4x + y^2 - 6y + 12 = 0.$$

18. For an introduction to other systems for locating points in a plane read the following articles on polar coordinates: Ref. 2, pp. 309–313; Ref. 42, pp. 129–142; and Ref. 72, pp. 563–565.

19. For a summary of Cartesian coordinate systems see Ref. 30, pp. 208–213.

11-5 COORDINATE GEOMETRY IN SPACE

If you were to write open sentences corresponding to sets of points in space, how many variables would generally be needed? We bypass any extensive discussion of three-dimensional coordinate geometry and investigate intuitively in the next exercises the use of three-dimensional coordinate systems for locating figures in space.

EXERCISE 11-5

1. (a) Draw a three-dimensional coordinate system and locate the following points: $A(1, 6, 1)$; $B(1, 1, 4)$; $C(1, 1, 1)$; and $D(3, 1, 4)$.
   (b) Draw $\overline{AB}$, $\overline{AC}$, $\overline{AD}$, $\overline{BC}$, $\overline{CD}$, and $\overline{DB}$. These segments appear to be the edges of what geometric figure?
2. Specify the coordinates of eight points in space such that they will be the vertices of a rectangular prism.
3. Describe the location of all points whose x-coordinate is 3.
4. Describe the location of all points whose x-coordinate is greater than 3.
5. How many coordinates would be needed to specify the location of a point in four-dimensional space? In n-dimensional space? See Ref. 50, pp. 306–309, for a discussion of n-dimensional coordinate geometry.

11-6 COORDINATES ON A SPHERE; SOME SPHERICAL GEOMETRY

Many adults are familiar with the system of locating points on the Earth's surface that uses the concepts of latitude and longitude. To understand this system you must realize the the Earth is often represented by a spherical solid with the surface of the Earth taken as a sphere, as in Fig. 11-25. A geometry quite unlike that with which we have been concerned applies where the points of a system occur on a portion of the surface known as a sphere. Such a geometry is called **spherical geometry**.

The first difference to be noticed between spherical and Euclidean plane geometry is in the interpretation of a line. In plane geometry a line is "straight"; in spherical geometry a line is "curved." In fact, lines on a sphere are particular circles, contained in the surface, called great circles.

What are great circles of a sphere? We first note that there are three possible cases for the intersection of a plane and a sphere: (1) the empty set [Fig. 11-26(a)], (2) a single point, in which case the plane is said to be **tangent** to the sphere [Fig. 11-26(b)], and (3) a circle [Fig. 11-26(c)]. If the

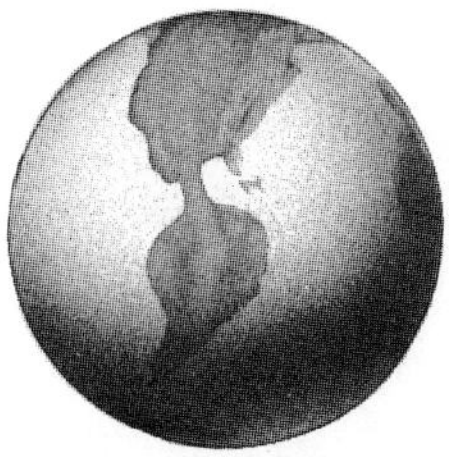

Figure 11-25

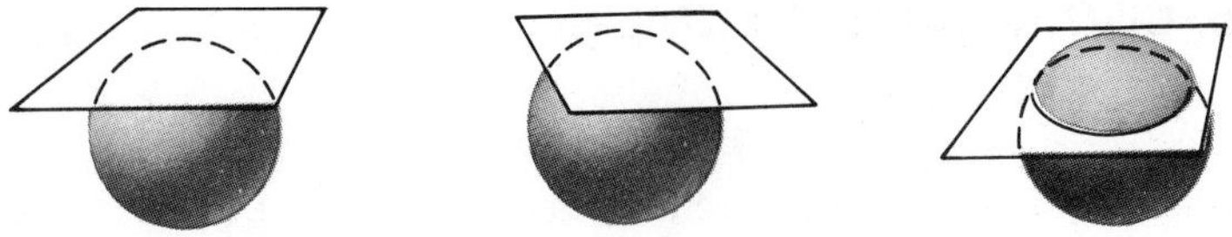

(a) (b) (c)

Figure 11-26

intersecting plane contains the center of the sphere, then the circle of the intersection is called a **great circle**. Otherwise, the circle of the intersection is called a **small circle**. (See Fig. 11-27.) How does the radius of a great circle of a sphere compare with the radius of the sphere? How does the radius of a small circle compare with the radius of the sphere?

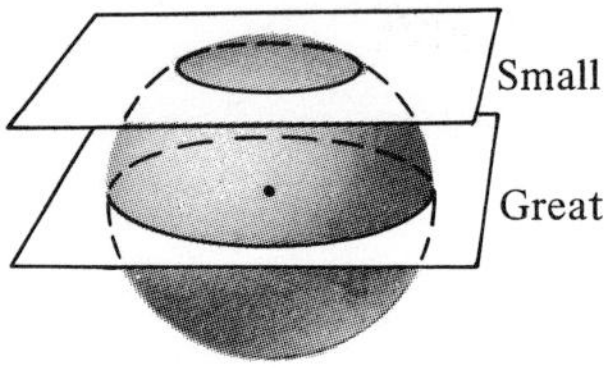

Figure 11-27

Why are the great circles taken to be the lines on a sphere? One of the reasons is that given any two points on a sphere, the shortest path on the sphere connecting the two points lies along a great circle containing the two points. This is similar to the situation for the shortest path on a plane connecting two points in the plane. In this case the shortest path lies along the straight line determined by the two points. In general, the shortest path on a surface between two points in the surface is called a **geodesic**. Straight lines contain the geodesics in a plane; great circles contain the geodesics in a sphere.

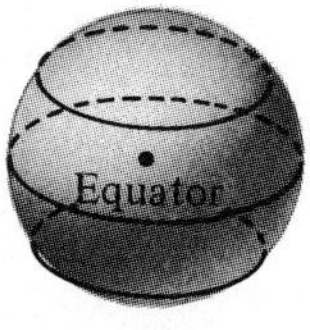

Figure 11-28

To return to the model of the Earth, the equator is a great circle. (An **equator** is a circle that separates a sphere into two hemispheres.) The **parallels of latitude** include the equator and the intersections of the sphere and planes parallel to the plane containing the equator. (See Fig. 11-28.) Are the parallels of latitude great or small circles?

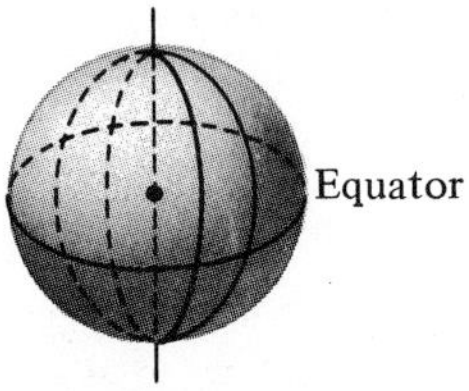

Figure 11-29

The **meridian circles**, used to determine longitude, are great circles of the sphere. They are the intersection of the sphere with planes that contain the center of the sphere and are perpendicular to the plane of the equator. (See Fig. 11-29.)

The intersection of all the planes that determine the meridian circles is the line called the **axis** of rotation of the Earth. The intersection of the axis (which is perpendicular to the plane of the equator and contains the center of the Earth) with the sphere determines the poles of the Earth.

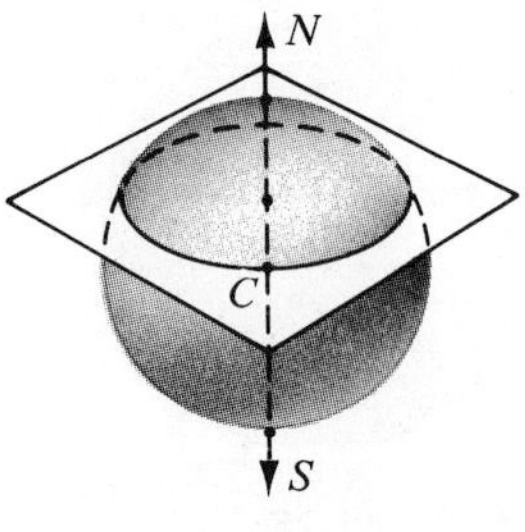

Figure 11-30

On a sphere every circle has an associated pair of **poles**. The poles, a pair of points, are the intersection of the sphere and a line containing the center of the sphere and perpendicular to the plane containing the given circle. (See Fig. 11-30.) What is the name of the segment that joins the two poles of a circle? In general, the endpoints of a diameter of a sphere are called **antipodal points.**

Exercise 11-6 will explore some other aspects of the geometry of a sphere.

We now investigate a system for coordinatizing a sphere. For the Earth, this leads to the concepts of latitude and longitude. Every point, P, in a sphere can be considered as the intersection of a parallel of latitude and a meridian circle, as in Fig. 11-31. To specify the location of a particular point it would seem that we need only specify the parallel of latitude and the meridian circle. Thus, a system of coordinatization similar to that for planes might be suggested. However, there is one slight difficulty. How many points are contained in the intersection of a parallel of latitude and a meridian circle? Why did you say two? We proceed with this difficulty in mind. Select a convenient parallel of latitude and a convenient meridian circle as reference lines. (For the Earth, these are respectively the equator and the meridian circle passing through the Greenwich Observatory, England.) One of the intersections of the two circles is called the origin, which we indicate as O in Fig. 11-32(a).

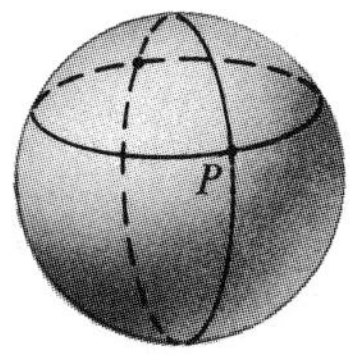

Figure 11-31

What set of numbers should we use to coordinatize the reference lines? The two planes e and p in Fig. 11-32(b), which are determined by the reference lines, are perpendicular and determine four regions on the sphere. We also locate the poles of the equator and call them N and S. The sphere is then coordinatized as follows. Every parallel of latitude intersects either $\widehat{ON}$ or $\widehat{OS}$. The point of intersection, X, and the origin, O, determine an arc. We coordinatize $\widehat{ON}$ and $\widehat{OS}$ by associating with each point X on the arc degree measure of $\widehat{XO}$. [See Fig. 11-32(c).] The parallels of latitude are then identified by the number associated with the intersection of the parallel of latitude and $\widehat{ON}$ or $\widehat{OS}$. Since this system would allow two different parallels of latitude to be associated with the same number (one above and one below the equator), some means of distinguishing the two sets is desirable. Thus, those parallels of latitude on the N side of e are called north and labeled N; those on the S side of e are called south and labeled S. The number and direction associated with a given parallel of latitude is called **latitude**. The latitude of circle A is $45N$. What is the latitude of circle B? What are the latitudes of the poles? Of the equator?

Every meridian circle intersects the equator at two points: one on the W side of p and one on the E side of p. [See Fig. 11-32(d).] Each of these points, Y, determines with the origin a minor arc. We coordinatize the two semicircles OWR and OER by associating with each point of the semicircle the arc degree measure of the minor arc determined by the given point and the origin. Each semicircle determined by the poles N and S on a meridian circle is called a **meridian**. Each meridian is identified by the number associated with the intersection of the meridian and the equator. To distinguish the meridians on the W side of p from those on the E side of p, the former are called west and labeled W while the latter are called east and labeled E. The number and direction associated with each meridian is

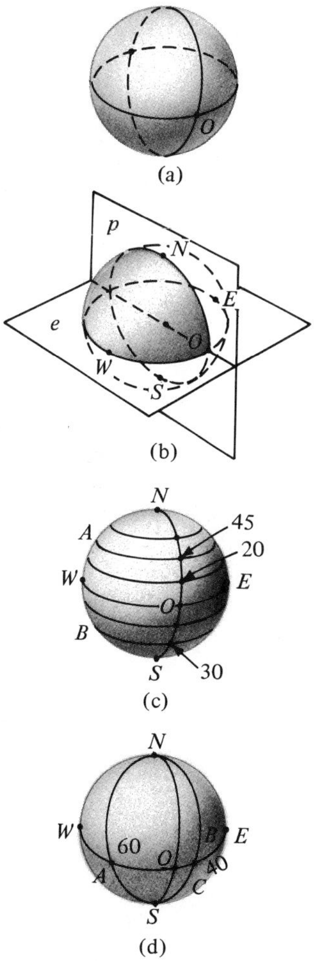

Figure 11-32

called **longitude**. The longitude of meridian *NBS* in Fig. 11-32(d) is 40*E*. What is the longitude of meridian *NAS*? Of meridian *NOS*? This meridian is called the **prime meridian**.

We can locate any point on the sphere by identifying the parallel of latitude and the meridian whose intersection is the point. We do so by giving the **spherical coordinates** of the point—an ordered pair consisting of the latitude and longitude associated with the point. Thus, in Fig. 11-33 we have *P*(20*N*, 10*E*).

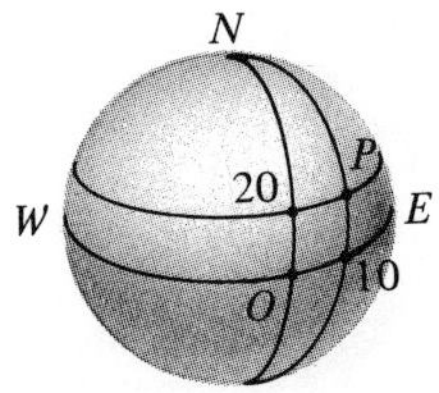

Figure 11-33

EXERCISE 11-6

1. On a model of a sphere (rubber ball, slate sphere, or the like):
   (a) Locate two points (not the endpoints of a diameter) and draw the great circle through the two points.
   (b) Draw a small circle through the two points.
   (c) Compare the length of the arc determined by the points on the great circle with the length of the arc determined on the small circle.

2. Why is it that ships normally do not sail along parallels of latitude when going from port to port?

3. What are the spherical coordinates of:
   (a) Your city or town. (b) New York City.
   (c) Melbourne, Australia. (d) Santiago, Chile.
   (e) Paris, France.

4. What cities have the following spherical coordinates?
   (a) 0, 0. (b) 56*N*, 38*E*.
   (c) 11*N*, 67*W*. (d) 23*S*, 43*W*.
   (e) 6*S*, 107*E*.

5. Many people think of a sphere as three-dimensional. In effect we have shown that it is two-dimensional. Explain.

6. You may wish to read or reread Ref. 8 at this time.

7. Do Exercise 1-7, 6.

8. On a spherical surface a **spherical angle** is sometimes defined as the union of two great-circle arcs that have a common endpoint. The common endpoint is called the vertex. The measure of the spherical angle is the measure of the dihedral angle (see Sec. 7-8) determined by the planes of the great circles that contain the defining arcs. If we confine ourselves to spherical angles whose vertices lie on the equator or poles of the Earth model, find the measure of the following angles:
   (a) An angle determined by any arc of the equator and a meridian arc one of whose endpoints is on the equator.
   (b) An angle whose vertex is the North Pole and whose sides are arcs on the 20*E* meridian and the 40*W* meridian.

9. A **spherical triangle** is sometimes defined to be the union of three points (in a hemisphere) and the minor arcs of the great circles determined by the three points.
   (a) From our work with plane triangles what might you guess is the sum of the measures of the three angles of a spherical triangle?
   (b) Consider the following spherical triangle: one vertex is the North Pole; vertex E is the intersection of the equator and the meridian whose longitude is 45E; vertex W is the meridian whose longitude is 45W. What is the measure of the angle at N? At E? At W?
   (c) What is the sum of the measures of the three angles of the spherical triangle?
   (d) Could this result be obtained with plane triangles? Does this suggest a different geometry from plane geometry?
10. In Sec. 3-8 it was suggested that there are geometries in which, through a given point, many lines are parallel to a given line; that there are geometries in which, through a given point, but one line is parallel to a given line; and that there are geometries in which, through a given point, no lines are parallel to a given line. If lines are defined to be great circles on a sphere, then which relationship about parallels to a given line holds for this spherical geometry? [*Hint:* Consider the equator. How many great circles of the sphere do not intersect the equator? How many lines are parallel to the equator?]
11. Read Ref. 105, pp. 457–461, and Ref. 103, pp. 559–563.

SUMMARY OF PRINCIPLES

Every point in a line can be located with one coordinate.

Every point in a plane can be located with two coordinates.

Every point in a sphere can be located with two coordinates.

Every point in three-dimensional space can be located with three coordinates.

If $P_1(x_1, y_1)$ and $P_2(x_2, y_2)$ are two points in a plane, then the distance d between the two points is given by

$$d = \sqrt{(x_2 - x_1)^2 + (y_2 - y_1)^2}.$$

On a spherical surface there is no line parallel to a given line.

The sum of the angles of a spherical triangle is greater than 180 in degree measure.

CHAPTER 12

Topological Considerations

12-1 INTRODUCTION

The geometry we have been discussing is really quite a special one. Specifically, we have considered problems of rigid motion (congruence transformations) and proportional shrinking or expanding (similarity transformations). There is a body of geometric subject matter that is more general and includes these previously mentioned ideas as subcases. The subject matter we are about to discuss comes under the general heading of topology and is closely aligned with our nonmetric ideas. However, one should not get the idea that there is no metric in a topology.

Basically, **topology** is concerned with properties of figures that are invariant under so-called elastic motion. Of concern to the topologist are such things as (1) points on, inside, or outside, (2) cutting, twisting, re-forming (seal ends together), (3) stretching, twisting (without cutting), traversability.

It turns out that all topological properties are geometric properties, but not all geometric properties are topological. Let us investigate some of these ideas.

12-2 INSIDE—OUTSIDE

Think for a moment of separation in a plane as it relates to closed curves. Consider a piece of string that is placed upon a card table as a model of a circle in a plane, as pictured in Fig. 12-1(a). Then suppose someone bumps against or gently jiggles the table, so that the string assumes a shape somewhat as suggested by Fig. 12-1(b). Are any properties of a circle preserved or invariant under this change? (The title of this section should give you a clue.) Suppose the string is in the shape of a square with knots tied at the vertices. Make sketches illustrating the possibilities before and after a jiggling of the table. List any properties of a square that might be maintained.

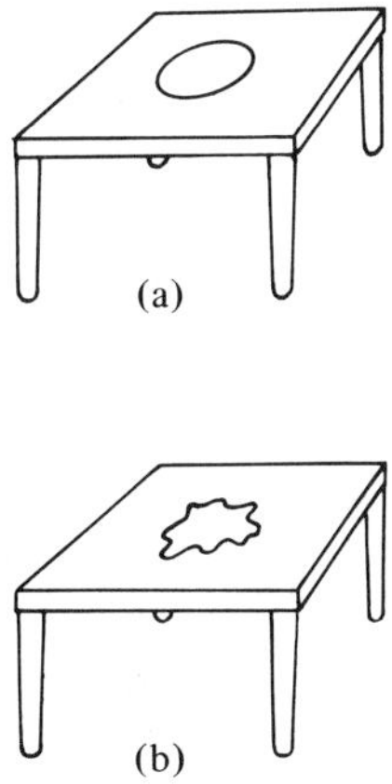

Figure 12-1

These situations illustrate some of the types of considerations a topologist investigates. The topologist says that the resulting figure in each of these two situations is topologically equivalent to the original figure. His argument would probably center around the idea of separation of a plane into an inside and an outside. Recall Exercise 3-4, 6. It is of special interest that neighborhoods[1] of points are maintained; that is, loosely speaking, points in the original figure that are "near" other points are still "near" the same points in the resulting figure. Also, there must be no creation of, or coalescing of, points. For example, consider only a string. Cut the string in the first illustration, loop it around in the shape of a knot, then seal the ends back together. Thus, as pictured in Fig. 12-2, although the string crosses itself, to the topologist it is "jumping over" itself—or at least the string does not fall upon itself at this point but passes overhead. Thus this resulting knotted piece of string is topologically equivalent to the original string circle.

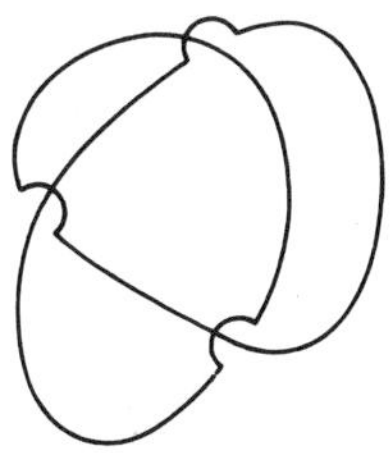

Figure 12-2

Notice that in all of these cases, distances, angles, straightness, all of which have concerned us earlier, are "destroyed" by the actions that are now allowed. We have been concerned, however, with separation. This illustrates the statement that topological properties are geometric, but not all geometric properties are topological.

What, then, of topological inequality? As an illustration of two topologically nonequivalent figures, consider two surfaces generated dynamically as follows. Start with a given circle in space.

(1) Rotate this circle about one of its diameters. What space figure is formed? [See Fig. 12-3(a).]

(2) Now think again of this same circle, but this time rotate it about a line that lies in the plane of the circle and does not contain any point of the circle. The surface generated appears somewhat as a doughnut, complete with hole. [See Fig. 12-3(b).] This is known mathematically as a **torus**, or **anchor ring**.

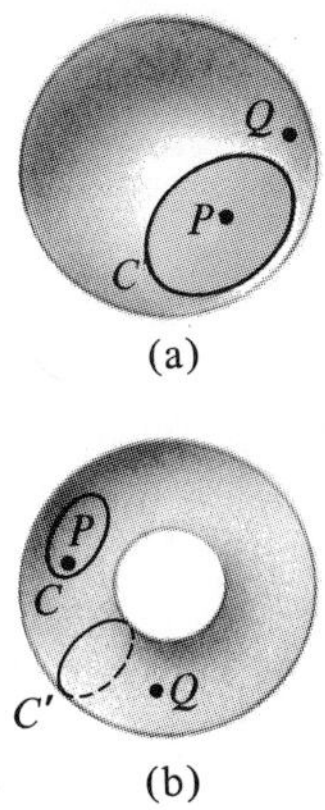

Figure 12-3

[1] For a discussion of neighborhood see Exercise 12-3, 9.

Now think of any circle drawn on the surface of the sphere, such as circle C in Fig. 12-3(a). [We say *on* the sphere to remind you of two things: (1) a sphere is a surface, and (2) we are dealing with a model of a sphere.] Does the circle separate this surface in such a way that there exist two points that cannot be joined by a curve on the surface without containing a point of this circle? Now think of circles on the surface of a torus in the same way. For example, consider circle C in Fig. 12-3(b). Does this same kind of separation occur? Now consider the generating circle of the torus in any one position, such as circle C' in Fig. 12-3(b). Can you find two points such that every path or curve that joins them must cross this circle in this position? Of course not. Thus, the topologist says that a sphere and a torus are not topologically equivalent.

More generally, existence of paths satisfying a set of conditions comes under the general heading of traversability.

12-3 TRAVERSABILITY

Perhaps one of the best known topological problems is the historic Königsberg bridge problem. This problem was solved by Leonard Euler in 1736, and ideas connected with it had a direct influence upon the field of topology.[2] Before considering this problem, let us investigate some fundamental ideas concerning the traversability of a network.

Recall the second situation above—that of the string in the form of a square, with knots at the vertices. The knots, after the distortion, allow us to distinguish the original vertices from the other points of the square. Topologically, vertices play a most important role.

Figure 12-4

Consider Fig. 12-4. Think of it as being a distortion of the original square. As with the square, the dots indicated are called **vertices** (singular: vertex), and the curves joining them are called **arcs** rather than sides. In fact, if we take a line segment and twist it in such a way that the two endpoints coincide, the figure we get is still an arc, but now it has only one endpoint, or one vertex. Is this last figure topologically equivalent to the original line segment from which it was distorted? Why not?

Returning to Fig. 12-4, since there are four vertices and four arcs in both the square and this distorted figure, and corresponding separation features, these two figures are topologically equivalent. Such figures are called networks. In fact, any combination of any finite (nonzero) number of arcs and vertices, planar[3] or nonplanar, is a **network**. We are interested at this point only in **connected networks**, which, as you might expect, come

[2] Much of the material in this chapter was suggested by Ref. 4.

[3] We have used "planar" with the meaning developed earlier in the text. In topology this term includes networks in a plane as well as networks that are topologically equivalent to networks in a plane. Thus, there is no contradiction in usage; our usage simply is less inclusive.

under our classification of a curve. In general the only points of the network that are of special interest are the vertices; all other points are merely parts of arcs. It does turn out that the arbitrary introduction of other points of arcs of a network as vertices does not affect the traversability of a network.

A **path** in a network is simply a set of one or more arcs and vertices that one can trace without lifting the pencil off the network and without retracing an arc of the network. A **closed path** is one for which the initial and terminal vertices are identical. This description resembles that for what previously studied figure? To determine whether a network is **traversable**, we need a method of concluding whether a single path can be found that includes all of the arcs of the network (there is no restriction on the number of trips through a vertex).

Notice that each vertex of Fig. 12-4 has two arcs connected to it. The vertex is said to have two **arc ends** and is called an *even vertex*. When would a vertex be called an *odd vertex*? Every arc end must be at a vertex; thus, we have no "stray" arc ends. The **order** of a vertex is the total number of arc ends at that vertex. What is the order of each vertex of Fig. 12-4?

We name a path in a network by listing each vertex of the path. In Fig. 12-5 find the following paths: (1) *ACB*, (2) *AC*, (3) *ADC*, (4) *ABCADC*, (5) *DACB*, (6) *AB*, (7) *BA*, and (8) *ACBA*. Notice that if the path is a closed path, the name of the terminal vertex is repeated even though it is the same as the initial vertex. Name the closed paths of Fig. 12-5.

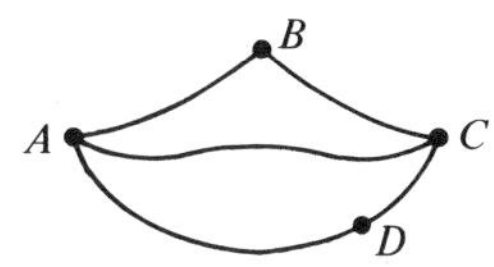

Figure 12-5

Equipped with vocabulary, symbolism, and these introductory notions, let us proceed with our analysis. It would seem that the total number of arcs and vertices could be any positive integer. However, consider the statement about no arc having a "stray" end. With this agreement, it follows that the number of arc ends is twice the number of arcs. Even the line segment with coincident endpoints, yielding only one vertex and one arc, has two arc ends. Thus the number of arc ends must be an even number. But the sum of arc ends in any network must be the sum of the orders of all vertices of the network, since the order of a vertex is simply the total number of arc ends at that vertex. Therefore, in any network the sum of the orders of all the vertices is an even number. Then from this we are able to conclude what about the total number of odd vertices? Explain why it must be an even number.

We wish now to investigate the possibility that a connected network contains more than two odd vertices, considering this from the point of view of traversability. Suppose we convince you that, with the possible exception of two vertices, each vertex of a traversable network is even. Would you, then, be willing to agree that if a connected network has more than two odd vertices, it cannot be traversed by a single path? (From Sec. 13-4 it can be seen that each of these statements is the contrapositive of the other, hence when either is true, the other is also.) Consider a vertex *P* as in Fig. 12-6. Let this be any vertex other than the initial or terminal vertices. Clearly, as one traverses the network over the chosen path, every vertex such as *P*, other than the inital and terminal vertices, must be an even vertex. Why? Since *P* is neither initial nor terminal, any path that enters *P* must leave *P* without retracing; hence, regardless of the number of times the path passes through *P*, it contributes two arc ends to the total number of arc ends at *P*; hence *P* is even. Since this is true of all vertices, with the possible exception of the terminal and initial vertices, if a connected network has more than two odd vertices, then it cannot be traversed by a single path; that is, it is not traversable.

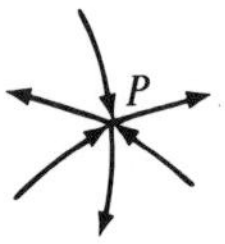

Figure 12-6

Let us now investigate two situations that will provide us sufficient information to conclude that under either of these conditions, a given network is traversable. Having excluded the possibility of more than two odd vertices in a traversable network, let us investigate the two remaining possibilities: no odd vertex, or exactly two odd vertices. Why can there not be a network with exactly one odd vertex?

We claim that if a connected network consists only of even vertices, then it is traversable. Further, in this case any vertex can be chosen as the initial vertex of the path and it will also be the terminal vertex. What name have we given to such a path? Think for a moment about the possible even vertices. Could not the arc ends at any such vertex be partitioned into two subsets such that any arc end in one set takes you into the vertex while any arc end in the other set takes you out of the vertex? Figure 12-7 shows four sets of even vertices. By using the idea formulated above, form connected networks with each of these sets. Try to find a combination that will not result in a traversable connected network; that is, try to end up with one arc end at some vertex without having some other arc end in the figure with which to join it. Indicate with arrows how you enter and how you leave each vertex. Also, indicate your initial vertex with *S* and your terminal vertex with *T*. By definition a network contains only a finite number of arcs. Therefore, does it not seem reasonable that such schemes as you have just employed could be applied, say, to a connected network with 24 even vertices? What about 23? What about *n* even vertices? (Actually, to prove this within a strict logical structure, it is necessary to employ a method known as mathematical induction. See Ref. 12, pp. 547–548; then see Ref. 4, pp. 13–18, 35, 39, problem 11.)

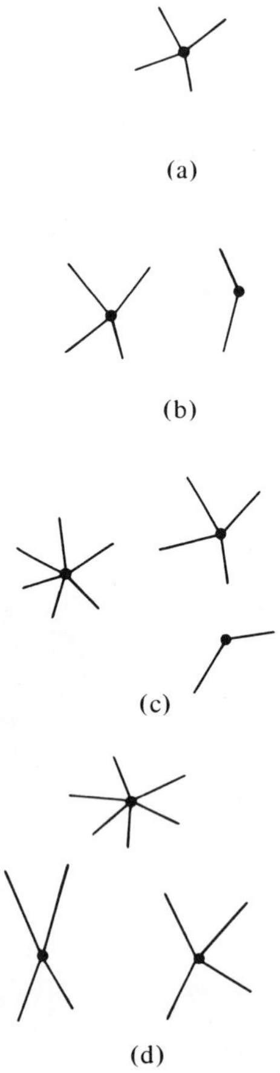

Figure 12-7

Our final claim regarding traversability is that if a connected network has exactly two odd vertices, it is traversable. Further, these two odd vertices are the initial and terminal vertices. Consider Fig. 12-7(d) again. Suppose we introduce into this figure two odd vertices, say *A* and *B*. Select one arc end of *A* and one arc end of *B*. Arbitrarily label an *s* on one of these arc ends and *t* on the other. Does not the remainder of the figure relate to the preceding paragraph? That is, starting the path along *s*, does not the remainder of the figure, with the exception of *t*, form a network with even vertices? Thus, if you select s at A, you leave A along s, complete the rest of the network, passing through A and B some even number of times; then you can complete the path by reentering B along t. Make a copy of Fig. 12-7(d); introduce vertex A with three arc ends, and vertex B with five arc ends. Then, as before, try to prevent completing a path involving these vertices. Also, as before, indicate direction along the arcs near the vertices and indicate the initial and terminal vertices. (Reminder: Each vertex must in some way be connected with the network. We are not allowing other than connected networks.)

We are now prepared to return to the Königsberg bridge problem. It originated in a puzzle regarding the seven bridges spanning the New Pregel River and the Old Pregel River, where there is an island at the point of confluence. Figure 12-8 shows the general layout of the mainland, island, and bridges. The puzzle is this: Is it possible to make a walking tour of this section of Königsberg, crossing each bridge exactly once? Although no one was able to arrange such a tour, it was Euler who first explained why it could not be done. His argument is based upon statements that we have just developed. In the next exercise you will be given some clues. In the meantime, use the statements above and try to explain why such a tour is impossible.

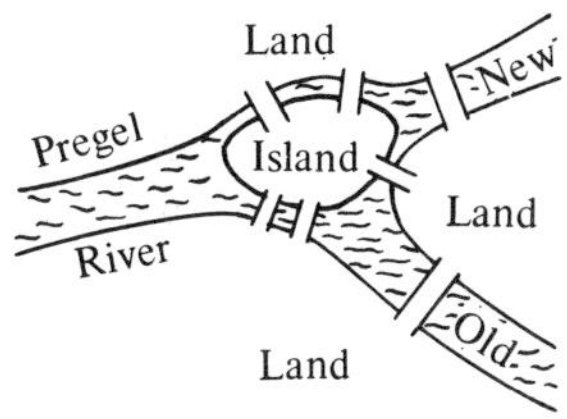

Figure 12-8

EXERCISE 12-3

1. For each of the following name a path that traverses the network. If this is not possible, explain.

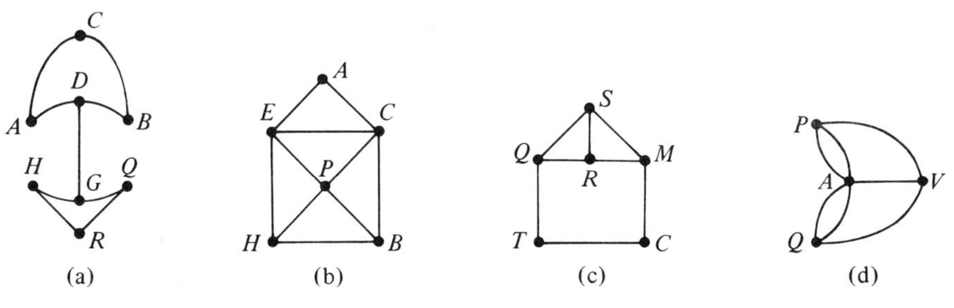

2. Think of the three rules we have for determining traversability; then answer the following questions.
   (a) If a given traversable network has an even number of points of its path added as vertices of the network, is the newly formed network traversable?
   (b) If a given network is traversable, is it still traversable if an odd number of points along the path are arbitrarily labeled as vertices?

3. On a sheet of elastic material, perhaps rubber, draw a triangle with a continuous stroke of a pencil—that is, in such a way that the vertices of the triangle receive no special emphasis over any other point of the triangle. Now stretch and twist the material, but do not tear it. Are you able to form a square? An irregular closed curve? A circle? Any configuration you make, provided you do not fold the triangle onto itself at any point, is topologically equivalent to any other configuration, even the original one. Perhaps for this reason topology is sometimes called the rubber sheet geometry.

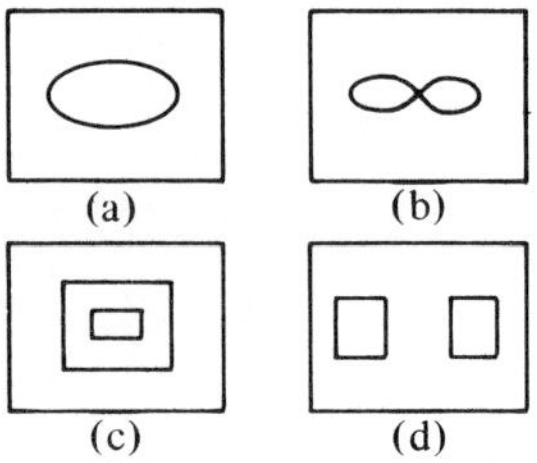

Figure 12-9

4. Two other properties of interest to the topologist are regional separation and one-sidedness.
   (a) Figure 12-9 shows four configurations. Using the concept of inside-outside separation, explain why no one of these is topologically equivalent to any other. (Rectangles depict planes.)
   (b) Perform the following experiments, which illustrate that a "two-sided" surface might really not be two-sided at all.
      (1) Cut a strip of paper with measurements 1 in. by 11 in. (An even longer piece makes the mechanics of this demonstration a little easier.) Label the four corners *R*, *S*, *T*, and *U* on both sides of the paper, as shown in Fig. 12-10(a). Label two points *A* and *B*, one on each side of this strip of paper. Take the two ends and paste them together as suggested by Fig. 12-10(b). Now with a pencil try to connect your two points *A* and *B*. Is this possible without drawing over the edge?
      (2) Now repeat the demonstration described in part (a), except that when you come to paste the ends together, give a half twist to one end and paste them as suggested in Fig. 12-10(c). Now try to connect your points *A* and *B* on this new strip. Is it necessary to cross the edge? Then how many surfaces does this strip of paper really have? Such a model is called a Möbius strip. See Ref. 21, pp. 12, 13, for an amusing anecdote concerning this.
   (c) See Ref. 109, pp. 118–122.

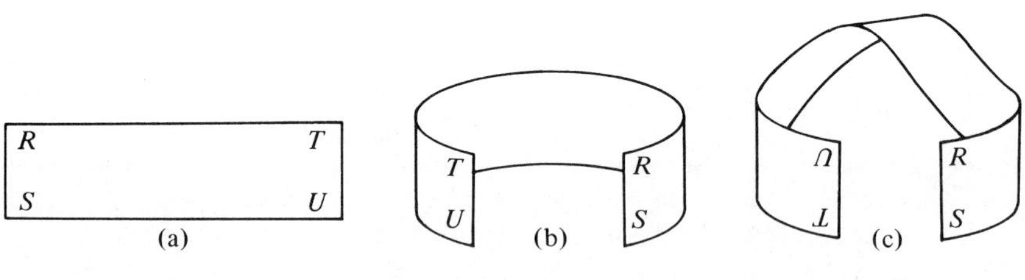

Figure 12-10

5. Let us return to the Königsberg bridge problem. Draw a model of this problem as follows. Consider the bridges as arcs in some network. Consider portions of land as vertices of this same network. If necessary, refer to problem 1(d) above. You should then have no difficulty completing an explanation.

6. Referring to the scheme outlined in problem 5, decide whether or not a person can complete the following "tours" as paths in a network. State which tours can be started and completed at the same point.

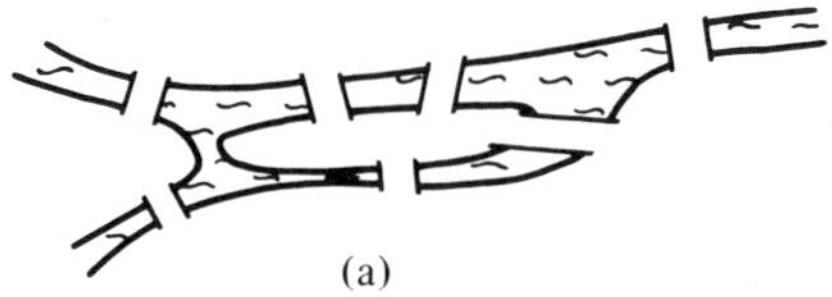

(a)

(b)

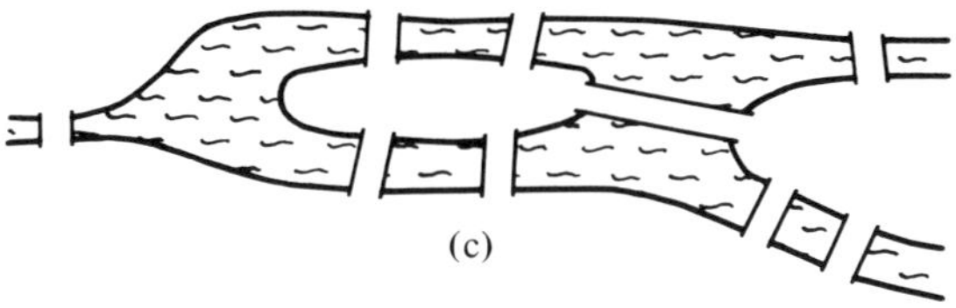

(c)

(d) The sketch for part (c) depicts the Pregel River as we understand the present situation. Of the two new bridges, however, the one on the extreme left is a railroad trestle. Since this might not be desirable as a walking bridge, try traversing the network with this bridge removed.

7. Figure 12-11 shows three squares and three circles. You are to connect each circle to each square with some curve. No curves are to cross (this means also that you are not permitted to run the curve through a square or circle).
(a) Is this possible if these figures are drawn on a sheet of paper?
(b) Is it possible if drawn on a sphere?
(c) Is it possible if drawn on a torus?

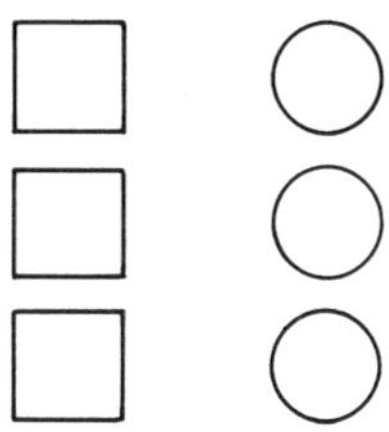

Figure 12-11

8. See Ref. 34, pp. 148–152, for general comments and pp. 153–158 for sample lessons for use in the elementary school.

9. (a) In Sec. 12-2 a neighborhood of a point is described as being the set of points "near" the given point. Actually, in a metric situation, a neighborhood of a point is often taken to be the set of points *within* a stated distance of the point.[4] With this understanding, describe:
(1) On a line, a neighborhood of a point on the line.
(2) In a plane, a neighborhood of a point in the plane.
(3) In space, a neighborhood of a point.

[4] Notice that, in a plane, this description of neighborhood results in a circular neighborhood of a point in the plane. There are instances when other polygonal regions are preferred, such as a rectangular or square region. In fact, in many topological discussions there is no metric involved. In these situations, the concept of neighborhood of a point is frequently left undefined, being characterized by the axioms and theorems of the system. See Sec. 13-2.

(b) A point, P, is a **boundary point** of a set if, and only if, every neighborhood of P contains (1) at least one point that is a member of the set, and (2) at least one point that is not a member of the set. The set of all boundary points of a given set is called the **boundary** of the set. With this understanding, describe:
   (1) In a line, the boundary of a ray.
   (2) In a plane, the boundary of a triangular region.
   (3) In a plane, the boundary of a circular region.
   (4) In space, the boundary of a space region that is 1 ft across throughout.

(c) Using the understandings of parts (a) and (b), we might describe the **interior** *of a set* S to be the set of points of S that have a neighborhood contained in S. How then should the exterior of a set be described? What is the boundary of the interior of a set? Compare these ideas with those previously developed concerning (1) the interior and exterior of a polygon, (2) the interior and exterior of a prism, (3) the interior and exterior of a closed prismatic surface.

(d) Topologically, a **closed set** is one that contains all of its boundary points. If a set contains none of its boundary points, it is usually called **open**. Use these terms as they apply to a ray. As they apply to a segment. If a set is not open, is it necessarily closed? Using point A as center, describe an open circular region of radius 2.

12-4 MAP COLORING

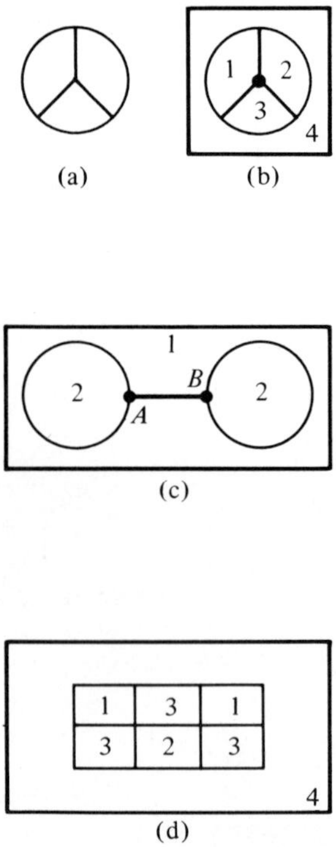

Figure 12-12

Another popular topological problem, perhaps as famous as that of the previous section, is the four-color map problem. Appearing in many guises, this puzzle may be stated simply as: How many colors do you need to color a map? The really puzzling thing is that no one knows for sure! Before attempting the solution, let us develop some basic vocabulary and understandings.

A **map** is merely a network, together with the surface containing the network. If we refer to a planar map, our discussion will center around planar regions and their boundaries. A **connected map** is one whose network is a connected network. In a map the regions of the surface partitioned by the network are called **faces.** The arcs and vertices of the network are **edges** and **vertices** of the map, respectively.

The rules in coloring a map are:

(1) Two faces with an edge in common must not be colored with the same color.

(2) Two faces with no border points in common, or with only vertices in common, may be colored with the same color.

(3) A new color is not introduced into a map if previously used colors can be used without violating the previous two rules. On occasion, in practice, this might require a rearrangement of those colors already present in the map. This is illustrated in note (3) below.

Notice three things:

(1) Figure 12-12(a) is a map. But, if you reread the previous material, the portion of the paper around the border is also a face. To remind you of this, we shall place our maps in rectangular regions. Then, all of the interior points of the region represent the totality of points of the surface in which the network is placed. Thus, we show the map of Fig. 12-12(a) as in Fig. 12-12(b). Different numerals indicate different colors. Thus, under the directions above, this map requires four colors.

(2) The edges in a map are usually edges of different faces. However, Fig. 12-12(c) pictures the edge AB, which is an edge of only one face. How many colors are required to color each of the other maps of Fig. 12-12? Explain why we are permitted to use the same color on both sides of edge AB in Fig. 12-12(c).

(3) With regard to rule (3) above, it would appear that the shaded region in Fig. 12-13 must be colored with a fifth color, since this region borders on regions numbered 1, 2, 3, and 4. Check this in the figure. However, if we change the "3" in region A to "4," then the shaded region can be numbered 3. If, in any map that you are coloring, you find yourself using a fifth color, recall this illustration before you proceed. Incidentally, what number can be used in the other face for which no color has been indicated?

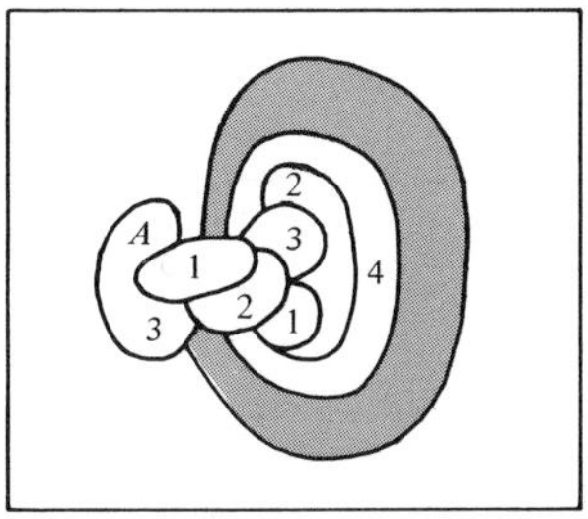

Figure 12-13

Recall the formula arrived at in answering problem 13 of the Exercises following Chapter 4. This formula is credited to Euler. It is valid not only for polyhedrons but for connected planar maps as well. Thus, in a connected map, with V, E, and F representing the numbers of vertices, edges, and faces, respectively,

$$\mathbf{V + F = E + 2.}$$

Euler's proof goes somewhat as follows.

By examining connected planar maps, we see that such maps can be built by starting with a single edge and proceeding with the following changes:

(1) Add a new edge joined at one end only.
(2) Add a new vertex in an existing edge.
(3) Add a new edge joined at both ends.

(At first glance it might seem that other changes can be made. Upon closer scrutiny, however, we find that these changes are but combinations or special applications of the three above.)

Figure 12-14 shows the two possible situations at the start. In either case, notice that $V + F = E + 2$. Now if we impose the first change above, we add one vertex and one edge; no face. Check this on a sketch. Hence, the formula obtains. Now impose the second change. Notice that we add

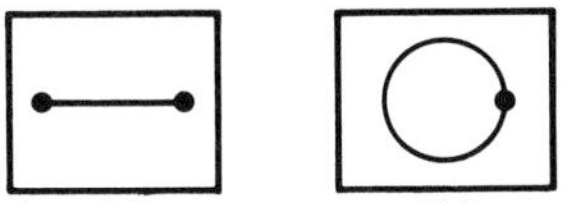

(a) (b)

Figure 12-14

one vertex and one edge; no face. If the third possible change is made in the original figures, we add one edge and one face; no vertices. Thus, none of these operations vitiates the formula. Then in the completed map this relationship continues to hold.

Although we omit the details, it can be shown on the foregoing basis that any regular connected planar map can be colored with six colors. (This should be interpreted as at most six colors—that is, six colors or less. Also, we should add that a **regular map** is a map each of whose vertices is of order three.) It can then be shown that any planar map can be colored with five colors.[5] Then why do we say in answering the original question that no one knows for sure how many colors are required? The reason lies in two facts: (1) no one has been able to find an example of a planar map that requires five colors; (2) no one has been able to prove that only four colors are required. In 1879 a "proof" was published by Kempe. However, in 1890 Heawood found an error in Kempe's reasoning. (See Ref. 10, p. 247.) The four-color statement has been proved for maps with less than thirty-eight regions. This fact indicates that if a map exists that requires five colors, it cannot be any "simple" example. Perhaps mathematics is not quite as "cut and dried" as some people seem to believe?

In Sec. 7-8 we mentioned that other proofs of the Euler formula (we did not call it Euler's formula than) exist using material from this chapter. Notice that we have not proved this formula for polyhedrons, but only for connected planar maps (using our meaning of planar). Using material from Secs. 12-3 and 12-4 along with extensions to (our) nonplanar networks, it is not difficult to prove the formula for polyhedrons. See Ref. 10, pp. 236–240, or Ref. 4, pp. 39–48.

Thus, once again we see that an investigation in one area of mathematics can illuminate quite a different area and provide a deeper understanding in both areas.

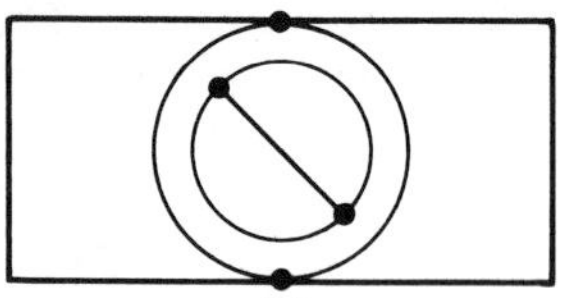
(a)

(b)

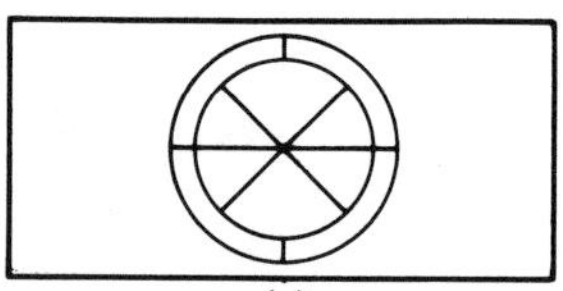
(c)

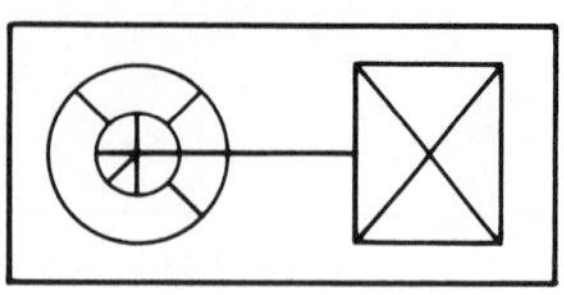
(d)

Figure 12-15

EXERCISE 12-4

1. Color each of the four maps in Fig. 12-15.
2. Draw and color a map in which no face has fewer than six edges.
3. Start with Fig. 12-14(a) and apply in succession the three possible changes that can be made in a map. (Start with the original figure, impose the first change, then impose the second change on this second figure. Then impose the third change upon this third figure.) Then check Euler's formula in this completed map.
4. Make an attempt to find a planar map that requires five colors. You might start by making changes in some maps pictured in this section.

[5] These details are expanded in Ref. 4, pp. 43–53.

The sum of the orders of all the vertices in a network is an even number.

A network has an even number of odd vertices, perhaps none.

If a connected network has more than two odd vertices, then it is not traversable.

If a connected network has no odd vertices, then it is traversable, and any vertex can be chosen as the initial and terminal vertex.

If a connected network has exactly two odd vertices, then it is traversable, and the two odd vertices are the initial and terminal vertices.

No planar map requires more than five colors.

In a connected planar map, with V, E, and F representing the number of vertices, edges, and faces, respectively, $V + F = E + 2$.

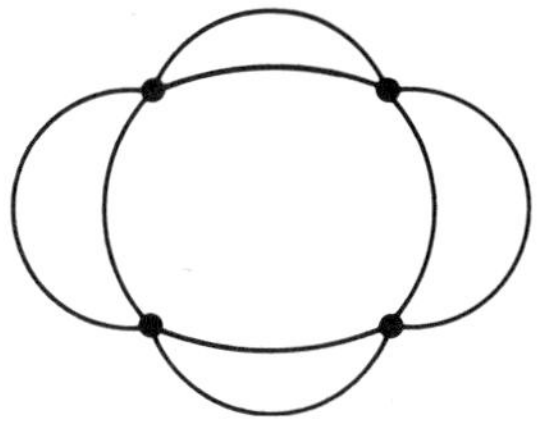

1

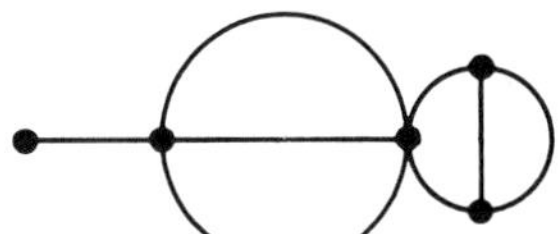

2

EXERCISES

1. (a) Which of the figures at the right are traversable networks?
(b) Design three traversable networks.
(c) Design three nontraversable networks.

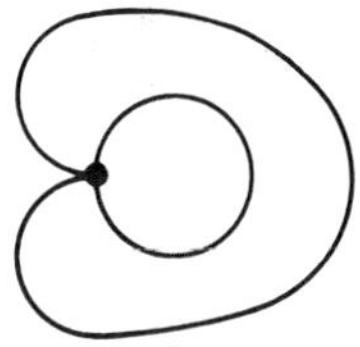

3

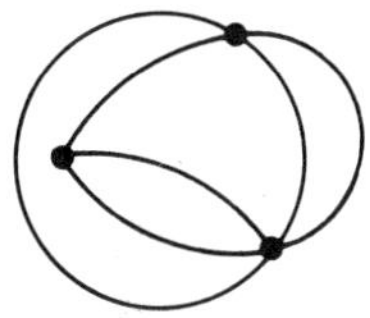

4

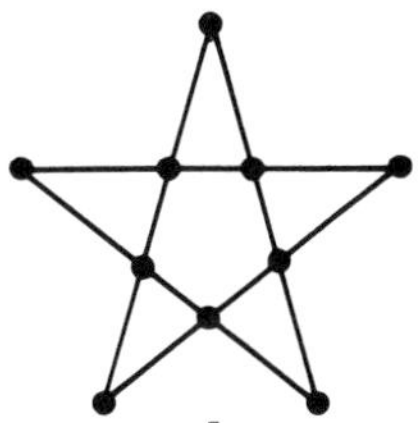

5

2. In Sec. 12-3 we restricted our discussion to connected networks. Why? In Sec. 12-4 we did not impose a corresponding restriction with respect to maps. Which maps in this chapter are not connected maps, hence illustrate other than connected networks?

3. You might raise the question concerning the three-dimensional analogue of the four-color map problem. Show by means of a sketch that for any (small) number of solids it is possible that they have a face in common. That is, show a method indicating that since this is always possible, there is no analogous space puzzle.

4. For more background material and practice exercises (many with solutions supplied) see: (a) Ref. 7, pp. 176–190 (perhaps the most interesting); (b) Ref. 21; (c) Ref. 34, pp. 159–163; (d) Ref. 51, pp. 379–393; (e) Ref. 28, pp. 338–356.

5. For some classroom applications see: (a) Ref. 109, pp. 118–122; (b) Ref. 92, pp. 503–507; (c) Ref. 34, pp. 153–158.

6. For more sophisticated treatments of topology see: (a) Ref. 36, pp. 306–318; (b) Ref. 107, pp. 465–474; (c) Ref. 121, pp. 18–24.

CHAPTER 13

To Build Further: Psychology, Logic, Structure

13-1 INTRODUCTION

The title of this final chapter suggests that our study of geometry is by no means completed.

The foregoing chapters have dealt extensively with what might be called the "content" of geometry—concepts and relationships. To a certain extent we have used different approaches to the study of geometry.

At times the approach has been quite inductive in that relationships or principles have been abstracted from particular situations. In Exercise 19 following Chapter 5, for example, the generalization concerning the congruence of alternate interior angles (in Euclidean geometry) is abstracted from a measurement situation. At other times the approach has been deductive: relationships or principles have been derived from other known relationships with recourse only to the logical connections between the relationships. In Exercise 18 following Chapter 6, for example, the same principle concerning the congruence of alternate interior angles formed by the transversal of two parallel lines is derived from the assumptions and principles concerning parallel lines.

Obviously, many geometric concepts are related to a wide variety of physical objects and relationships encountered in the "everyday world." We have relied heavily on such common experiences in introducing and developing many geometric concepts.

Certainly you realize by now that there is more than one *geometry.* Geometries are sometimes classified according to the methods used in describing and deriving the geometric relationships (examples: analytic geometry, coordinate geometry, and synthetic geometry). Sometimes the classification is according to the concepts of major interest (examples: topology and projective geometry). Sometimes the classification is according to the nature of the fundamental assumptions (examples: Euclidean geometry and non-Euclidean geometries). (Note that these classifications overlap.)

We could suggest many directions for further study in geometry; in this chapter we emphasize but two. The first is the organization of geometric concepts and relationships as mathematical structures. It is most remarkable that the seeming plethora of concepts and relationships in a given geometry can be derived by logical deduction from a few basic terms —left undefined—and a few basic statements whose truth value is assigned.

A second avenue of exploration is the psychology of geometric concepts. It is worth noting especially that young children come to school having already formulated ideas about many geometric situations.

Some other directions we might have suggested for further study include: the various kinds of geometry, applications of geometry in science and industry, and the history of the development of geometry. The Exercises and References in this book provide some guidance for further reading in these areas.

EXERCISE 13-1

1. For a brief overview of the evolution of geometry, read Ref. 27, pp. 203–207.
2. For a description of some of the various types of geometry, read Ref. 49, pp. 164–179.

13-2 LOGIC AND STRUCTURE

To consider the organization of geometric concepts let us now briefly investigate the role of logic and structure.

Recall that in earlier chapters you often encountered phrases such as:

it would seem that . . . ,
it appears that . . . ,
would you be willing to agree that . . . ,
does it not seem reasonable to assume that . . . ,
let us agree that . . . ,

and so on. Later you were asked to show, verify, justify, or demonstrate certain relationships. Soon you were asked to "prove" a statement. Just what is meant by *prove*? Does this mean something different from, or more than, show or demonstrate? Occasionally, you will hear a question to the effect: "You have shown it true for this case. Now, can you prove that it is always true?" What does a person mean when he asks someone to prove that it is always true? Surely he means that an explanation is to be supplied that conforms to some familiar and established structure.

Suppose, for purposes of comparison, that someone has asked: "Who won the Broad Jump event in the 1936 Olympic Games?" How would you go about deciding who won? Surely you would check the records. What records? The records of some committee or organization in charge of overseeing this event. How does the committee decide the winner? Clearly, a carefully detailed *structure* or set of rules is used. Terms are carefully *defined*, certain agreements are reached, then conclusions follow. The structure of geometry, and of almost any other mathematical structure, is rather similar. Specifically, what is this structure?

If you look up the directions for a game or contest, you will find that, although technical terms are carefully explained, many other terms are not. This is basic to any such structure. You simply cannot define every term employed. Likewise with certain agreements. You cannot *prove* every fact in a given mathematical system. There must be some fundamental facts that one assumes to be true and upon which the remainder of the structure rests. In geometry the basic agreements are referred to as **axioms** or **postulates**. (Although former usage assigned these terms slightly different meanings, present usage treats them synonymously.) Thus, our basic structure consists of certain undefined terms and a set of postulates. Given these basic elements, what are the rules of procedure? In geometry they are simply the rules of logic. In the remainder of this chapter we discuss some other components of a logical geometric structure.

13-3 DEFINITION

What do we mean when we say a certain term or concept is defined? Usage varies. Pedagogically, two usages are important: operational definitions and classical definitions.

An **operational definition** might best be described as a working definition, which merely allows one to recognize a certain concept and apply the proper name. This type of definition is often employed at an introductory and elementary level. Specifically, students in the elementary school should develop descriptions of geometric concepts to the extent they are able to apply the vocabulary in an accurate manner. On the other hand, there is no real cause for alarm if a student is unable to provide a really precise definition. Thus, a student might say, "A parallelogram has opposite sides parallel." Even with this imprecise definition, the student must be prepared to decide which figures with opposite sides parallel are, indeed, parallelograms; which are not (for example, a regular hexagon); and, perhaps, which are special parallelograms. It may not be necessary *at this point* for the student to realize that although every rectangle is a parallelogram, not every parallelogram is a rectangle. This is a matter for the instructor to decide. However, at some stage the student's understanding of a given concept should approach a classical definition.

A **classical definition** must satisfy at least the following five requirements:

(1) It must name the concept.
(2) It must characterize to the point of reversibility.
(3) If classification is possible, it should place the concept in its nearest, or smallest, class or set.
(4) If it does classify, it must distinguish this concept from other members of the same class.
(5) The terms used should be "simpler" than the terms defined. That is, only previously defined terms, or more simple and, perhaps, previously encountered, undefined terms should be employed. (This creates a real problem until some basic vocabulary is established upon which to build.)

Items (1) and (2) are, perhaps, the most important, since they indicate the two main uses of a definition: (1) knowing certain characteristics—apply a name; (2) knowing a name—these same characteristics are present. This is what most authors mean when they say a definition must be reversible. An illustration of items (3) and (4) is the following. In Sec. 6-6 we define a rectangle as a parallelogram with one right angle. This is a classical definition. If we defined a rectangle as a quadrilateral with opposite sides parallel and containing one right angle, this definition would satisfy (4) but not (3). However, notice that we have defined a quadrilateral with opposite sides parallel to be a parallelogram. This means that every rectangle is a parallelogram. Hence, to satisfy (3) we must classify a rectangle as a parallelogram rather than as a quadrilateral.

1. Do the definitions given in this text for the following terms possess the characteristics of a classical definition?
   (a) Rectangle—see Sec. 6-6. (b) Polygon—see Sec. 3-4.
   (c) Polyhedron—see Sec. 4-8. (d) Space—see Sec. 1-4.
2. Euclid wrote (as pointed out in Ref. 5, p. 21): "A point is that which has no part." Comment on this as a classical definition of *point*. How have we defined *point*? For further discussion of Euclid's definitions, axioms, and theorems, see Ref. 5, pp. 15–22.

13-4 ELEMENTARY LOGIC AND PROOF

We return to the question: What is proof? Proof in geometry really refers to consistency within a certain framework. When we say that a certain fact has been proved, what we really are saying is that, based upon our agreements concerning undefined terms and postulates, and following consistent rules of logic, this fact may be concluded. Basic to all of this is an understanding of an implication.

An **implication** is merely a statement of the form: If (a statement) is true, then (a statement) is true. Let us use lowercase letters $p, q, r, \ldots,$ to represent statements. Then an implication may be abbreviated:

If p, then q.

If p represents the statement, "It is warm," and q represents the statement, "$2 + 1 = 7$," then the abbreviation above represents the implication, "If it is warm, then $2 + 1 = 7$." (In logic statements are often compounded when there is no evident relationship between the parts; however, in geometry we usually look for and study related statements.) In logic an implication is symbolized:

$$p \longrightarrow q \qquad \text{or} \qquad p \Longrightarrow q.$$

We use the former.

If we wish to prove a given implication true, we must start by assuming that the statement in the "if" part—called the **antecedent** or **hypothesis**—is given to be true. Then, by using only the basic undefined terms, postulates, definitions, and previously proved implications, along with common rules of logic, we reach the statement in the "then" part—called the **consequent** or **conclusion**. See how these last two sentences apply to the proof of problem 19 of the Exercises following Chapter 6. Actually, this paragraph describes a **direct proof**. An indirect proof is illustrated in problem 7 of the Exercises at the end of Chapter 3. (More on indirect proof shortly.)

Having thus proved a given implication, we can use it in subsequent proofs. An implication for which a proof is sought within a given structure is often called a **proposition.** Once the proposition is proved it becomes a part of the structure and is often called a **theorem**. (Some authors treat "proposition" and "theorem" synonymously, leaving context to indicate the state, proved or unproved, of the implication.) If a proposition is *easily* proved on the basis of some theorem within the structure—that is, the truth of the proposition follows almost immediately from this theorem—the proposition is often called a **corollary** *to* or *of* this theorem.

Let us make two important observations:

(1) When an implication is proved, we prove not that q is true, but only that q is true *if p is true.*

(2) When an implication is proved, it is proved only in the particular structure being considered. (Of course—and this is the real value of structuring mathematics—any other structure that possesses all the elements of the proof could also include the given implication among its theorems.)

Related to any given implication are three other implications: the converse, the inverse, and the contrapositive. Symbolically,

given an implication	$p \longrightarrow q$,
the **converse** is	$q \longrightarrow p$,
the **inverse** is	p not true $\longrightarrow q$ not true,
the **contrapositive** is	q not true $\longrightarrow p$ not true.

Basically, logic may be considered as a study of patterns of thinking. In geometry two patterns of thinking are most often employed: inductive and deductive. Deductive thinking is exemplified above by what we described as a direct proof. In general, thinking from postulates and definitions is called *postulational thinking*, *deductive thinking*, or simply **deduction**. Thinking from appearances, or special cases, is often called *intuitive thinking*, *inductive thinking*, or simply **induction**. (Not to be confused with the method of mathematical induction mentioned in Chapter 12.)

Deductive proofs are based upon several important facts of logic. Perhaps the most used is one called the *law of the syllogism*:

If [if p then q, and, if q then r], then [if p then r].

This law says: given the two implications, "if p then q" and "if q then r," the implication, "if p then r" follows. This is often abbreviated:

If p implies q, and q implies r, then p implies r;

and may be symbolized:

$$[(p \longrightarrow q) \text{ and } (q \longrightarrow r)] \longrightarrow (p \longrightarrow r).$$

Then a deductive proof can be written as a series of syllogisms. A **syllogism**, or *syllogistic argument*, consists of three parts:

(1) A general statement—theorem, postulate, definition, or the like.

(2) A particular statement—a specific statement that declares the truth of the hypothesis of a general statement in a given situation.

(3) A conclusion—another specific statement, one that relates the conclusion of the general statement to the given situation.

In the proof of problem 19 of the Exercises following Chapter 6, the explanation of step (1) is the general statement of the first syllogism; the facts given form the particular statement; and step (1) is the conclusion. Explain step (3) of this same proof in terms of the three parts of a syllogistic argument. Do the same for step (6).

An indirect proof is based upon the *law of the contrapositive*. This law states that an implication and its contrapositive are logically equivalent.

For two statements to be **logically equivalent**, they must have the same truth value. That is, when one statement is true, the other statement is true; and when one is false, the other is false.

Now let us investigate the law of the contrapositive. Suppose it is known that a given implication, $p \longrightarrow q$, is true, and that q is not true. Can we conclude anything from these two facts? Do you think that p is true? If p is true, then what of q? Can q be both true and not true at the same time? (The statement that this is not possible is known in logic as the *law of contradiction*.) We are led to conclude that p is not true. Collecting all of this information, we see that for this case we have verified that if we know $p \longrightarrow q$ is true, then (not q) $\longrightarrow$ (not p) is also true. By a similar argument we can verify the converse of this previous statement. Hence, the two implications are equivalent, and the law of the contrapositive is verified for this case. (In logic this law is easily proved for all cases using the method of truth tables.) In the next exercises you will investigate a relationship between the converse and inverse statements.

What has this to do with indirect proof? Refer to problem 7 of the Exercises following Chapter 3. Write the statement of the problem in if-then form. Then write the contrapositive of this implication. Now read through the steps of the proof. Do you see that these steps prove the contrapositive of the given statement? Thus, an **indirect proof** is often a deductive proof of the contrapositive of the given statement. We have just agreed that in all cases an implication and its contrapositive are logically equivalent; hence, in proving the contrapositive we have proved the given implication. We also see that an indirect proof is another example of deductive proof.

EXERCISE 13-4

1. Given the following:

p: A glop is a dlir.
q: A tib is red.

(a) Write symbolic and verbal statements that represent the implication, p implies q, and the converse, inverse, and contrapositive statements.

(b) Take the converse statement of part (a) as a new given implication and write in verbal form the converse, inverse, and contrapositive statements of this new implication.
(c) Compare the statements in parts (a) and (b) and then decide what relationship exists between the converse and inverse of a given implication.
(d) If the given implication in part (a) is true, what other statement is known to be true?
(e) In part (b) if the given implication is false and its converse is true, what other statement is true?

2. (a) For a discussion of "proof" see Ref. 115, pp. 483–484, and Ref. 36, pp. 88–94.
(b) For more complete discussions of symbolic logic, see: (1) Ref. 2, pp. 1–8, 12–31; (2) Ref. 12, pp. 234–250.
(c) For other readings related to logic and reasoning, see: (1) Ref. 117, pp. 457–458; (2) Ref. 10, pp. xv–xix; (3) Ref. 9, pp. 6–11, 14–18; (4) Ref. 27, pp. 177–195.

13-5 CONSISTENCY AND INDEPENDENCE[1]

Suppose you are interested in establishing the truth of a certain theorem. How do you proceed? First, you consider how it might follow from other theorems. This means that you must prove these other theorems, which you do by going back to others, and so on. Eventually you will reach a statement that it is not possible to prove. That is, the statement must be taken as a postulate, or axiom. Some people may say about the statement, "This is a 'self-evident' truth," or "Clearly, this is so." No idea of proof is involved—the truth of the statement is simply assumed. With definitions, too, you state one definition in terms of other objects, until you reach a point at which you can no longer use defined terms; hence, undefined terms enter. Thus, the entire structure rests upon assumed postulates and undefined terms. Perhaps this is what Bertrand Russell was referring to when he said: "Mathematics may be defined as the subject in which we never know what we are talking about, nor whether what we are saying is true."[2]

The question is this: Is the choice of postulates and undefined terms completely arbitrary? The answer is an emphatic yes! However, although theoretically there are no restrictions on the postulational structure, in practice certain characteristics are highly desirable. Two of these are consistency and independence of the postulates.

[1] Much of the material in this section was suggested by Ref. 45. See Exercise 13-5, 1.

[2] Bertrand, Russell, "Recent Work on the Principles of Mathematics," *International Monthly*, **4** (1901), p. 84.

A system is **consistent** if the postulates and the theorems derived from them do not contradict each other. Why is consistency important? Surely, if one postulate contradicts another, the system is quite uninteresting. After all, if we study a system in which a given statement can be proved both true and false, this means that the law of contradiction does not hold in this system. It might not be obvious from the list of postulates that an inconsistency exists in a structure; thus we might study a given system without knowing that it contained an inconsistency. But once an inconsistency is discovered, the whole structure becomes uninteresting. Thus, in setting up the postulational structure we try to avoid any inconsistency.

How can you show that a given set of postulates is consistent? One relatively easy way is to build a model. When you build a model, you must specify what you mean by the undefined objects. In a geometric system it is necessary to establish what you mean by "point," "line," and "on." Then, the postulates become verifiable theorems in the model. This process should become clear in the next exercises, in which you are asked to construct such a model and verify a set of postulates to be consistent.

A set of postulates is **independent** if no one postulate can be derived from the others. Why is independence desirable? (Notice the choice of words: we state that consistency is "important" and that independence is "desirable"—neither is required.) Consistency is rather pressing, since the system collapses if discovered inconsistent; independence is not as critical. Pedagogically, it may be much easier if the postulates are not independent (that is, there are more postulates than necessary), but mathematically, it is often important that the fewest possible restrictions be placed upon the structure. This gives the system a wider application.

It is not practical to require students in their first exposure to formal geometry to prove every theorem that is provable. In the usual high school geometry course, for instance, it is quite proper to accept as postulates many statements that can be proved on the basis of other statements. An example of this appears in the usual section on congruence theorems. Most texts at this level take *SSS*, *SAS*, and *ASA* statements as postulates. Actually, it is necessary[3] to take only one statement as a postulate, often the *SAS*; the others can then be proved.

How is independence of a postulate proved? The most common way is to show that a system exists (build a model) in which the other postulates hold and in which the questioned postulate does not hold. Historically, independence is of interest, since this question was raised about Euclid's parallel postulate. In the attempt to prove this postulate from the others, it was found that it could even be denied. That is, a completely consistent system could be established in which the fifth postulate did

[3] Still only pedagogically "necessary"; see Exercise 7-6, 4.

not hold. This development clearly showed the independence of this postulate and led to the study of non-Euclidean geometries.

EXERCISE 13-5

1. (a) Accept "point" and "line" as undefined terms, accept the relation "on," and build a model to show the consistency of the following six postulates.

P-1: There is at least one point and one line.
P-2: For two distinct points there is at least one line on both of them.
P-3: For two distinct points there is at most one line on both of them.
P-4: For two distinct lines there is at least one point on both of them.
P-5: There are at least three points on any one line.
P-6: Not all points are on the same line.

[*Hint:* Select seven letters and let these be the points of the system. Then select triples of these letters to be the lines. Thus, you might select the first letters of the alphabet as the points—A, B, C, D, E, F, and G. Then you might select the following triples as lines—$\{A, B, C\}$, $\{A, D, E\}$, $\{A, F, G\}$, $\{B, D, G\}$, $\{B, E, F\}$, $\{C, D, F\}$, and $\{C, E, G\}$. By "on" then we mean "is an element of." Thus, D is on $\{A, D, E\}$, $\{B, D, G\}$, and $\{C, D, F\}$ and no others. Now, check each of the six postulates in these two lists. (You might find the checking easier if you refer to Fig. 13-1. But we remind you that the "lines" in the figure are not the lines of this system. A line in this system is nothing more than a triple of letters, one of the seven listed.)]

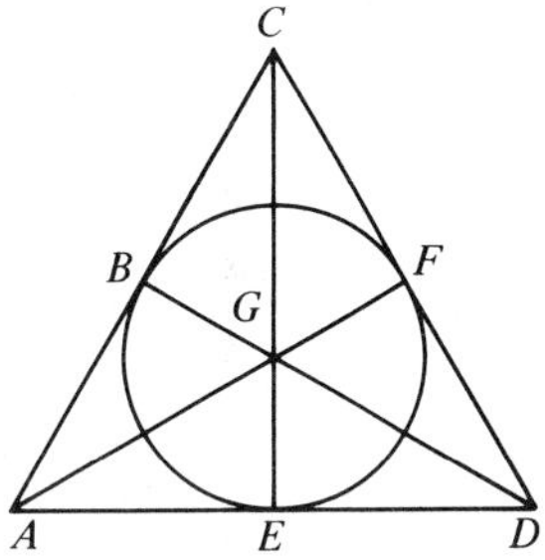

Figure 13-1

(b) The postulates listed in part (a) are from Ref. 45, p. 43, and can be taken as axioms of a projective plane. Compare these with those for an affine plane listed on p. 52 of this same reference. See also pp. 42–43, 49–55 for an excellent discussion of axiomatic structure in general, including consistency and independence of the axioms.

(c) For further comments on general foundations of geometry see Ref. 26, pp. 1–20.

2. (a) For examples of finite geometries see Ref. 30, pp. 36–38, and Ref. 52, vol. I, pp. 1–7.

(b) For comments on non-Euclidean geometries, see: (1) Ref. 103, pp. 559–563; (2) Ref. 105, pp. 457–461.

(c) For an interesting analysis of Euclidean organization of geometry see Ref. 106, pp. 578–584.

(d) For general comments on teaching geometry see Ref. 119, pp. 287–295, and Ref. 28, pp. 324–337.

3. (a) For general readings on "other" geometries see Ref. 13, Preface and pp. 1–29.

(b) For a formal and rigorous development of affine and Euclidean geometries see Ref. 23, pp. 14–47. Although you may find some of the discussion rather detailed, the general comments would prove valuable.

(c) For a development of incidence geometry see Ref. 43, pp. 84–89, and for a detailed discussion of the independence of the incidence postulates see Ref. 102, pp. 269–277.

13-6 PSYCHOLOGY AND TEACHING

Our discussion of the psychology of geometric concepts centers on the work of Jean Piaget and his colleagues who have made extensive investigations into the understandings of geometric concepts possessed by children and the differences that exist between these conceptions at various ages. Our discussion of the teaching of geometry is limited to an indication of places in which you can begin to find more information on this topic.

Two major results of the work of Piaget are of interest to us. In studying what he calls the spontaneous geometry of the child—that is, the geometric concepts the child formulates for himself without adult teaching—Piaget has found that children at various age levels differ in the concepts they possess and also in the way in which they know these concepts.

Historically, the study of geometric concepts was formalized in the following order. Euclidean geometry (which deals with figures, angles, parallels, measurement, and so on) was organized by Euclid around 300 B.C. Projective geometry (which has its roots in the problems of perspective) was organized in the seventeenth century. Topology (which deals with general qualitative relations such as separation, open and closed figures, interior and exterior points, connectedness, and so on) is the most recently organized.

Piaget has found that for the young child the order of appearance of spatial and geometric concepts is not that of the historical development of organized geometry but rather the reverse. Young children seem to acquire the topological notions first. Projective and Euclidean notions come later, developing simultaneously and interdependently. This later development does not seem to occur unless the topological concepts are already present. However, once the projective and Euclidean concepts develop, they become dominant, masking the fact that the child has already developed the topological concepts.

According to Piaget's interpretation of his investigations, the development of geometric concepts seems to occur as follows.[4] In the earliest stages of development each object and distinct pattern is associated with its own space, which is totally unrelated to the space associated with any other object. Coordination of the various spaces begins to occur when the child begins to use projective operations. The elementary projective concepts have their basis in the same mental operations as topological concepts but with the addition of a "viewpoint." The Euclidean measurement relations depend on the projective coordination of viewpoints for the development of coordination of shape and coordination of size (distance). Piaget notes that the last two stages seem to be very interdependent.

[4] The discussion in this paragraph is suggested by Ref. 38a, pp. 467–468, 476.

Note, however, that the order in which a young child develops the basic spatial and geometric notions does not necessarily mean that he understands topology, projective geometry, or Euclidean geometry as abstract mathematical structures. This suggests a second major result of Piaget's work: the way in which children know and can operate with their conceptions of space.

The child's mental operations appear to progress through a series of stages. It also appears that the higher and later stages of thought depend upon the adequate development of the lower and earlier stages.

The child's potential development of thought, Piaget feels, is closely tied to his actions.[5] The conceptual development—topological through Euclidean—presupposes a growing number of increasingly complex coordinations between different actions. In fact, "spatial concepts are considered to be internalized actions."[6] These actions are internalized in a series of stages, each stage marked by the way in which a child knows and can operate with concepts. The first stage (up to age 4 or 5) is characterized by thought that reproduces action as a result of a direct tie to perception. The child does not think of any action but that which he perceives. With growing coordination of physical action (4 or 5 to 7 or 8 years) an accompanying internal coordination of schematic outlines occurs. The child anticipates potential action little by little on a trial-and-error basis. Piaget suggests that this represents the beginning of the formation of "trains of ideas." Out of this process "arises the stage of concrete operations (7–8 to 11–12 years) when the schemata are coordinated sufficiently to be combined, and consequently, for each one to be mentally explored in alternate directions."[7] It is not until the child can think in terms of alternate directions (to action) that his thought is no longer dependent on his perceptions. This Piaget calls an "operational system." The child's thought before this stage Piaget often calls "preoperational." "With the further development of operational coordination it becomes possible to conceive of several systems simultaneously. Thus the child now disposes (about 11–12 years) of abstract operations, capable of being expressed in propositional form. This marks the final stage in the development of intuition and the beginning of a type of thought which whilst being the culmination of the continuous internalizing of actions, at the same time prepares the ground for the elaboration of abstract spatial concepts."[8] This final stage is often called the stage of "formal operations."

Notice that according to the preceding description, the child is not initially capable of understanding the deductive nature of abstract mathe-

[5] The discussion, in this paragraph is suggested by Ref. 38a, pp. 454–455.

[6] Ref. 38a. p. 454.

[7] Ref. 38a. p. 454.

[8] Ref. 38a. p. 455.

matical structures (see Sec. 13-4) until the stage of concrete operations, and he is not fully capable of such understanding until the stage of formal operations.

We have selected Piaget's work as suggestive of further study because he and his colleagues have conducted such extensive investigations of children's conceptions of geometry and because they have based their investigations on both psychological and mathematical structures. Of course, theirs is not the only possible approach, but it provides a point of departure for considering the educational question: What geometry content and what instructional methods should be selected for the students in our schools?

EXERCISE 13-6

1. For a brief introduction to Piaget's thoughts on the learning of mathematical concepts by young children read Ref. 120, pp. 74–79.

2. For a more complete exposition of Piaget's investigations into the child's conception of space and geometry read Refs. 38a and 38b. The first reference is concerned mainly with topological and projective concepts (nonmetric) but includes ideas on the transition from projective to Euclidean space. The second reference deals with Euclidean concepts, giving special attention to metric geometry.

3. In this exercise you are asked to carry out investigations similar to those used by Piaget in formulating his ideas about the nature of the child's conception of space and geometry. We suggest that you give each task to children of various ages from 3 to 12 years. Observe and note carefully what each child does and says. Do not try to teach the child! The following tasks are suggested in Ref. 120.

(a) A topological problem: Ask a child to copy a square or a triangle.

(b) A projective construction: Attach a piece of modeling clay to one end of seven or eight straws so that each straw can stand vertically. Set two of these straws about 15 in. apart. Ask the child to place the remaining straws in a straight line between these two.

(c) Euclidean measurement task: Show a child a tower of wooden blocks on a fairly low table. Then ask him to build a second tower of blocks (using another set of blocks) the same height as the first but placed on a table somewhat higher or lower than the first tower.

4. This exercise lists items from the References that give illustrations of geometric concepts and instructional techniques suggested for various grade levels. The listing is, of course, by no means exhaustive.

(a) For elementary school—primary levels:

(1) Ref. 16. (2) Ref. 61, pp. 356–358.
(3) Ref. 63, pp. 191–192. (4) Ref. 64, pp. 229–230.
(5) Ref. 67, pp. 422–426. (6) Ref. 69, pp. 193–194.
(7) Ref. 71, pp. 32–36. (8) Ref. 84, pp. 453–459.
(9) Ref. 85, pp. 450–454.

(b) For elementary school—upper division:
(1) Ref. 48. (2) Ref. 74, pp. 144–148.
(3) Ref. 80, pp, 443–449.

(c) For elementary school—all levels, general:
(1) Ref. 57, pp. 441–447. (2) Ref. 58, pp. 460–467.
(3) Ref. 62, pp. 645–646. (4) Ref. 68, pp. 673–682.
(5) Ref. 77, pp. 3–10. (6) Ref. 79, pp. 119–128.

(d) For junior high school:
(1) Ref. 35. (2) Ref. 48.

(e) For senior high school:
(1) Ref. 35. (2) Ref. 90, pp. 226–238.
(3) Ref. 91, pp. 165–169. (4) Ref. 94, pp. 765–771.
(5) Ref. 96, pp. 466–471. (6) Ref. 98, pp. 645–646.

5. If you are interested in the actual geometry content of present mathematics programs, we suggest that you obtain complete sets of commercial or experimental programs and analyze the content from kindergarten through twelfth grade.

13-7 CONCLUSION

You will leave this book with, we hope, at least two lasting impressions: (1) your study of geometry has only begun, and (2) nonetheless, you have learned much about the concepts, methods, and language of geometry.

As a final note we return to the question raised in Chapter 1: *What is geometry*? In that chapter we gave a very simplified answer. Let us now rephrase the question: What do you think geometry is? Each one of us, from the young child to the adult, has his own conception of geometry. The authors' ideas are partly expressed in this book. What are yours?

SUMMARY OF PRINCIPLES

Law of the contrapositive: An implication and its contrapositive are logically equivalent.

Law of the syllogism: If p implies q, and q implies r, then p implies r.

Law of contradiction: It is not possible that the statement, p, and the statement, not p, be true simultaneously.

There are no restrictions on the postulates and undefined terms of a postulational system.

APPENDIX A

Review of Set Concepts

A-1 SETS, ELEMENTS, NOTATION

By **set** we usually refer to some collection or class of things. That is, for some reason we want to consider a group of things together: a set of children, a set of states, a set of numbers, a set of geometric figures. In order that a set be useful or well defined, the reader must be able to determine whether any particular object belongs in it.

Each of the objects included in a set is called an **element** or **member** of the set. For example, the number three is an element of the set of whole numbers less than ten. Jim is a member of Miss Smith's third-grade class.

If names for all the elements in a set are listed, braces are often placed around the list to indicate that the collection constitutes a set. Notice that the items in the list are separated by commas and that the order of the items makes no difference. For example, "{penny, nickel, dime, quarter, half-dollar}" is a name for the set of coins currently used in the United States. The set of even whole numbers less than ten is named by "{0, 2, 4, 6, 8}." The set of objects on my desk is named by "{pen, paper, dictionary, paper clip, glass}." This third example shows that the objects in a set might have only one thing in common—that they are members of a particular set.

Sometimes it is inconvenient to list all the elements of a set. Set-builder notation helps to eliminate this problem. For example, consider the following name for the set of numbers between zero and one hundred: "$\{x \mid x$ is a number greater than zero and less than one hundred$\}$." This is certainly more convenient than listing all the elements! The braces indicate a set; the letter x functions as a variable for the elements in the set; the vertical stroke is read "such that." Thus, "$\{x \mid x$ is an animal$\}$" is read "the set of all x such that x is an animal" or "the set of all animals."

It is also inconvenient to write out names for sets, using either lists or set-builder notation, each time the set is to be mentioned. Sets are often named with capital letters or other special symbols (as in geometry). Once we list or otherwise indicate the elements of a set, we can agree that a certain letter (or other symbol) also names the set. For example, $A =$ {Jack, John, Bob} or $N = \{x \mid x$ is a natural number$\}$. Note that we are using "$=$" to mean that the two names represent the same set.

The symbol "$\in$" is used to indicate that a given object is an element of a particular set. "Jack $\in A$" is read "Jack is an element of the set A," and "$3 \in N$" means that three is a natural number. The notation "$\notin$" means that an object is not an element of a named set. For example, "Fred $\notin A$" is read "Fred is not a member of set A."

EXERCISE A-1

1. Name the elements in the set of New England states and place the list in braces.
2. Name the members of the set of whole numbers less than twenty but greater than fifteen and place the list in braces.
3. Write a name for the following set using set-builder notation:

   {Alabama, Alaska, Arizona, . . . , Wisconsin, Wyoming}.
4. Given that $A = \{1, 3, 2\}$ and $B = \{2, 4, 6\}$, tell whether each statement below is true or false.

   (a) $2 \in A$.
   (b) $3 \in B$.
   (c) $6 \notin B$.
   (d) $2 \in A$ and $2 \in B$.
   (e) $1 \in A$ or $1 \notin B$.

A-2 NULL SET, UNIVERSAL SET

The set that contains no elements is often called the **empty set, null set,** or **void set.** The symbols commonly used to represent the empty set are "{ }" and "$\varnothing$."

Note that "{0}" and "{∅}" are *not* symbols for the empty set. The first symbol, "{0}", indicates a set with one element—the number zero. The second symbol, "{∅}", indicates a set with one element—the empty set.

In mathematics we are not usually concerned with sets whose elements are just anything whatever. (Although we could be.) Rather, we are usually concerned with sets whose elements belong to a more restricted class of things—for example, sets whose elements are numbers, or sets whose elements are geometric figures. For any given discussion in which sets are used, the class of objects from which elements of the sets are selected is called the **universal set** or **universe of discourse**. At times, I or U is reserved to indicate the universal set. We prefer U. Thus, every element of any set in a particular discussion must belong to U for that discussion.

EXERCISE A-2

1. Which of the following is a null set?
   (a) The set of all female past Presidents of the United States.
   (b) The set of all living organisms on Mars.
   (c) The set of all whole numbers less than ten but greater than nine.
2. Which of the following statements are true?
   (a) "{ }" and "{∅}" mean the same thing.
   (b) "0" and "{ }" name the same things.
   (c) "∅" and "{ }" name the same sets.

A-3 FINITE AND INFINITE SETS

Sets may be distinguished by the number of elements they contain. Without getting too technical, we may characterize a **finite set** as one for which the number of elements can be counted (at least in the imagination) and a counting number or zero assigned to the set. Even if the number of elements is very large, the end of the counting process is conceivable. The set of blades of grass in a well-kept lawn is finite. An **infinite set** is one for which the end of the counting process is not possible; no matter how much counting you do, there is always at least one more element in the set. The set of natural numbers is an infinite set.

Obviously, it is not possible to list all the elements in an infinite set. For those cases where some sort of list is desired, several objects in the set are listed and this list is followed by the ellipsis. For example, A is the set of even whole numbers. $A = \{0, 2, 4, 6, \ldots\}$. Notice that the set-builder notation would also be appropriate here: $A = \{x \mid x \text{ is an even whole number}\}$.

EXERCISE A-3

Which of the following are infinite sets and which are finite sets?
(a) Set of all grains of sand on the earth.
(b) Set of all natural numbers greater than one and less than two.
(c) Set of all rational numbers between zero and one.
(d) Set of all locations in ordinary three-dimensional space.

A-4 VENN DIAGRAMS

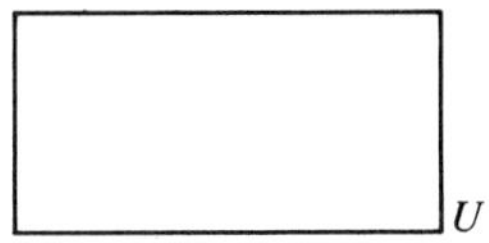

Figure A-1

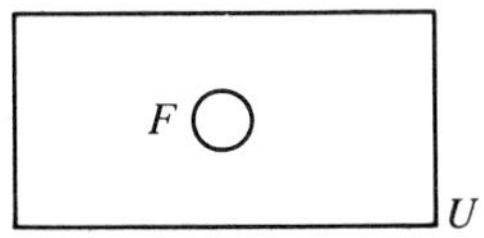

Figure A-2

Individual sets are often represented by diagrams, sometimes called Venn diagrams or Euler circles. The universal set is often represented by a rectangular region. Figure A-1 represents a universal set. Other sets are represented by circles and their interior points. Since every element of a given set is in the universal set, the circle for the set is drawn in the interior of the rectangle for the universal set. Figure A-2 shows the Venn diagram for set F.

We shall consider next some relationships that exist between sets.

A-5 EQUAL SETS

If two sets have exactly the same elements, then the sets are said to be **equal**. The set of even numbers and the set of whole numbers that are multiples of two are equal sets, since they have identical elements. The equal sign is used to indicate that two names represent the same set.

EXERCISE A-5

Which sets are equal?
(a) $A = \{1, 2, 3, 4, 5\}$.
(b) $B = \{\frac{1}{1}, \frac{4}{2}, \frac{6}{2}, \frac{8}{2}, \frac{10}{2}\}$.
(c) $C = \{0, 1, 2, 3, 4\}$.
(d) $D = \{x \mid 0 < x < 6$ and x is a whole number$\}$.
(e) $E = \{3, 4, 5, 2, 1\}$.

A-6 ONE-TO-ONE CORRESPONDENCE, EQUIVALENT SETS

If two sets can be matched, using all elements of both sets, so that for each element of the first set there is a unique element of the second set and vice versa, then the match is said to be a **one-to-one correspondence** between the two sets. The symbol "$\leftrightarrow$" is sometimes used to indicate the pairs of elements that are matched. For example, if $A = \{a, b, c\}$ and $F = \{1, 2, 3\}$, then the match $a \leftrightarrow 1$, $b \leftrightarrow 2$, and $c \leftrightarrow 3$ is a one-to-one correspondence.

Note that if $G = \{1, 2, 3, 4\}$, there is no one-to-one correspondence between A and G. For each element in A we can find a match in G—for example, a with 1, b with 2, and c with 3. But if we must use every element in G, then one of the elements in A is used twice. That is, one of the elements in A does not have a unique match in G.

Any two sets that can be put in a one-to-one correspondence are said to be **equivalent sets.** The symbol "$\sim$," called "tilde," is used to indicate that two sets are equivalent. For example, "$A \sim B$" is read "A is equivalent to B." Notice that for finite sets this is tantamount to saying that the two sets have the same number of elements.

EXERCISE A-6

1. Show that there is a one-to-one correspondence between $A = \{m, n, o, p\}$ and $B = \{1, 2, 3, 4\}$.

2. Show another way to match the elements of A and B in problem 1 that will also show there is a one-to-one correspondence between the sets.

3. Which sets are equivalent?

(a) $M = \{0, 1\}$. (b) $N = \{1, 2, 3\}$.
(c) $P = \{1, 0\}$. (d) $Q = \{3, 4, 5\}$.
(e) $R = \{a, b, c\}$. (f) $\varnothing$.

4. Which sets in problem 3 are equal?

5. Which statements below are true?

(a) Any set is equivalent to itself.
(b) If a given set is equivalent to a second set, then the second set is equivalent to the given set.
(c) If a given set is equivalent to a second set, and the second set is in turn equivalent to a third set, then the given set is equivalent to the third set.
(d) Any two sets that are equivalent are also equal.
(e) Any two sets that are equal are also equivalent.

6. Write the statements of problem 5(a), (b), and (c) in symbolic notation.

A-7 SUBSETS

If every element of one set is an element of a second set, the first set is said to be a **subset** of the second set. The second set is called the **superset** of the first set. Thus, if A is the set of girls in Miss Smith's class and B is the set of children in Miss Smith's class, then A is a subset of B. The symbol "$\subset$" is often used to indicate the subset relation. For example, "$A \subset B$" is read "A is a subset of B." The symbol "$\supset$" is often used to indicate the superset relation. For example, "$B \supset A$" is read "B is the superset of A."

Note that every set is a subset of itself. This will be the case in our example if Miss Smith's class contains only girls.

Note that the empty set is a subset of every set. This will be the case in our example if Miss Smith's class contains only boys.

If a set contains at least one element not in a given subset of that set, the subset is called a **proper subset** of the set. This will be the case for our example if Miss Smith's class contains both girls and boys.

EXERCISE A-7

1. Which of the following are subsets of $K = \{$John, Jack, Jim$\}$?
   (a) {John}.
   (b) {John, Jim}.
   (c) {John, Jack, Jim}.
   (d) $\varnothing$.
   (e) {John, Jack, Sam}.
   (f) {David, Linda, Jim}.
2. Which sets are proper subsets of K in Problem 1?
3. List all the subsets of $\{1, 2\}$.
4. Given $A = \{0, 1, 2\}$, $B = \{0\}$, $C = \{2, 3\}$, and $D = \{2, 1, 0\}$, which of the following are true?
   (a) $B \subset A$.
   (b) $C \subset A$.
   (c) $D \subset A$.
   (d) $\varnothing \subset A$.
   (e) $A \subset B$.
   (f) $D \supset B$.

A-8 DISJOINT SETS

If two sets have no elements in common, the sets are said to be **disjoint.** If Miss Smith's class contains both boys and girls, the subset of all the boys and the subset of all the girls are disjoint. The set of boys who are ten years old and the set of boys who are eleven years old are disjoint sets.

EXERCISE A-8

Which sets are disjoint?

(a) $P = \{a, b\}$.
(b) $Q = \{b, c\}$.
(c) $S = \{M, n\}$.

A-9 COMPLEMENTS, PARTITIONS

Given a particular set and two subsets of this set, if the two subsets are disjoint and if every element of the given set is in one or the other of the two subsets, the subsets are said to be **complements** of each other in the given set. In general, the universal set is separated by any set into two disjoint sets, the set and its complement in the universal set. In Miss Smith's class, the set of boys and the set of girls are complements. The complement of A is written $\bar{A}$, $\sim A$, or A'. We use A'.

Given a particular set and two or more subsets of this set, if the subsets are all disjoint one from another and if every element of the given set is in one or another of the subsets, the given set is said to be **partitioned.** We could partition Miss Smith's class into subsets of children of the same age. As long as every member of the class is in one and only one of the subsets, we have a partitioning of the class.

EXERCISE A-9

1. Given $W = \{$Diane, Carla, Dolores, Ingrid$\}$ and $C = \{$Diane, Carla$\}$, what is the complement of C in W?
2. What is the complement of the universal set?
3. Find a partition of $E = \{2, 4, 6, 8, 10, 12, 13\}$ into four subsets.

For purposes of analysis it is often convenient to think of a set as being derived from two or more sets. We consider four of the many possible set operations: union, intersection, difference, and Cartesian product.

A-10 SET UNION

If a set can be thought of as being the collection of all elements of two sets, then the set is said to be the **union** of the other two sets. The symbol for union is "$\cup$" and is called "cup." Thus, "$A \cup B$" means the set of elements whose elements are also members of A or members of B. If $A = \{a, b, c\}$ and $B = \{b, c, d, e\}$, then $A \cup B = \{a, b, c, d, e\}$. Note that names of elements that are in both sets A and B are not repeated in the listing of the set that is the union of the two sets.

EXERCISE A-10

The following sets are to be used in problems 1–7.

$$A = \{a, b, c, d, e\}, \qquad B = \{f, g, h\}, \qquad C = \{b, c\},$$

$$D = \{d, e, f, g\}, \qquad U = \{x \mid x \text{ is a letter}\}.$$

1. $A \cup B =$ ____________. **2.** $B \cup A =$ ____________.

3. $A \cup C =$ ____________. **4.** $A \cup D =$ ____________.

5. $A \cup A =$ ____________. **6.** $A \cup \varnothing =$ ____________.

7. $A \cup U =$ ____________.

8. What is the union of the set of even whole numbers and the set of odd whole numbers?

9. What is the union of the set of even whole numbers and the set of whole numbers?

10. What is the union of the nonnegative rational numbers with the set of nonnegative irrational numbers?

A-11 SET INTERSECTION

If a set can be thought of as being the set of elements common to two sets, the set is said to be the **intersection** of the other two sets. The symbol for intersection is "$\cap$" and is called "cap." Thus, "$A \cap B$" means the set whose elements are members of A and also members of B. If $A = \{a, b, c\}$ and $B = \{b, c, d, e, f\}$, then $A \cap B = \{b, c\}$. If $C = \{d, e, f\}$, then $A \cap C = \varnothing$.

Note that the symbol for intersection is often replaced by a dot or omitted, so that $A \cap B = A \cdot B = AB$.

EXERCISE A-11

1. What is the intersection of the set of odd numbers and the set of even numbers?

2. What is the intersection of the set of even numbers and the set of prime numbers?

3. What is the intersection of the set of even numbers and the set of whole numbers?

The following sets are to be used in problem 4–11.

$$F = \{0, 1, 2, 3, 4\}, \qquad G = \{2, 3, 4, 5\}, \qquad H = \{1, 2\},$$

$$I = \{5, 6, 7\}, \qquad J = \{9, 0\}, \qquad U = \{x \mid x \text{ is a whole number}\}.$$

4. $F \cap G =$ ____________. **5.** $G \cap F =$ ____________.

6. $F \cap H =$ ____________. **7.** $F \cap I =$ ____________.

8. $F \cap J =$ ____________. **9.** $F \cap F =$ ____________.

10. $F \cap \varnothing =$ ____________. **11.** $F \cap U =$ ____________.

A-12 SET DIFFERENCE

If a set can be thought of as being formed by deleting from a given set the elements that are also contained in a second set, the set is called the **difference** of the two sets. The symbol for difference is a minus sign and is called "minus." If $F = \{1, 2, 3, 4, 5\}$ and $E = \{2, 4, 6, 7\}$, then $F - E = \{1, 3, 5\}$.

EXERCISE A-12

1. Given that $M = \{\alpha, \beta, \gamma, \delta\}$ and $N = \{\beta, \gamma, \pi\}$, find each of the following sets: (a) $M - N$, (b) $M - M$, (c) $M - \varnothing$.
2. Suppose that C is a subset of A. $A - C$ is the ____________________ of C in A.
3. Given $U = \{0, 1, 2, 3, 4, 5, 6, 7, 8, 9, 10\}$, $A = \{1, 2, 3\}$, and $B = \{0, 1, 2, 3, 4, 5\}$, find each of the following sets:
   (a) $B - A$.
   (b) The complement of A in U.
   (c) The intersection of B with the set found in (b).
4. What relationship is there between the sets found in problems 3(a) and (c)?

A-13 CARTESIAN PRODUCT

If a set can be thought of as being formed by the pairing of each element of one set with every element of a second set, the set is called the **Cartesian product** of the two sets. Notice that the members of the Cartesian product are pairs. Each pair has one element from each of two sets. The symbol "$\times$" is often used to indicate the operation of Cartesian product and is read "cross." Thus, "$A \times B$" means the Cartesian product of A and B. If $M = \{a, b, c\}$ and $N = \{0, 1\}$, then $M \times N = \{(a, 0), (a, 1), (b, 0), (b, 1), (c, 0), (c, 1)\}$.

EXERCISE A-13

1. Write the Cartesian product for each pair of sets.
   (a) $A = \{2, 3\}$, $B = \{5, 6, 7\}$. (b) $C = \{1\}$, $D = \{8, 9, 10\}$.
   (c) $E = \{\text{Jack, Jim, John}\}$, $F = \{\text{Mary, Sue, Carol}\}$.
2. Using A as in problem 1(a), find $A \times A$.
3. Defining A and C as in problem 1:
   (a) Find $A \times C$. (b) Find $C \times A$.
   (c) Is $A \times C = C \times A$? (d) Is $A \times C \sim C \times A$?

A-14 MULTIPLE OPERATIONS

If A and B are sets, then $A \cup B$, $A \cap B$, $A - B$, and $A \times B$ are sets in themselves. Thus, the analysis and symbolism is easily extended to situations in which three or more sets are used or more than one operation is used. In these cases parentheses are normally inserted to indicate the operations to be performed first. Thus, $A \cup (B \cap C)$ means find $B \cap C$ first, then find the union of A with this new set.

EXERCISE A-14

Use the following sets in this exercise.

$$A = \{a, b, c, d\}, \quad B = \{b, c, d, e\}, \quad C = \{d, e, f\}, \quad D = \{b, c, d\}.$$

1. $(A \cup B) \cup C$.

2. $A \cup (B \cup C)$.

3. $(A \cap B) \cap C$.

4. $A \cap (B \cap C)$.

5. $(A \cup B) \cap C$.

6. $(A \cap C) \cup (B \cap C)$.

7. $(A \cap B) \cup C$.

8. $(A \cup C) \cap (B \cup C)$.

9. $(A - D) \cap B$.

A-15 VENN DIAGRAMS AGAIN

Some of the concepts presented above are easily represented by Venn diagrams.

$A \subset B$:

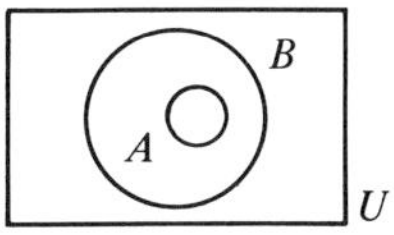

$C = D$:

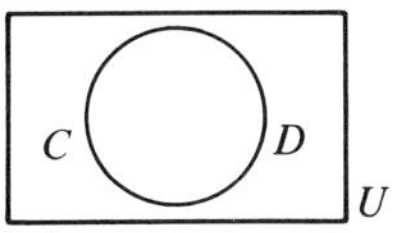

E is the complement of F in G:

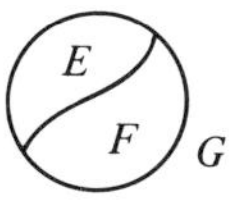

The complement of A in U (shaded portion):

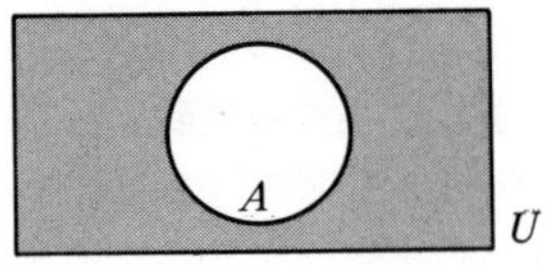

X is partitioned into L, M, N, and P:

R and S are disjoint sets:

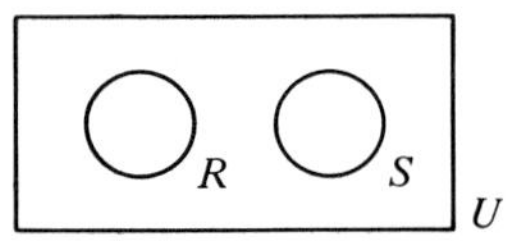

$S \cap T$ (the shaded portion):

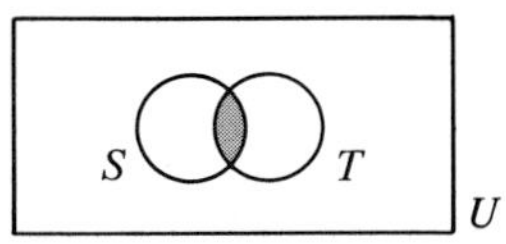

$H \cup I$ (the shaded portion):

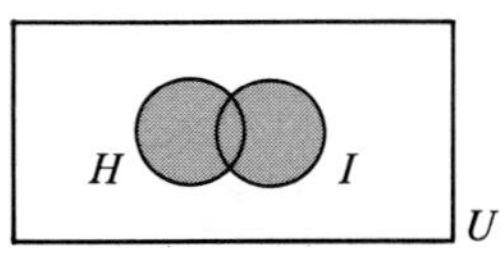

$J - K$ (the shaded portion):

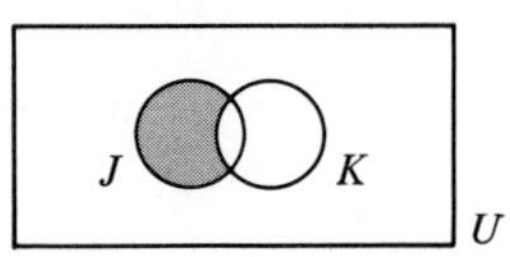

$(A \cap B) \cap C$ (the shaded portion):

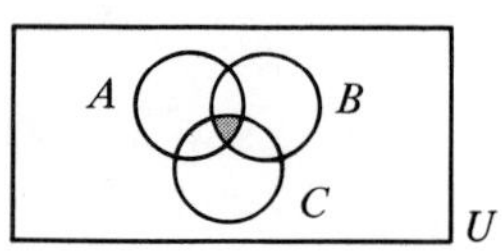

EXERCISE A-15

Draw Venn diagrams to illustrate the situations in problems 1–7.

1. The union of two sets, one of which is a subset of the other.
2. The union of two disjoint sets.
3. The intersection of two sets is empty.
4. The intersection of two sets, one of which is a subset of the other.
5. The union of three sets.
6. The complement of the union of two sets.
7. The intersection of one set with the complement of a second set.
8. For further reading on set concepts consult Ref. 37, pp. 1–47; Ref. 19; Ref. 25, pp. 71–82; and Ref. 47, pp. 1–6, 11–14, 21–30.

APPENDIX B

The Point of Confusion

Figure B-1

Figure B-2

With a view toward updating the geometric content in the elementary and secondary schools, one of many changes has been the introduction of set notation and vocabulary. Among the goals, of course, is clarity and precision in use of this notation and vocabulary. However, there is other than complete agreement in the literature with regard to the answer to the question: In Fig. B-1 what is $\overline{AB} \cap \overline{CD}$?

Since intersection is an operation on sets, the answer must be a set. In Fig. B-2, for instance, if $R = \{1, 2, 3\}$ and $S = \{3, 5, 7, 9\}$, then "clearly" the answer must be $\{3\}$, surely not 3. In similar fashion, then, $\overline{AB} \cap \overline{CD} = \{P\}$. However, it is not uncommon to find this written: $\overline{AB} \cap \overline{CD} = P$.

The issue raised may seem trivial until we reflect that one of the justifications for "all this new stuff" is clarification and removal of possible "points of confusion" or ambiguity. Let us observe another illustration, which brings out why this is not mere hairsplitting.

Using conventional notation, let $\overleftrightarrow{GH}$ symbolize line GH, $\overline{GH}$ symbolize segment GH, and $\angle GHJ$ symbolize angle GHJ with vertex at H. Then, in Fig. B-3, what is $\overleftrightarrow{GJ} \cap \angle HGJ$? Surely, $\overleftrightarrow{GJ} \cap \angle GHJ = \{G, J\}$, since this set contains the points G and J, which are the only points in common to the line GJ and the angle GHJ. In this same figure, what is $\overline{GJ} \cap \overline{HG}$? Since the point G is the only point in common to the segments GJ and HG,

$$\overline{GJ} \cap \overline{HG} = \{G\}.$$

But this is exactly the situation illustrated in Fig. B-1. Then how can the answer be listed:

$$\overline{AB} \cap \overline{CD} = P?$$

To agree with the situation above we must write:

$$\overline{AB} \cap \overline{CD} = \{P\}.$$

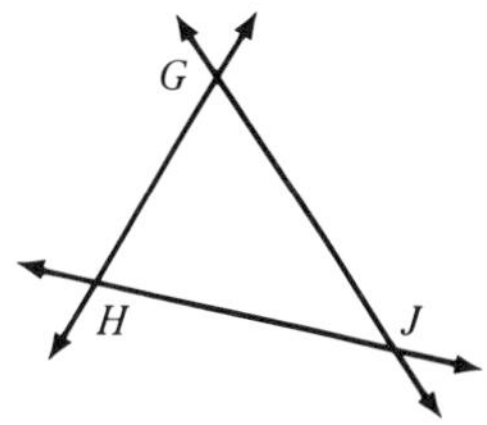

Figure B-3

"P" in this context is the name of a point; thus to give this answer is to claim that a point is a subset of a segment. But to say that a point is a subset is to confuse "subset" and "member of." The confusion seems to lie in the fact that the symbol "P" is used in two different, hence ambiguous, ways: (1) as the name of a set and (2) as the name of an element of a set. The point is not a set but is a *member* of a set, one of many subsets of the segment. This use of "P" in itself is no error, *provided* this usage is clear to those using it in this manner and clear to those reading the material. It is quite within reason that something be an element of a set and yet be a set in its own right. (A familiar example appears in the next paragraph.) However, it would seem that in such a situation different names should be used: one for the element, another for the set.

Thus in Fig. B-1 we have the set of points that we refer to as line segment AB, with three of its elements named, the points A, B, and P. Then each of these points, and combinations of them, constitute some of the subsets of $\overline{AB}$. Using the usual subset notation:

$$\{A\} \subset \overline{AB},$$
$$\{B\} \subset \overline{AB},$$
$$\{P\} \subset \overline{AB},$$
$$\{A, P\} \subset \overline{AB}.$$

Notice, also, that $\{P\} \subset \overline{CD}$. Now, the intersection of two sets is defined to be the set that contains exactly those elements common to both sets. This requires that the intersection set be a subset of each of the given sets. Therefore, $\overline{AB} \cap \overline{CD} = \{P\}$, since of all the original sets, this is the one containing all the elements of each set that are common to both sets, and no other elements. Now then, if the set that contains the point P is defined to be E—that is, let $\{P\} = E$—then

$$\overline{AB} \cap \overline{CD} = \{P\} = E.$$

The confusion in situations of this type resembles that which some students experience when confronted with the symbols $\varnothing$, $\{\ \}$, 0, $\{0\}$, and $\{\varnothing\}$. The first two symbols name the empty set, the third represents the number zero, the fourth names a set containing the number zero, and the last names a set that is not empty since it contains the empty set as an element. Notice that the first and last symbols illustrate that something can be an element of a set and yet be a set in its own right.

Perhaps those who would write concerning Fig. B-1, "$\overline{AB} \cap \overline{CD} = P$," should do one of three things:

(1) Point out that they are saying $P = \{P\}$.

(2) Introduce another symbol, clearly defined, such as "Let $E = \{P\}$," then $\overline{AB} \cap \overline{CD} = E$.

(3) Admit the ambiguity (to the reader) and write "$\overline{AB} \cap \overline{CD} = \{P\}$."

Perhaps the first of these would be most readily acceptable in view of the usage in some texts. This is similar to saying "area of a triangle" means "area of the surface bounded by a triangle"—a convention which many textbooks now point out and subsequently use. Surely, however, either of the other two suggestions is more precise, hence mathematically more desirable.

APPENDIX C

Patterns

On the next few pages you will find patterns for some polyhedrons and miscellaneous items. For best results you should draw these on heavier paper and have available (1) a razor blade, (2) scissors and/or a paper cutter, (3) a ruler with a sharp edge, (4) rubber cement or white glue.

Basically you should find these patterns rather easy to develop into reasonably satisfactory models. For best results:

(1) Cut on solid lines; use the paper cutter for longer cuts.

(2) Score dashes with the razor blade and fold back away from the printed side. The ruler or other sharp straight edge should improve the fold.

(3) Fit the model together without glue to check alignment.

(4) Spread glue on top of tabs—only a few at a time.

(5) If you wish to color the model, it might be advisable to color before gluing.

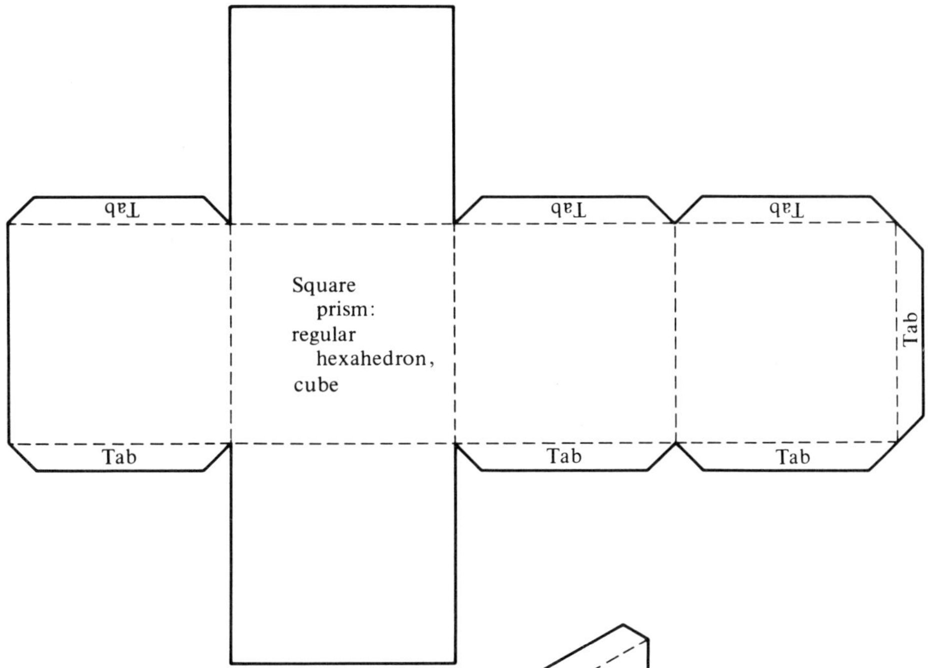
Tab
Tab
Tab
Square
prism:
regular
hexahedron,
cube
Tab
Tab
Tab
Tab

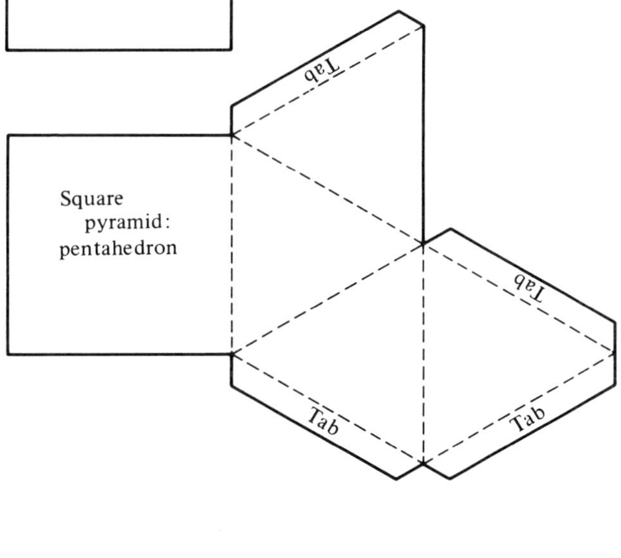
Tab
Square
pyramid:
pentahedron
Tab
Tab
Tab

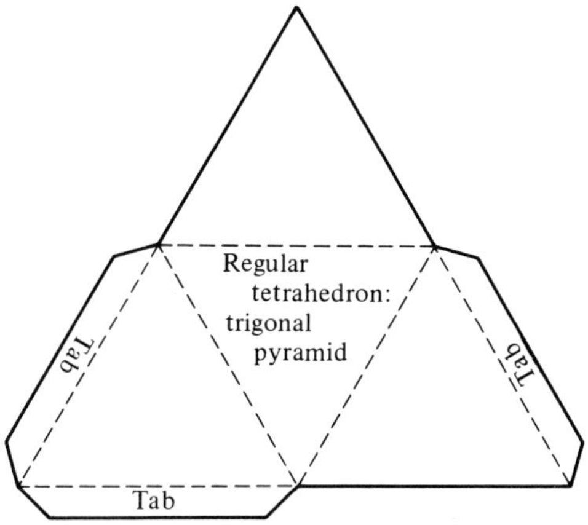

Regular
tetrahedron:
trigonal
pyramid
Tab
Tab
Tab

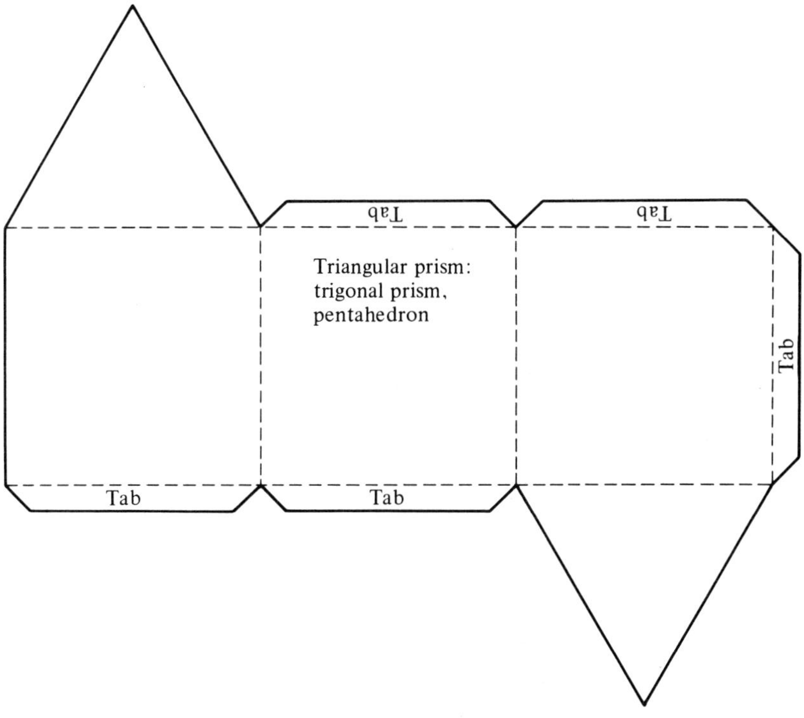

Triangular prism:
trigonal prism,
pentahedron
Tab
Tab
Tab
Tab
Tab

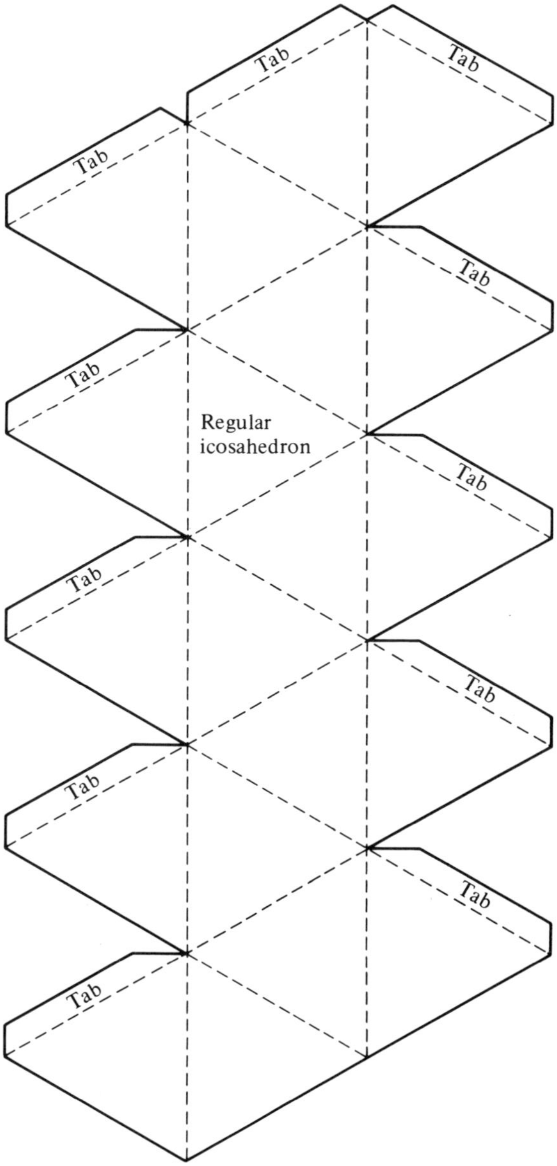
Tab
Tab
Tab
Tab
Regular
icosahedron
Tab
Tab
Tab
Tab
Tab
Tab

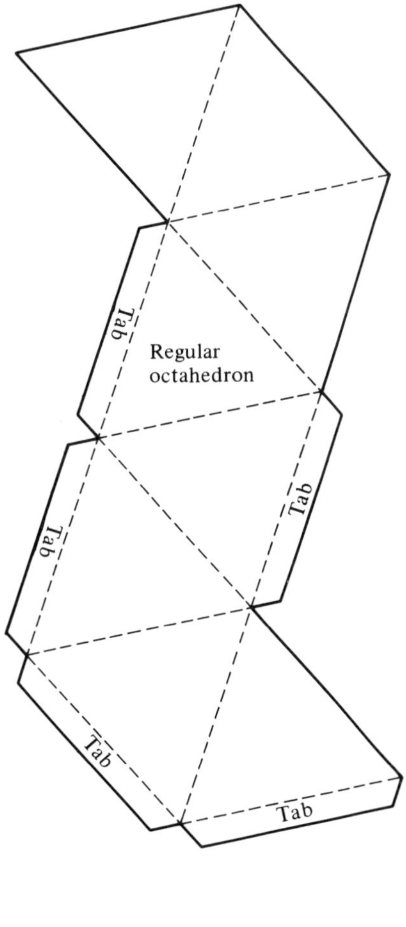
Tab
Regular
octahedron
Tab
Tab
Tab
Tab

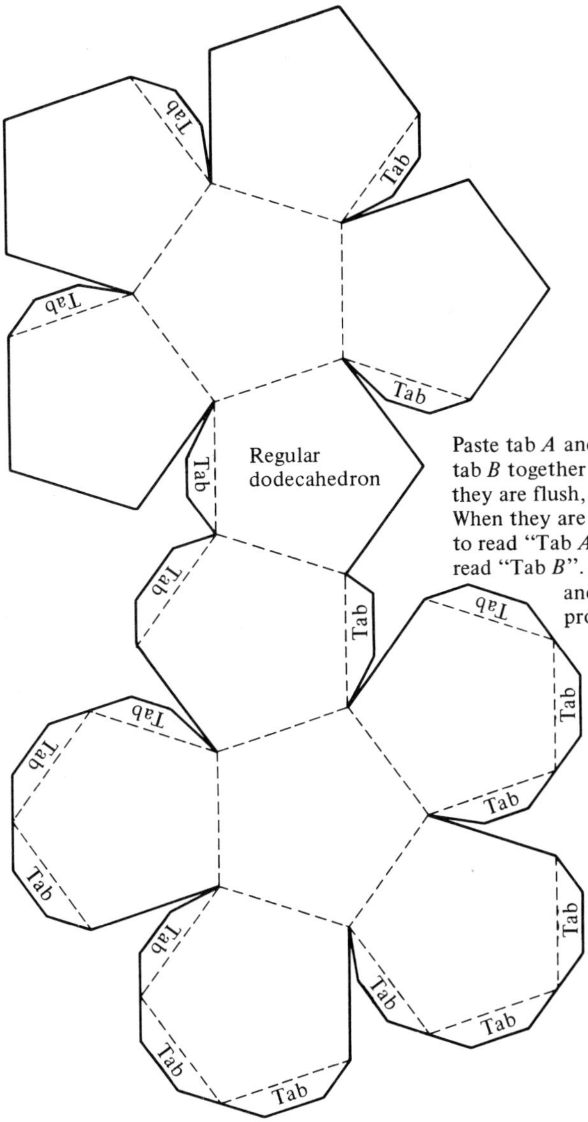

Paste tab *A* and tab *B* together so that they are flush, and "Tab *B*" is against "Tab *A*". When they are together you should not be able to read "Tab *A*", and should not be able to read "Tab *B*". This should be done and dried before proceeding with other steps.

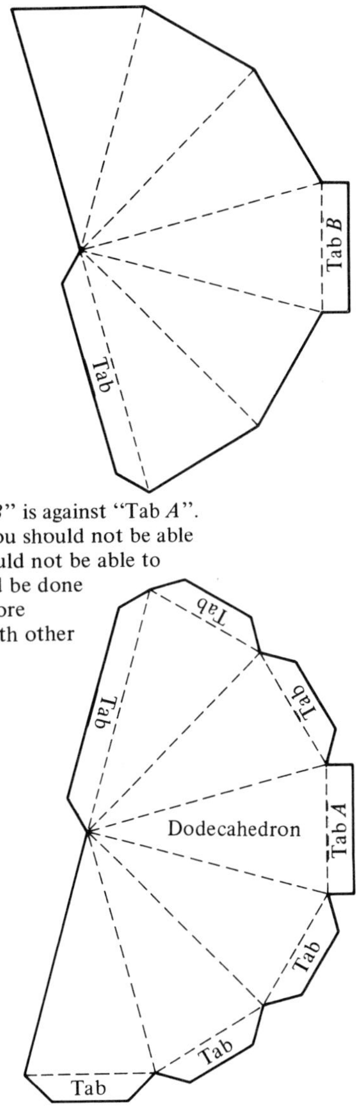

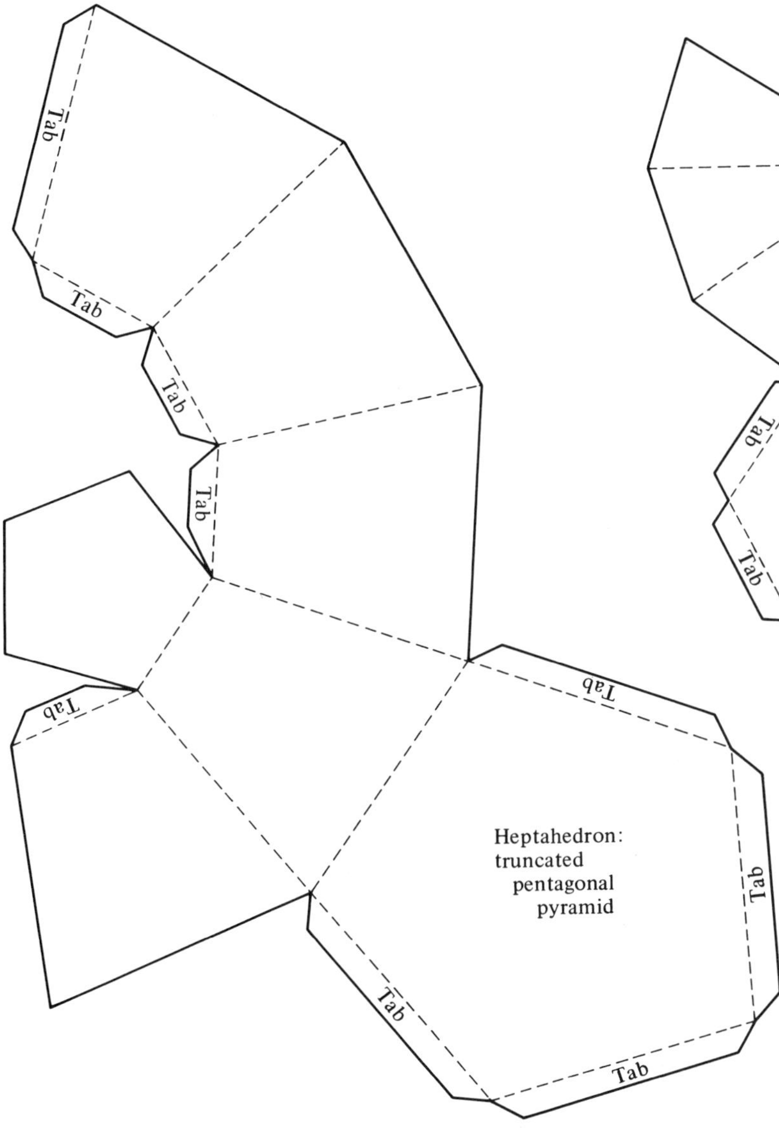
Tab
Tab
Tab
Tab
Tab
Tab
Tab
Tab
Heptahedron:
truncated
pentagonal
pyramid

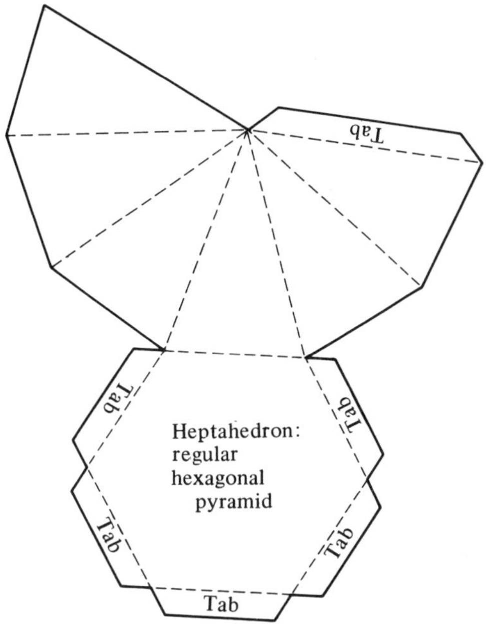
Tab
Tab
Tab
Heptahedron:
regular
hexagonal
pyramid
Tab
Tab
Tab

Note: Fold *forward* on code line. Do not cut.

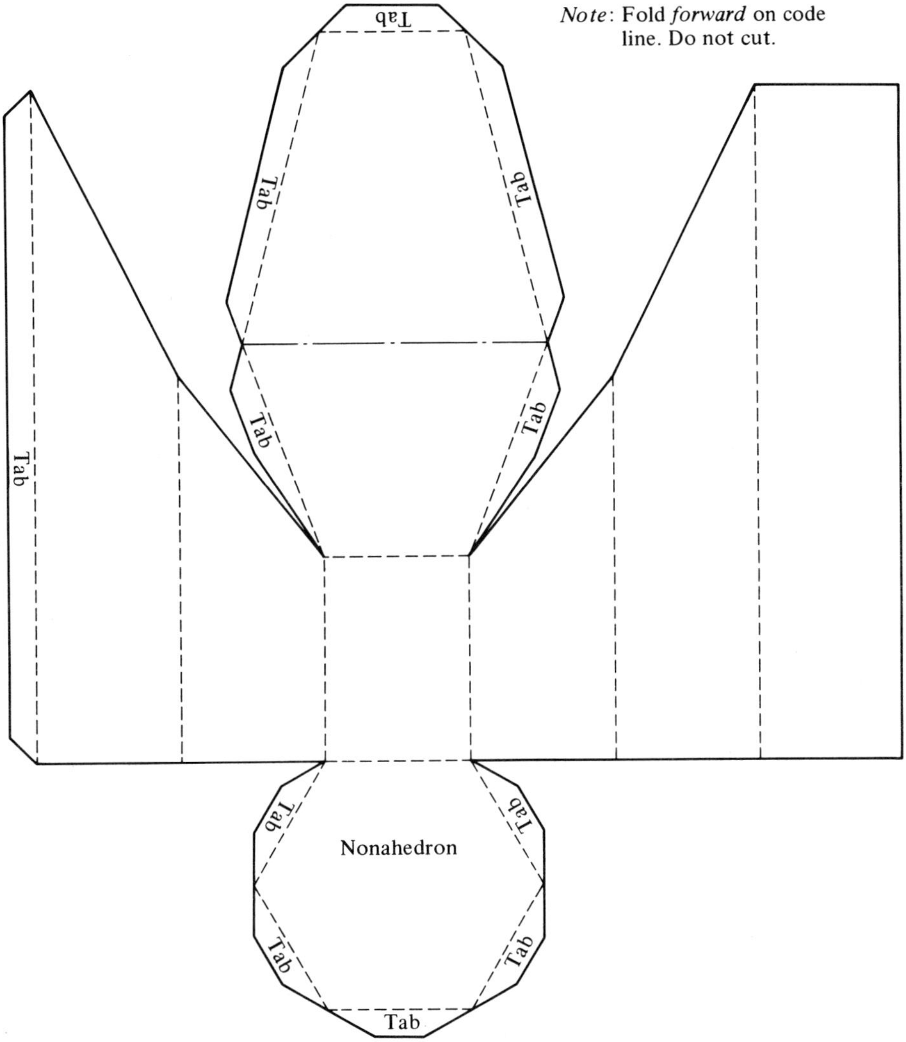

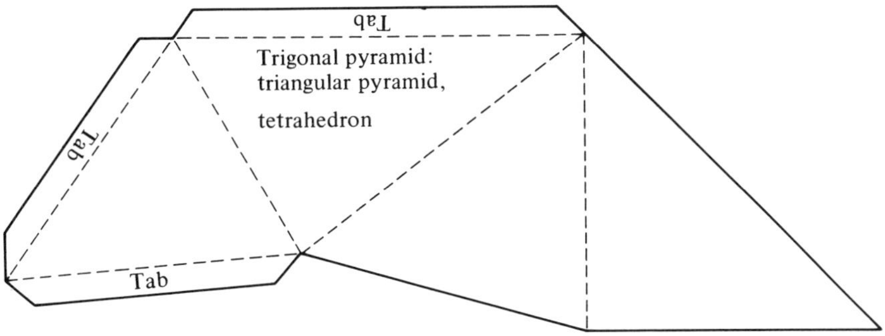
Tab
Trigonal pyramid:
triangular pyramid,
tetrahedron
Tab
Tab

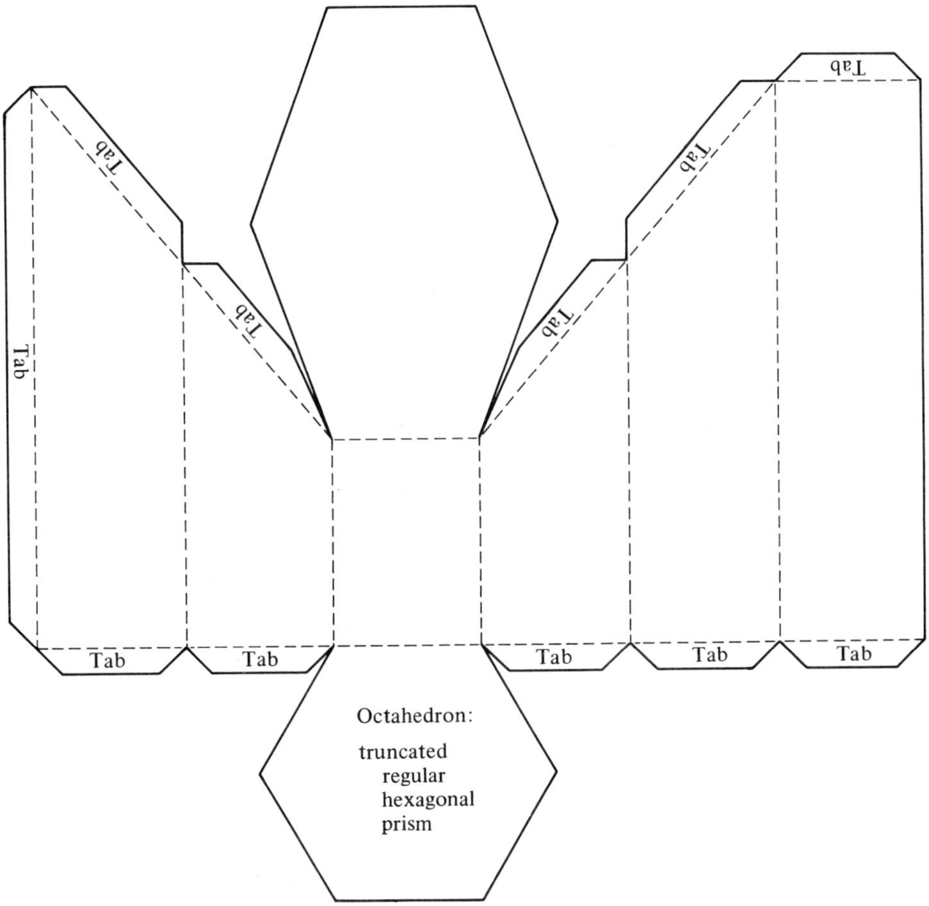
Tab
Tab
Tab
Tab
Tab
Tab
Tab
Tab
Tab
Tab
Tab
Octahedron:
truncated
regular
hexagonal
prism

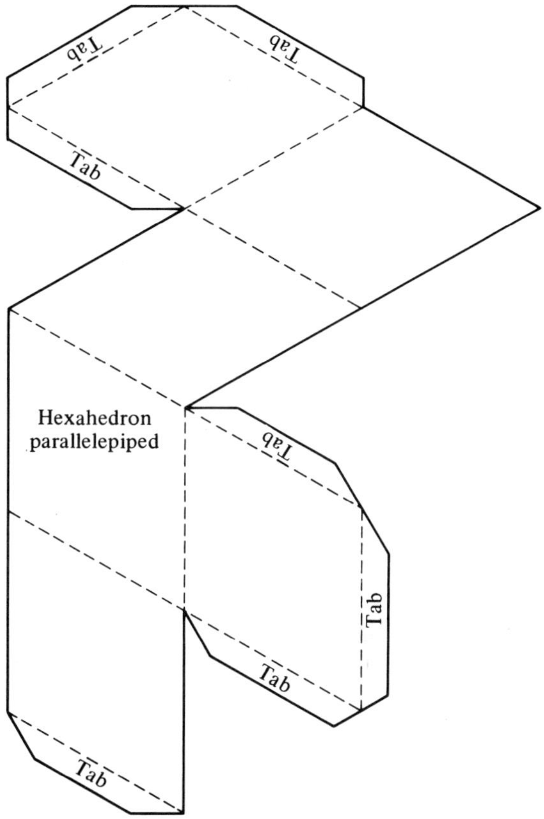
Tab
Tab
Tab
Hexahedron
parallelepiped
Tab
Tab
Tab
Tab

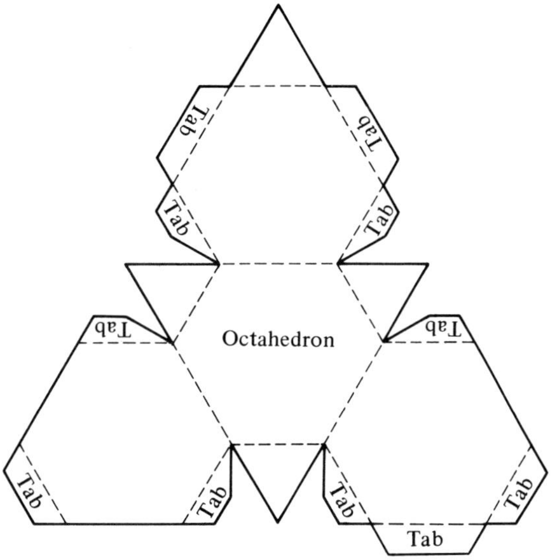
Tab
Tab
Tab
Tab
Tab
Octahedron
Tab
Tab
Tab
Tab
Tab
Tab

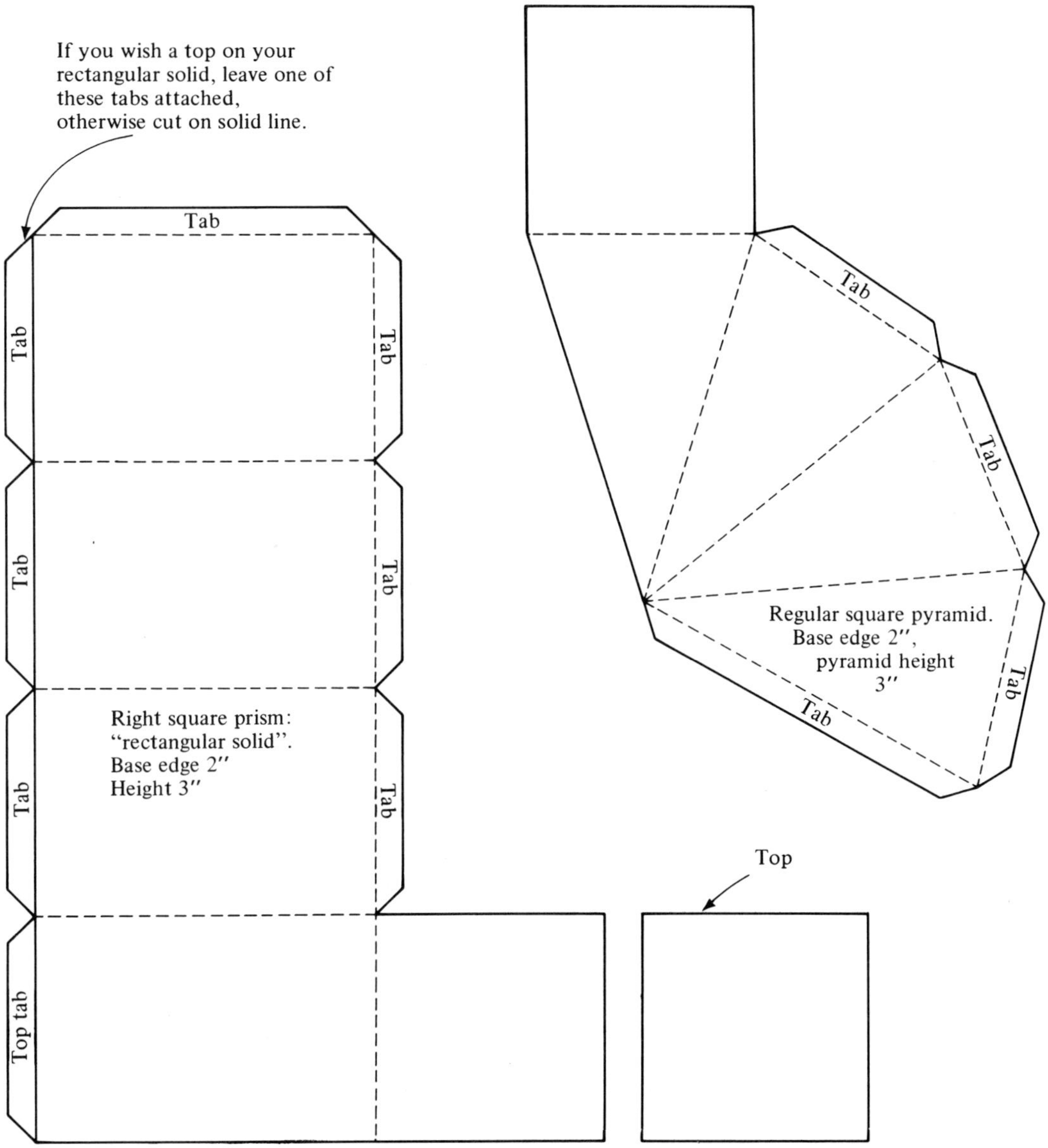
If you wish a top on your
rectangular solid, leave one of
these tabs attached,
otherwise cut on solid line.
Tab
Tab
Tab
Tab
Tab
Tab
Tab
Top tab
Right square prism:
"rectangular solid".
Base edge 2″
Height 3″
Tab
Tab
Tab
Tab
Regular square pyramid.
Base edge 2″,
pyramid height
3″
Top

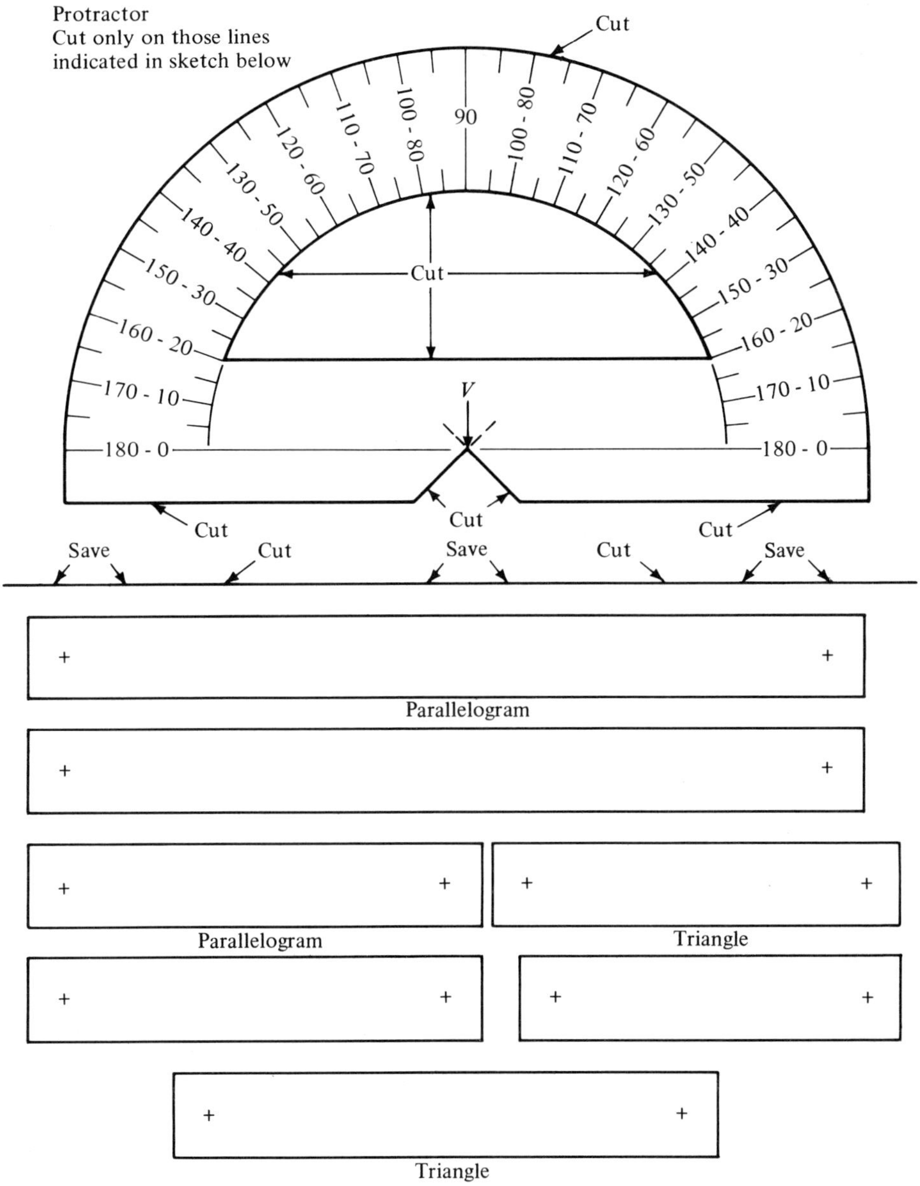

Cut on solid lines.
Cross marks indicate pinholes.

References

BOOKS AND PAMPHLETS

1. Abbott, Edwin A. *Flatland: A Romance of Many Dimensions.* New York: Dover Publications, Inc., 1952.

2. Allendoerfer, Carl B., and Cletus O. Oakley. *Principles of Mathematics,* 2d ed. New York: McGraw-Hill Book Company, 1963.

3. Anderson, Richard D. *Concepts of Informal Geometry,* Studies in Mathematics, vol. V (preliminary ed.). New Haven: Yale University Press, 1960.

4. Arnold, B. H. *Intuitive Concepts in Elementary Topology.* Englewood Cliffs, N.J.: Prentice-Hall, Inc., 1962.

5. Barker, Stephen. *Philosophy of Mathematics.* Englewood Cliffs, N.J.: Prentice-Hall, Inc., 1964.

6. Barry, Edward H. *Introduction to Geometric Transformations.* Boston: Prindle, Weber & Schmidt, Inc., 1966.

7. Bergamini, David. *Mathematics.* New York: Time, Inc., 1963.

8. Burger, D. *Sphereland.* New York: Thomas Y. Crowell Company, 1965.

9. Cooley, Hollis R., and Howard E. Wahlert. *Introduction to Mathematics.* Boston: Houghton Mifflin Company, 1968.

10. Courant, Richard, and Herbert Robbins. *What Is Mathematics?* New York: Oxford University Press, 1941.

11. Coxeter, H. S. M. *Introduction to Geometry*. New York: John Wiley & Sons, Inc., 1961.

12. Denbow, Carl H., and Victor Goedicke. *Foundations of Mathematics*. New York: Harper & Row, Publishers, 1959.

13. Dorwart, Harold L. *The Geometry of Incidence*. Englewood Cliffs, N.J.: Prentice-Hall, Inc., 1966.

14. Durell, Fletcher, and E. E. Arnold. *New Solid Geometry*. Columbus, Ohio: Charles E. Merrill Books, Inc., 1932.

15. Glenn, William H., and Donovan A. Johnson. *The Pythagorean Theorem*. St. Louis: Webster Publishing Division, McGraw-Hill, Inc., 1960.

16. Hawley, Newton, and Patrick Suppes. *Geometry for Primary Grades*, Books One and Two. San Francisco: Holden-Day, Inc., 1960.

17. Hogben, Lancelot. *The Wonderful World of Mathematics*. New York: Doubleday & Company, Inc., 1955.

18. Johnson, Donovan A. *Paper Folding for the Mathematics Class*. Washington, D.C.: National Council of Teachers of Mathematics, 1957.

19. _______ and William H. Glenn. *Sets, Sentences, and Operations*. St. Louis: Webster Publishing Division, McGraw-Hill, Inc., 1960.

20. _______. *The World of Measurement*. St. Louis: Webster Publishing Division, McGraw-Hill, Inc., 1961.

21. _______. *Topology: The Rubber Sheet Geometry*. St. Louis: Webster Publishing Division, McGraw-Hill, Inc., 1960.

22. Keedy, Mervin L., and Charles W. Nelson. *A Modern Introduction to Geometry*. Reading, Mass.: Addison-Wesley Publishing Company, Inc., 1965.

23. Levi, Howard. *Foundations of Geometry and Trigonometry*. Englewood Cliffs, N.J.: Prentice-Hall, Inc., 1960.

24. _______. *Topics in Geometry*. Boston: Prindle, Weber & Schmidt, Inc., 1965.

25. Marks, John L., C. Richard Purdy, and Lucien B. Kinney. *Teaching Elementary School Mathematics for Understanding*, 2d ed. New York: McGraw-Hill, Inc., 1965.

26. Meserve, Bruce E. *Fundamental Concepts of Geometry*. Reading, Mass.: Addison-Wesley Publishing Company, Inc., 1955.

27. _______ and Max A. Sobel. *Introduction to Mathematics*. Englewood Cliffs, N.J.: Prentice-Hall, Inc., 1964.

28. _______. *Mathematics for Secondary School Teachers*. Englewood Cliffs, N.J.: Prentice-Hall, Inc., 1962.

29. Modenov, P. S., and A. S. Parkhomenko. *Geometric Transformations*, vol. I, trans. Michael B. P. Slater. New York: Academic Press, Inc., 1965.

30. Moise, Edwin E. *Elementary Geometry from an Advanced Standpoint*. Reading, Mass.: Addison-Wesley Publishing Company, Inc., 1963.

31. _______ and Floyd L. Downs, Jr. *Geometry*. Reading, Mass.: Addison-Wesley Publishing Company, Inc., 1964.

32. Mold, Josephine. *Circles*. New York: Cambridge University Press, 1967.

33. _______. *Solid Models*. New York: Cambridge University Press, 1967.

34. National Council of Teachers of Mathematics. *Enrichment Mathematics for the Grades*, 27th Yearbook. Washington, D.C.: The Council, 1963.

35. _______. *Geometry*, Experiences in Mathematical Discovery, 4. Washington, D.C.: The Council, 1966.

36. _______. *Insights into Modern Mathematics*, 23rd Yearbook. Washington, D.C.: The Council, 1959.

37. _______. *Topics in Mathematics for Elementary School Teachers*, 29th Yearbook. Washington, D.C.: The Council, 1964.

38. (a) Piaget, Jean, and Bärbel Inhelder. *The Child's Conception of Space*, trans. F. J. Langdon and J. L. Lunzer. New York: Humanities Press, Inc., 1967, and London: Routledge and Kegan Paul Ltd., 1956.
(b) _______ and Alina Szeminska. *The Child's Conception of Geometry*, trans. E. A. Lunzer. London: Routledge and Kegan Paul Ltd., 1960.

39. Plato. *Meno*, trans. Benjamin Jowett. Indianapolis: The Bobbs-Merrill Company, Inc., 1949.

40. Ransom, William R. *Can and Can't in Geometry*. Portland, Me.: J. Weston Walch, Publisher, 1960.

41. Ranucci, Ernest R. *Points, Lines, and Planes: An Introduction to Geometry in Two Dimensions*. New York: The Macmillan Company, 1963.

42. Rees, Paul K. *Principles of Mathematics*. Englewood Cliffs, N.J.: Prentice-Hall, Inc., 1965.

43. Ringenberg, Lawrence A. *Informal Geometry*. New York: John Wiley & Sons, Inc., 1967.

44. School Mathematics Study Group. *Intuitive Geometry*, Studies in Mathematics, vol. VII (preliminary ed.). Stanford, Calif.: Stanford University Press, 1961.

45. Seidenberg, A. *Lectures in Projective Geometry*. Princeton, N.J.: Van Nostrand-Reinhold Company, 1962. One of a series of texts entitled "The University Series in Undergraduate Mathematics."

46. Smart, James R. *Introductory Geometry: An Informal Approach*. Belmont, Calif.: Brooks/Cole Publishing Company, 1967.

47. Smith, Seaton E., Jr. *Explorations in Elementary Mathematics*. Englewood Cliffs, N.J.: Prentice-Hall, Inc., 1966.

48. Spooner, George. *Mathematics Enrichment Programs, A, B, C: Part 2, Geometry*. New York: Harcourt, Brace & World, Inc., 1962.

49. Stanton, Ralph G., and Kenneth D. Fryer, eds. *Topics in Modern Mathematics*. Englewood Cliffs, N.J.: Prentice-Hall, Inc., 1964.

50. Stein, Sherman K. *Mathematics: The Man-Made Universe*. San Francisco: W. H. Freeman and Company, 1963.

51. Van Engen, Henry, Maurice L. Hartung, and James E. Stochl. *Foundations of Elementary School Arithmetic*. Glenview, Ill.: Scott, Foresman and Company, 1965.

52. Veblen, O., and J. W. Young. *Projective Geometry*, vols. I and II. Boston: Ginn & Company, 1910.

53. Wenninger, Magnus J. *Polyhedron Models for the Classroom*. Washington, D.C.: National Council of Teachers of Mathematics, 1966.

54. Whitehead, Alfred North. *An Introduction to Mathematics*. New York: Oxford University Press, 1958.

55. Wolfe, Harold E. *Introduction to Non-Euclidean Geometry*. New York: Holt, Rinehart and Winston, Inc., 1945.

ARTICLES IN PERIODICALS

The Arithmetic Teacher

56. Bourne, H. N. "The Concept of Area," XV, No. 3 (March 1968), 233–243.

57. Brune, Irvin H. "Some K–6 Geometry," XIV, No. 6 (October 1967), 441–447.

58. Buck, Charles. "Geometry for the Elementary School," XIV, No. 6 (October 1967), 460–467.

59. Buckland, G. T. "The Prismoidal Formula," VI, No. 1 (February 1959), 44–45.

60. Denmark, Tom. "An Intuitive Introduction to the Euclidean Concept of Betweenness," XV, No. 8 (December 1968), 683–686.

61. Felder, Virginia. "Geometry Concepts in Grades K–3," XII, No. 5 (May 1965), 356–358.

62. Giddings, Marie. "Being Creative with Shapes," XII, No. 8 (December 1965), 645–646.

63. Goldmark, B. "Geometry in the Primary Grades," X, No. 4 (April 1963), 191–192.

64. Heard, Ida Mae. "Developing Geometric Concepts in the Kindergarten," XVI, No. 3 (March 1969), 229–230.

65. Helgren, Fred J. "The Metric System in the Elementary Grades," XIV, No. 5 (May 1967), 349–353.

66. Hilire, Paul A. "The Volume of a Sphere," VII, No. 5 (May 1960), 268.

67. Inskeep, James E., Jr. "Primary Grade Instruction in Geometry," XV, No. 5 (May 1968), 422–426.

68. Krause, Eugene F. "Elementary School Metric Geometry," XV, No. 8 (December 1968), 673–682.

69. Lamb, Pose M. "Geometry for Third and Fourth Graders," X, No. 4 (April 1963), 193–194.

70. Mahlenbruck, D. E. "Points and Lines," XI, No. 4 (April 1964), 247.

71. Mastain, Richard K., and Bernice C. Nossoff. "Mathematics in Kindergarten," XIII, No. 1 (January 1966) 32–36.

72. Needleman, Joan R. "Discovery Approach—Polar Coordinates in Grade Seven?" XIV, No. 7 (November 1967), 563–565.

73. Overholt, Elbert D. "From Number Lines to 2-D Space Concepts," XIII, No. 2 (February 1966), 107–109.

74. Parker, Robert. "Graph Paper: A Versatile Aid," XVI, No. 2 (February 1969), 144–148.

75. Peskin, Anne S. "Geometric Representation of Binomial by Binomial—Laboratory Style," XV, No. 1 (January 1968), 40–44.

76. Pray, Richard H. "The Metric System IS Simple," VIII, No. 4 (April 1961), 179.

77. Robinson, G. Edith. "The Role of Geometry in Elementary School Mathematics," XIII, No. 1 (January 1966), 3–10.

78. Sganga, Francis T. "A Bee on a Point, a Line, and a Plane," XIII, No. 7 (November 1966), 549–552.

79. Shah, Sair Ali. "Selected Geometric Concepts Taught to Children Ages Seven to Eleven," XVI, No. 2 (February 1969), 119–128.

80. Skypek, Dora Helen. "Geometric Concepts in Grades 4–6," XII, No. 6 (October 1965), 443–449.

81. Smart, James R., and John L. Marks. "Mathematics of Measurement," XIII, No. 4 (April 1966), 283–287.

82. Strangman, Kathryn B. "Grids, Tiles, and Area," XV, No. 8 (December 1968), 668–672.

83. Vigilante, Nicholas J. "The Address of a Point," XV, No. 8 (December 1968), 689–693.

84. _______. "Geometry for Primary Children: Considerations," XIV, No. 6 (October 1967), 453–459.

85. _______. "Why Circumvent Geometry in the Primary Grades?" XII, No. 6 (October 1965), 450–454.

86. Walter, Marion. "An Example of Informal Geometry: Mirror Cards," XIII, No. 6 (October 1966), 448–452.

87. _______. "Some Mathematical Ideas Involved in the Mirror Cards," XIV, No. 2 (February 1967), 115–125.

88. Woods, Dale, and William E. Hoff. "Introducing Models for *n*-Dimensional Geometry in the Elementary School," XIII, No. 1 (January 1966), 11–13.

89. Wray, D. Eileen. "You and the Metric System," XI, No. 8 (December 1964), 576–580.

The Mathematics Teacher

90. Adler, Irving. "What Shall We Teach in High School Geometry?" LXI, No. 3 (March 1968), 226–238.

91. Allendoerfer, Carl B. "The Dilemma in Geometry," LXII, No. 3 (March 1969), 165–169.

92. Auclair, Jerome A., and Thomas P. Hillman. "A Topological Problem for the Ninth-Grade Mathematics Laboratory," LXI, No. 5 (May 1968), 503–507.

93. Berman, Roslyn M., and Martin Berman. "Concave Polygons," LVI, No. 6 (October 1963), 403–406.

94. Birkhoff, George D., and Ralph Beatley. "A New Approach to Elementary Geometry," LXI, No. 8 (December 1968), 765–771.

95. Boe, Barbara. "A Study of the Ability of Secondary School Pupils to Perceive the Plane Sections of Selected Solid Figures," LXI, No. 4 (April 1968), 415–421.

96. Buck, Charles. "What Should High School Geometry Be?" LXI, No. 5 (May 1968), 466–471.

97. Carver, Margaret L. "Graphing Pictures," LII, No. 1 (January 1959), 41–43.

98. Davis, Edward. "The First Days of Geometry," LVI, No. 8 (December 1963), 645–646.

99. Dye, David L. "What Is a Trapezoid?" LX, No. 7 (November 1967), 727–728.

100. Eves, Howard. "A Geometry Capsule Concerning the Five Platonic Solids," LXII, No. 1 (January 1969), 42–44.

101. Hinkel, Wilfred H. "The Trapezoid—And Area," LIII, No. 2 (February 1960), 106–108.

102. Huffman, David C. "Independence of the Incidence Postulates," LXII, No. 4 (April 1969), 269–277.

103. Kattsoff, Louis O. "Problems in Presenting Non-Euclidean Geometries to High School Teachers," LIII, No. 7 (November 1960), 559–563.

104. Keedy, M. L. "What Is a Trapezoid?" LIX, No. 7 (November 1966), 646.

105. Maiers, Wesley W. "Introduction to Non-Euclidean Geometry," LVII, No. 7 (November 1964), 457–461.

106. Meder, A. E., Jr. "What is Wrong with Euclid?" LI, No. 8 (December 1958), 578–584.

107. Meserve, Bruce E. "Topology for Secondary Schools," XLVI, No. 7 (November 1953), 465–474.

108. _______ and R. E. Pingry. "Some Notes on the Prismoidal Formula," XLV, No. 4 (April 1952), 257–263.

109. Pederson, Jean J. "Dressing Up Mathematics," LXI, No. 2 (February 1968) 118–122.

110. Ranucci, Ernest R. "A Tiny Treasury of Tessellations," LXI, No. 2 (February 1968), 114–117.

111. Rucker, Isabelle P. "The Metric System in Junior High Schools," LI, No. 8 (December 1958), 621–623.

112. Ryoti, Don E. "What Is an Isosceles Trapezoid?" LX, No. 7 (November 1967), 729–730.

113. Sanders, Walter J., and J. Richard Dennis. "Congruence Geometry for Junior High School," LXI, No. 4 (April 1968), 354–369.

114. Sitomer, Harry, and Howard F. Fehr. "How Shall We Define Angle?" LX, No. 1 (January 1967), 18–19.

115. Smith, Stanley A. "What Does a Proof Really Prove?" LXI, No. 5 (May 1968), 483–484.

116. Troyer, Robert. "Rotations, Angles and Trigonometry," LXI, No. 2 (February 1968), 123–129.

117. Wiseman, John D. "Chains of Reasoning in Geometry," LII, No. 6 (October 1959), 457–458.

118. Wright, Frank. "The SMSG Angle Is a Good One," LX, No. 8 (December 1967), 856–857.

119. Young, J. W. A. "The Teaching of Mathematics," LXI, No. 3 (March 1968), 287–295.

Scientific American

120. Piaget, Jean. "How Children Form Mathematical Concepts," CLXXXIX, No. 5 (November 1953), 74–79.

121. Tucker, A. W., and H. S. Bailey, Jr. "Topology," CLXXXII, No. 1 (January 1950), 18–24.

Index *

* Numerals within brackets refer to a problem in an Exercise on the page listed.

INDEX OF SYMBOLS